Lecture Notes in Energy

Volume 80

Lecture Notes in Energy (LNE) is a series that reports on new developments in the study of energy: from science and engineering to the analysis of energy policy. The series' scope includes but is not limited to, renewable and green energy, nuclear, fossil fuels and carbon capture, energy systems, energy storage and harvesting, batteries and fuel cells, power systems, energy efficiency, energy in buildings, energy policy, as well as energy-related topics in economics, management and transportation. Books published in LNE are original and timely and bridge between advanced textbooks and the forefront of research. Readers of LNE include postgraduate students and non-specialist researchers wishing to gain an accessible introduction to a field of research as well as professionals and researchers with a need for an up-to-date reference book on a well-defined topic. The series publishes single- and multi-authored volumes as well as advanced textbooks.

Indexed in Scopus and EI Compendex The Springer Energy board welcomes your book proposal. Please get in touch with the series via Anthony Doyle, Executive Editor, Springer (anthony.doyle@springer.com)

More information about this series at http://www.springer.com/series/8874

Machiel Mulder

Regulation of Energy Markets

Economic Mechanisms and Policy Evaluation

 Springer

Machiel Mulder
Faculty of Economics and Business
Centre for Energy Economics Research
University of Groningen
Groningen, The Netherlands

ISSN 2195-1284   ISSN 2195-1292   (electronic)
Lecture Notes in Energy
ISBN 978-3-030-58318-7   ISBN 978-3-030-58319-4   (eBook)
https://doi.org/10.1007/978-3-030-58319-4

This Springer imprint is published by the registered company Springer Nature Switzerland AG
The registered company address is: Gewerbestrasse 11, 6330 Cham, Switzerland

Preface

As energy is a crucial commodity in modern societies, it's important to understand how to realize its optimal allocation across alternative utilisations. In principle, when energy markets would function according to microeconomic textbooks, there is no need to intervene. After all, well-functioning markets automatically find the optimal allocation and, hence, generate the maximum welfare. Energy markets, however, are quite different from this theoretical benchmark of perfectly functioning markets. Energy markets generally suffer from a large number of fundamental shortcomings, which are called market failures. These shortcomings include the presence of information asymmetry between consumers and producers in retail energy markets, presence of natural monopolies in the transport through networks, the public-good character of reliability of energy supply, negative externalities related to the consumption of fossil energy, and the ability of suppliers to behave strategically and raise market prices.

Because of these market failures, policy intervention in energy markets may be helpful to improve the allocation. This intervention is called sector-specific regulation. This regulation includes measures such as rules regarding information provision to residential consumers, rules on the maximum tariffs that network operators may charge, incentives for network users to pay attention to the consequences of their behaviour for the balance of energy systems, incentives for producers of energy to use more renewable energy, and measures to enlarge the market in order to mitigate the ability of dominant firms to abuse market power.

The search for the optimal regulation of energy markets is the topic of this textbook. This book is the result of more than 20 years of experience in the field of the economic analysis of energy markets. At 1 January 2000, I became head of the Energy department of the Netherlands Bureau for Economic Policy Analysis (CPB), after a career in the field of environmental, agricultural and small-business economics. Since then, encouraged by its complexity and societal relevance, I have worked (more than) full-time on the application of energy economics to policy making. I am grateful that I was able to do this at various positions, not only at the CPB, but also at the environmental consultancy CE-Delft and the Authority for Consumers and Markets (ACM), which is the competition authority and regulator of, amongst others, the Dutch energy markets. This experience has provided my in-depth insight in the actual functioning of energy markets as well as the actual political decision making in practice. The list of people I have worked with in this period and who contributed to my pleasure of working in and knowledge of this field is extensive. I am grateful to all of them.

In all my previous positions, I worked on the bridge between academic research and practical application in policy making. Gradually, however, my focus shifted fully to the academic world. In 2013, I became, initially part-time next to my position at the ACM, full professor of Regulation of Energy Markets at the Faculty of Economics of Business of the University of Groningen. I am still grateful to both the board of the ACM and the board of the university for their support. A year later, I started teaching my M.Sc. course called Economics of Regulating Markets. For both my students and me, this was a new experience. Students were less used to get a course that was not directed at developing theoretical concepts, but that applies these concepts to actual problems in the field of energy. According to the course evaluations, students did like the course, which motivated me to further develop the course materials. Initially, these materials mainly consisted of my lecture slides, notes on the weekly assignments and a number of scientific articles. Already in the first year of this course, students asked for a textbook in which all the topics were systematically and carefully described. Although many textbooks in the field of energy economics have been published, a textbook on the regulation of energy markets combining theoretical economic concepts and practical applications did not exist. Therefore, I decided to write it myself. Though, it took some time before I really started, because of many other activities, such as the organisation of the International Conference of the International Association for Energy Economics (IAEE) in Groningen in June 2018 and the organisation of the multidisciplinary full-time Energy Minor at the University of Groningen. At the IAEE conference in Groningen, however, I shared my plan with Barbara Fess of Springer, who responded very positively. She introduced me to her colleagues Johannes Glaeser and Judith Kripp, who were equally positive about the idea. They have guided me throughout the process of writing the book in a highly pleasant way, for which I am very grateful. I also thank Sudhany Karthick of Springer Publishing Services for carefully managing the production process.

The manuscript of this textbook has been used by about 50 students of my M.Sc. course in Spring of 2020. During the (online) lectures and tutorials, I received useful feedback on formulations, figures and tables which were less clear. Despite the draft character of the manuscript, these students appreciated it highly according to the course evaluations. I thank these students for accepting the manuscript as sole course material, including the several typos and grammatical errors, and for their valuable feedback and suggestions.

Besides these students, a large number of experts in the field have commented on the manuscript, which I appreciate very much. I thank Mathieu Fransen and Paul Giesbertz for the extensive and very helpful discussions on the description of electricity systems and markets. For the texts on coal, oil, gas, hydrogen and/or heat systems, markets and regulation, I thank Kick Bruin, Robert Haffner, Jann Keller, Peter Perey and Ton Schoot Uiterkamp. I thank Peter Dijkstra, Lennard Rekker and Edwin Woerdman for their feedback on the texts on tariff regulation and/or environmental regulation. I also thank Anneke Boelkens, Daan Hulshof and Jose Luis Moraga for carefully reading the texts on microeconomic concepts, information asymmetry in retail markets and/or distributional effects. Finally, I am grateful to two anonymous reviewers for their valuable suggestions. Of course, I remain fully responsible for any remaining error or shortcoming in this book.

Finally, I am very grateful to my wife, Mieke, for accepting my long office days, facilitating me in all respects and her willingness to listen to my stories about the progress. Without her love and care, it would not have been possible to write this book.

Groningen/The Hague, The Netherlands   Machiel Mulder
July 2020

Contents

About the Author

Machiel Mulder (1960) is Professor of Regulation of Energy Markets and director of the Centre for Energy Economics Research at the Faculty of Economics and Business of the University of Groningen in the Netherlands since 2013. He has a broad experience in the field of regulation of energy markets, both as academic and as practitioner. As academic, he has published many articles on energy economics and regulation of energy markets in various scientific journals and he has developed and taught undergraduate, graduate and executive courses on energy policy and regulation of energy markets. Until 2019, he was also president of the Benelux Association for Energy Economics (BAEE). In that position, he organised the 41st International Conference of the International Association for Energy Economics (IAEE) in Groningen in June 2018.

Until 2016, he combined his academic position with the role of Specialist Regulatory Economics at the Netherlands Authority for Consumers and Markets (ACM), who regulates, amongst others, the Dutch energy market. In this position, he was responsible for the economic analysis behind the regulatory decisions regarding the Dutch energy market, while he also participated, among others, in European working groups directed at the regulation of energy markets at EU level. Earlier, he was head of the Economics department of the environmental consultancy CE-Delft, head of the Energy department of the Netherlands Bureau for Economic Policy Analysis (CPB), scientific researcher at the Dutch Agricultural Economics Research Institute (LEI), and scientific member of staff of the Council of Small and Medium-sized firms (RMK). In all these positions, he worked on the bridge between academic research and practical application in policy making, which resulted in a large number of research reports on specific policy issues, such as the design of environmental tax systems, security of energy supply measures, costs and benefits of unbundling of the energy industry, design of fiscal measures for gas exploration, effectiveness of tariff regulation for energy networks, and monitoring market power in electricity markets.

In his current position, his main teaching responsibilities consist of coordinating the multidisciplinary Energy Minor for Bachelor students, the M.Sc. course Economics of Regulating Markets, and several executive Energy courses offered by the

University of Groningen Business School. In addition, he teaches in several other programmes, such as in the European Master in Sustainable Energy System Management of the Hanze University of Applied Sciences.

List of Figures

List of Tables

List of Boxes

Economic Analysis of Energy Markets: An Introduction

1.1 Introduction

This chapter explains why the economic analysis of energy markets is relevant for current societal debates and how the optimal regulation of these markets can be determined. Energy is an essential commodity for economic growth, and, therefore, its availability at efficient prices is crucial for societies. Using energy, however, has significant negative environmental effects, as it is still mainly based on fossil energy. The challenge for policy makers is, hence, to foster an efficient supply of energy, while mitigating negative environmental effects. This chapter introduces the public-interest approach as a tool to find the optimal regulation to realize the societal objectives regarding energy markets. This approach departs from the theoretical benchmark of perfectly functioning markets and analyzes how fundamental shortcomings can be overcome.

1.2 Background

1.2.1 Importance of Energy for Economic Development

Energy is a crucial input for today's societies. The high living standards in many parts of the world have only been possible thanks to the wide availability of energy. Industries need oil and gas for the production of all kind of commodities, transportation of goods and people require energy in the form of motor fuels or electricity while for heating of houses and buildings, energy is needed to produce the heat. Consequently, a strong relationship exists between all human activities and energy use. The past decades have witnessed that economic growth has coincided with a growth in energy consumption, (see Fig. 1.1). Since 2000, global GDP has increased by 64%, while the use of the coal, oil and gas increased by 64%, 42% and 24%, respectively. The global consumption of coal has increased from 3.3 to 5.4

© Springer Nature Switzerland AG 2021
M. Mulder, *Regulation of Energy Markets*, Lecture Notes in Energy 80,
https://doi.org/10.1007/978-3-030-58319-4_1

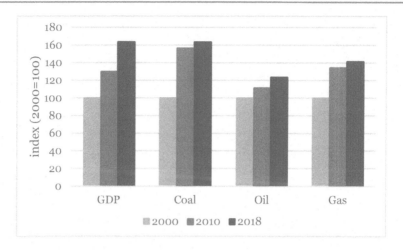

Fig. 1.1 Global GDP and use of coal, oil and natural gas in 2000, 2010 and 2018 (index: 2000 = 100). *Source* GDP: Worldbank; consumption of coal, oil and gas: IEA

billion ton of coal equivalents (Mtce) since 2000, the daily global consumption of oil has increased from 80 million barrels a day to almost 100 in 2019, while the global consumption of natural gas grew from 2500 to 4000 billion cubic metre (bcm).[1]

The strong increase in the use of energy is mostly realized by the increase in the production of fossil energy, but the production of energy from renewable sources, in particular from wind and solar, has grown (in relative terms) even stronger. The global production of electricity by wind turbines and solar panels, for instance, has increased to 1619 TWh in 2017, which was 6% of global electricity production. Because of this strong growth, the share of fossil energy in the total energy mix has decreased to 82% in 2017, which is, however, still a large portion (IEA 2019).

Overall, the growth in energy consumption has been smaller than the growth in income, which implies that the energy intensity, i.e. the energy use per unit of income, has gone down. This reduction in energy intensity results from higher efficiency in the use of energy (e.g. more energy-efficient engines) and changes within the economic structure (i.e. lower shares of energy-intensive industries and higher shares of services industries, which are generally less energy intensive). Despite this strong decline in energy intensity, the average energy consumption per capita has increased over the past decades because of the growth in income per capita. Since the population has also increased, the global total energy use has grown as well.

---

[1]For the definition of units, see Appendix 1.

1.2.2 Societal Consequences of Energy Use

The importance of energy for economic activities implies that the availability as well as the prices of energy carriers are important factors for economic development and social well-being. If firms and people don't have good access to energy, they are hampered in their activities. For instance, in several regions, in particular in Africa and Asia, only a fraction of the population is connected to an electricity grid. In Europe and North–North America, everyone is connected to an electricity grid, but not all of them are always able to pay the bills of energy companies, indicating that access to affordable energy is also an issue in more developed countries.[2]

Because the developed countries are highly dependent on the availability of energy, they are also more vulnerable to interruptions in supply. Geopolitical events occasionally result in negative shocks to oil and gas markets which lead to surging energy prices, while technical incidents in electricity networks may deprive large number of people and business temporarily from access to power (see Bhattacharyya 2019).

In addition, the continuing growth in the consumption of fossil energy results in growing emissions of CO_2 (carbon dioxide), CH_4 (methane) and N_2O (nitrous oxide) which are greenhouse gases and, hence, contribute to climate change. The use of fossil energy contributes to about 75% of the total global emissions of greenhouse gases (see Fig. 1.2). Moreover, the emissions of ultrafine particles harm, in particular, the quality of air and, hence, the environment for life. Although the carbon intensity of energy use has been more or less stable on a global level, the continuing growth in the level of consumption of fossil energy results in an ongoing increase in the level of global carbon emissions. Since 1990, the global emissions have increased by about 50% (IEA 2019).

From this brief overview, it appears that energy is an important contributor to economic development and well-being of human beings, but limited availabilities and high prices of energy can result in social and economic problems, while the growing use of fossil energy has dramatic environmental effects.

1.3 Energy Markets, Climate Policy and Energy Transition Policy

1.3.1 Energy Markets

Because of the societal importance of energy, governments in many countries have formulated policy targets to promote the availability, the affordability and the sustainability of energy. These three objectives together are often called the 'energy trilemma' (see e.g. WEC 2018). Strictly speaking, these three objectives do not

---

[2]Sources: EU Energy Poverty Observatory: https://www.energypoverty.eu/. Worldbank: https://databank.worldbank.org/source/world-development-indicators.

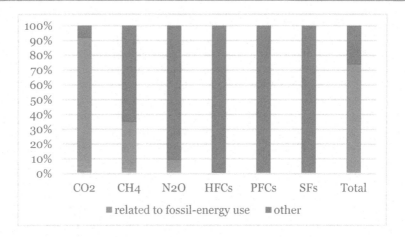

Fig. 1.2 Contribution of fossil energy consumption to global greenhouse gas emissions, 2017 (in %). *Source* IEA (2019), CO_2 Emissions from fuel combustion

refer to a trilemma in which only one of the objectives can be realized as generally all three objectives have to be realized, although governments (societies) make different decisions regarding the relative importance of the separate objectives. In order to realize these objectives, governments implement various types of regulatory measures, such as direct intervention, public ownership and financial support systems. A key condition, however, to realize the three energy objectives is the presence of well-developed energy markets, because well-functioning markets are generally seen as fundamental requirements for the economic development of societies (see e.g. Kay 2004).

Energy markets have evolved strongly over the past decades. The term energy market represents a complex system in which producers, transport operators, traders and consumers interact with each other in order to realize their individual objectives. Markets for coal and oil exist already for many decades, as these markets do not depend on the presence of physical transport infrastructures. Gas and electricity systems, however, were centrally coordinated until the roll-out of competition which started off in many countries about 30 years ago. Investments in and the deployment of traditional power plants, for instance, were initially coordinated at a national level while electricity prices were collectively determined based on the costs of the entire production, transport and distribution systems (EC 2015). Connections with neighbouring countries were created to realize a more efficient deployment of, in particular, coal and hydroelectric power stations in various European countries and to support each other in securing the system frequency (UCTE 2010).

The organization of electricity and gas systems has changed drastically with the introduction of market principles. Central management and coordination have been replaced by decentralized decision-making regarding investments in production facilities, deployment of power plants, exploration of gas fields, and utilization of gas storages. Note that decentralized decision-making must not be confused with the decentralization of energy systems. The latter concept means that energy is more frequently produced within distribution networks, whereas decentralized decision-making implies that decisions on investments, among other things, are no longer coordinated, but are made by a number of decision-making units (e.g. firms, households) independently of each other. Uncoordinated decentralized decision-making is an important precondition for efficient markets, because only then can competition can develop. Therefore, coordination between firms which jointly constitute a significant part of the market is generally not allowed in countries that have liberalized their markets. In this textbook, we will not discuss this general policy to foster competition in an economy, but we focus on sector-specific regulation to reduce bottlenecks for competition where this is possible (such as in the case of international trade in energy) and to introduce regulation where competition cannot exist (such as in the case of the management of gas transport and distribution networks and electricity transmission and distribution networks).

1.3.2   Climate Policy and Energy Transition Policy

The promotion of markets in energy systems over the past decades is accompanied with intensified interventions to make energy systems less dependent on fossil energy in order to realize climate-policy objectives. The general objective of this policy is to reduce greenhouse gas emissions in order to mitigate climate change. As carbon emissions coming from the use of fossil energy are a major contributor to the overall rise in greenhouse gas emissions, as we have seen in Fig. 1.2, one of the key aims of climate policy is to substitute fossil energy consumption by renewable energy consumption.

Many countries and regions have formulated ambitious targets for this substitution. The European Union, for instance, pursues a 20% share of renewable energy in the total energy consumption in 2020 and a 32% share in 2030, while the greenhouse gas emissions have to be reduced by 40% in 2030.[3] At present, the percentage of renewable energy in the EU is approximately 13. As electricity offers greater opportunities for renewable energy than, for example, fuels, these high ambitions imply that, in 2030, the share of renewably generated electricity in the total power consumption in the EU must be more than 50%. This increased portion of renewably generated electricity will have to be realized while total energy consumption will also increase further, among other things because of electrification of transport and house heating.

This switch to an energy system which is based on renewable energy is called energy transition. Energy transition is a more comprehensive concept than just a change in the composition of energy consumption. Energy systems are in a constant state of flux, often as a result of changing relative prices. Energy transition can be

---

[3]Source: https://ec.europa.eu/clima/policies/strategies/2030_en.

defined as a change in the composition of energy consumption which is intentionally pursued by means of all sorts of societal interventions. Energy transition in the framework of the above-mentioned decentralized energy markets means that governments aim to influence the decisions made by the decentralized units (i.e. firms, consumers) in such a way that, at the end of the day, energy produced by renewable means will represent a greater portion of the national energy consumption.

In order to induce market participants to make other decisions than they would otherwise do, governments are taking numerous measures to make renewable energy more appealing or even, in some cases, mandatory. These options vary from direct regulation and information provision to providing financial incentives. Improving incentives for market participants is the most efficient solution. Examples of these measures include subsidies for renewable energy, taxes on the use of fossil energy, mandatory closure of coal-fired power stations, a cap-and-trade system for CO_2 emissions, a system of guarantees of origin for renewably generated power, supplier obligations to offer a fixed percentage of renewable energy and network operator obligations to give priority to the supply of renewable power (regardless of network congestions). Besides that, governments in a number of countries are taking measures in order to raise the flexibility of the power system, such as promoting hydrogen production through electrolysis, in order to be able to deal with increasing amounts of fluctuating supply of renewable energy and increased volatility in electricity consumption due to, among others, growing use of heat pumps for heating houses of residential consumers.

1.3.3 Energy Markets and Climate Policy

From this overview of various types of climate-policy measures directed at the energy sector, it is clear that the political ambition to significantly increase the role of renewable energy within the framework of energy markets brings two policy fields together: developing and promoting competition in energy markets on the one hand, while maintaining the reliability of supply, and realizing climate-policy objectives on the other.

The promotion of market forces in the energy sector is directed at realizing a reliable energy supply at a higher efficiency. The objective of this policy is, therefore, that energy is produced as cost-effectively as possible (which is called productive efficiency), that consumers don't pay a higher price than is necessary (which refers to allocative efficiency) and that innovation is promoted (which is called dynamic efficiency). The goal of climate policy is to reduce greenhouse gas emissions, among other things by realizing a transition in the energy sector. This requires a more intensive intervention of governments in the decisions of energy producers and consumers which could be seen as inconsistent with the concept of liberalized markets based on free decentralized decision-making.

Since energy transition focusses on encouraging businesses and consumers to make alternative choices, i.e. using less energy and more renewable energy, it is in itself at odds with the basic principles of a free market in which all players decide for themselves what type of energy they prefer to use and how much. Furthermore, market participants are encouraged to instal renewable technologies, the characteristics of which are different from those of the conventional technologies, such as increased dependency on weather and fewer possibilities to control production. It is often argued that because of these different characteristics, energy transition should lead to an adjustment of the energy market design (see e.g. Vries and Verzijlbergh 2018).[4] The question is, though, to what extent this energy transition will negatively affect the way in which the energy market operates (see e.g. Baldwin et al. 2012). If it has an impact, one has to analyze what measures will be required to safeguard both the energy transition and the well functioning of markets. These topics will be addressed in this textbook.

1.4 Approaches to Analyze Regulation

1.4.1 Public-Interest Approach

In order to determine the appropriate role of governments in energy markets, we follow the public-interest approach. Based on this approach, we want to analyze how governments can improve welfare by intervening in energy markets. In an ideal society, politicians are fully directed at making decisions which foster the well-being and welfare of society as a whole. This viewpoint on government interventions is called the public-interest approach, which can be seen as a normative theory (Veljanovski 2012). According to this approach, intervention in society is only needed if it raises welfare, which may only happen in case of market failures.

Market failures are fundamental shortcomings in markets which make it impossible for a market to realize the maximum possible level of welfare. An example of market failures is the existence of externalities, which are costs (such as carbon emissions) which are not taken into account by economic agents and, as a result, they produce or consume too much from a social-welfare point of view. The public-interest theory states that governments may implement measures like subsidies or taxes in order to internalize such externalities. Another market failure is, for instance, market power of one or more firms which enable them to charge too high prices, which are prices that are higher than needed to recoup the costs of an efficient firm. The public-interest theory says that in such situations, governments should prevent firms to acquire market power or, otherwise, they should regulate firms with market power by capping the prices at the level of costs of an efficient firm or by imposing regulatory rules on the behaviour of these firms. In Sect. 4.2.5, we discuss the various types of market failures, while in Sect. 3.4.2, we will discuss the various types of government interventions in energy markets.

---

[4]Also in the EU, market rules are changing in order to facilitate the energy transition. See e.g. https://ec.europa.eu/commission/priorities/energy-union-and-climate/fully-integrated-internal-energy-market_en.

To what extent policy measures should be implemented to address market failures also depends on their effectiveness and efficiency. Here, it is important to realize that not only markets can fail, but also governments can fail to implement the most appropriate measures. These costs and risks of regulatory intervention, which can be seen as regulatory failures, are discussed in Sect. 4.2.6.

The choice to depart from the perspective of market failures also implies that policy objectives can be assessed. What this means becomes clear when we look at the policies regarding, for instance, energy transition. Energy transition is, as defined above, the ambition of a society to drastically change the nature of an energy system. Even though there may be many actors in society who share this ambition, this target is generally set by the government and does not necessarily reflect the optimal social outcomes. When policy targets are ambitious, for instance, regarding the share of renewable energy, one may assume that they are likely to be achieved less quickly than intended, but this does not necessarily mean that markets fail. Energy markets only fail when there are flaws which prevent market participants from entering into transactions which would be beneficial to either of them or the society in general. Hence, it is important to realize that the public-interest approach is not taking the objectives of governments for granted, but that it tries to determine the optimal policies from a social-welfare point of view.

1.4.2 Other Approaches

The normative public-interest approach to government intervention may, however, not explain how government policies are actually being determined. In order to find this explanation, a positive approach instead of a normative approach is required. These positive approaches include theories like the capture theory and the economic theory of regulation (Veljanovski 2012; Viscusi et al. 2005). According to the capture theory, government interventions in society result from lobbying behaviour of interest groups. This theory states that, among others, the better a group is organized, the more influence it may have on government decisions. As smaller groups are generally more coherent than larger groups, this theory predicts that government intervention is often biased towards small interest groups. Another positive theory on regulation is the economic theory of regulation which departs from the interest of politicians. This theory states that politicians are permanently searching for political support in order to become re-elected and, therefore, their decisions are meant to maximize this support. One of the predictions of this theory is that regulation may not only be implemented in industries dominated by a few firms in order to protect consumers from too high prices, but also in industries where firms compete heavily with each other resulting in low profits. In the latter case, politicians may protect firms from intensive competition by implementing regulatory measures which, for instance, restrict entry of new firms.

Although the public-interest approach is the default perspective in this book, occasionally we will also pay attention to the actual design of regulation and to what extent this may deviate from the optimal regulation. The focus of the book is, however, on the design of regulation from a social-welfare point of view.

1.5 Organization of the Book

1.5.1 Target Group

The book is in particular meant for final year undergraduate as well as graduate students from various disciplines who want to learn more about the economics behind energy systems and policies as well as the economic analysis of regulation of energy markets. For students having a solid background in economics, the application of economic concepts to the field of energy markets and policy may still be new and relevant. For other students, like those from physics or other social sciences, the economic concepts will be clearly explained in separate sections which will enable them to understand how these concepts can be applied to the regulation of energy markets.

1.5.2 Learning Objectives

After studying the material offered in this textbook, students should be able to apply the concept of market failures in discussing the role of governments in energy markets. In addition, students should understand the key characteristics of the functioning of energy markets, in particular electricity and gas markets. Moreover, students should understand the various ways regulators can intervene in energy markets, in particular how regulators can regulate the tariffs of network operators and how the various types of market failures in the wholesale and retail markets can be addressed.

1.5.3 Outline

Using the public-interest approach, this textbook extensively analyzes the various kinds of market failures in the different components of the energy supply chains. This analysis is supported by providing ample empirical evidence on the actual functioning of these markets as well as actual regulatory measures taken by governments. The textbook also pays attention to the impact of energy transition on the functioning of these markets and discusses the question to what extent the designs of these markets have to be adapted in order to facilitate the energy transition.

This textbook on the regulation of energy markets consists of 12 chapters. Each chapter starts with presenting the outline and ends with providing some exercises to enable readers to apply the lessons learned. After this introductory chapter, on the background and scope of the book, Chap. 2 introduces the basic physical properties of various energy carriers and the various types of activities within energy supply chains, while Chap. 3 introduces the basics of energy markets and energy policies. Chapter 4 introduces the microeconomic perspective and the economic concepts which are needed to think systematically about the optimal regulation of energy markets. This chapter explains how the theoretical concept of well-functioning markets with perfect competition is used as a benchmark for the assessment of actual markets.

In the consecutive chapters, the various market failures in energy markets are discussed and how regulation can be used to overcome these failures. This analysis starts, in Chap. 5, with the problem of information asymmetry in retail energy markets and how this market failure can be addressed. Chapter 6 discusses the regulation of the natural monopolies in transmissions and distribution networks. The semi-public-good character of the infrastructures which affect the reliability of energy supply is the topic of Chap. 7. Chapter 8 discusses externalities related to the supply and use of fossil energy and how these can be addressed by regulators. Chapter 9 explains why market power is a fundamental problem in energy markets and how this can be dealt with, while Chap. 10 analyzes market inefficiencies due to international trade barriers. In Chap. 11, the attention is directed to distributional effects and equity and fairness concerns and the options for regulators to address these issues. Finally, Chap. 12 summarizes the lessons learned.

Exercises

1.1 Explain why in developed countries many people may have problems with the availability of energy while everyone is connected to an electricity or gas network.

1.2 What is meant by a decentralized organization of energy systems?

1.3 What makes that the public-interest approach to regulation is a normative approach?

1.4 Explain the tension between policies to foster energy markets and policies directed at energy transition.

1.5 What is the difference between energy intensity and energy efficiency?

References

Baldwin, R., Cave, M., & Lodge, M. (2012). *Understanding regulation; Theory, strategy and practice* (2nd ed.). Oxford University Press.
Bhattacharyya, S. C. (2019). *Energy economics; Concepts, issues, markets and governance* (2nd ed.). Berlin: Springer.
European Commission (EC). (2015b). Renewable energy progress report. Brussels.
IEA. (2018). Electricity information.

IEA. (2019). World energy outlook.

Kay, J. (2004). *Culture and prosperity; The truth about markets—Why some nations are rich, but most remain poor.* Harper Business.

Schavemaker, P., & van der Sluis, L. (2017). *Electrical power system essentials* (2nd ed.). New York: Wiley.

Smets, A., Jäger, K., Isabella, O., van Swaaij, R., & Zeman, M. (2016). *Solar energy: The physics and engineering of photovoltaic conversion technologies and systems.* UIT Cambridge Ltd.

UCTE. (2010). The 50-year success story; Evolution of a European interconnected grid. Brussels.

Veljanovski, C. (2012). Economic approaches to regulation. In Baldwin, R., Cave, M., & Lodge, M. (Eds.), *The Oxford handbook of regulation.* Oxford University Press.

Viscusi, W. K., Harrington, J. E., & Vernon, J. M. (2005). *Economics of regulation and antitrust.* The MIT Press.

Vries, L. J., & Verzijlbergh, R. A. (2018). How renewable energy is reshaping Europe's electricity market design. *Economics of Energy & Environmental Policy, 7,* 31–50.

World Energy Council (WEC). (2018) World Energy Trilemma Index 2018. London, United Kingdom.

Energy Carriers and Supply Chains

2

2.1 Introduction

Knowledge of the physical characteristics of energy and the various types of activities within energy supply chains is necessary before being able to analyse the optimal regulation of energy markets. Therefore, this chapter introduces the various types of energy carriers, distinguishing primary from secondary carriers and non-renewable from renewable sources (Sect. 2.2). This chapter also discusses the concept of the energy balance, which describes the origins of supply and the destinations of use of energy within an economy. The chapter proceeds by describing the various layers in the supply chains, starting from exploration of energy resources, production, conversion, storage, transport until consumption of energy (Sect. 2.3). This chapter concludes by discussing how these energy supply chains of activities can be organized (Sect. 2.4).

2.2 Sources and Use of Energy

2.2.1 Primary Sources and Conversion to Final Energy

In physical sense, energy refers to the potential to produce work.[1] Work refers to the force causing a displacement of an object. This force is measured in Joule, which is the amount of energy transferred to an object which is displaced. When we speak about energy, we generally do not refer to this work, but to the carriers of this energy, i.e. the energy commodities. In physical terms, the energy content of these

---

[1] See https://www.physicsclassroom.com/Class/energy/u5l1e.cfm.

© Springer Nature Switzerland AG 2021
M. Mulder, *Regulation of Energy Markets*, Lecture Notes in Energy 80,
https://doi.org/10.1007/978-3-030-58319-4_2

physical commodities can be expressed as Joule per unit of the carrier (e.g. kg and m³). For instance, 1 tonne of crude oil is set equal to 42 GJ and 1 m³ of natural gas is set equal to 35.17 MJ.[2]

Some commodities include the energy themselves, and others don't but can be used to carry energy (Fig. 2.1). The first type of commodities is called primary energy carriers or energy sources. The conventional primary energy carriers include coal, oil and gas. Besides these fossil energy carriers, renewable energy sources exist like wind speed and sunshine. The total supply of primary energy carriers in an economy is called primary energy supply.

An example of a commodity that can be used to carry energy is water. By adding energy to water molecules, they carry (more) energy: this energy can be heat. Energy can also be added to electromagnetic fields which results in the commodity electricity. Because the energy content is added to these commodities, electricity and heat are called secondary energy carriers. Other secondary energy carriers are, for instance, gasoline, which is based on oil as primary energy source, and hydrogen, which can be produced through processes like Steam Methane Reforming (SMR), based on natural gas and steam, and electrolysis by which water is split into hydrogen and oxygen by means of electricity.

The process of adding energy to other commodities is called conversion. Since the process of conversion requires work, this process uses also energy itself. This use of energy during the conversion process is called energy loss, as this energy is converted into another form of energy which is not used (generally in the form of heat).[3] The lower the energy loss, the higher the efficiency of the conversion. Hence, the efficiency of conversion (η) is the ratio between the energy content of the useful output (E^o) and the energy content of the input (E^i):

$$\eta = \frac{E^o}{E^i} \tag{2.1}$$

See Table 2.1 for the efficiencies of various types of conversion. These efficiencies play a crucial role in the competitive position of the various technologies in energy markets.[4] The efficiency of conversion affects the marginal costs and, hence, the position of technologies in the merit order, as will be discussed in Sect. 3.3.4. These marginal costs (mc) are related to the unit price of the primary energy (p_f) and

---

[2]GJ is 10^9 J and MJ is 10^6 J. See Appendix 1 for the definition and units of energy. As crude oil and natural gas are natural products, the precise contents and energy value vary across types of oil and gas. The energy contents mentioned here are defined by convention. See for further conversions: https://www.iea.org/statistics/resources/unitconverter/.
[3]Note that physically, energy always remains in one or the other form.
[4]As efficiency is defined as the ratio between the energy content of the useful output versus the energy content of the input, it is key to define what is defined as useful. In this respect, one has to define how the energy content of residuals is treated. In the case of heating, the distinction between Lower Heating Value (LHV) and Higher Heating Value (HHV) is relevant. LHV refers to the energy value of the useful output not taken into account the energy content of the water vapour and other gases like carbon dioxide (CO_2) resulting from the conversion process, while HHV includes this energy content as well.

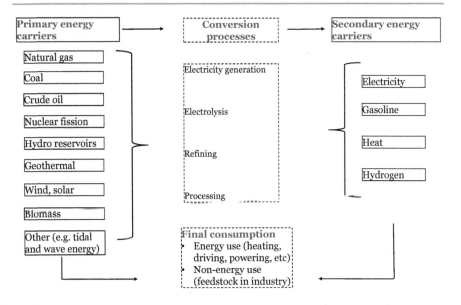

Fig. 2.1 Main primary and secondary energy carriers, conversion processes and final consumption

the conversion efficiency. Ignoring other variable costs of production, the marginal costs of a power plant i can be determined by the ratio between these two variables:

$$mc_i = \frac{p_f}{\eta_i} \tag{2.2}$$

In some cases, more than one type of output is utilized. This occurs, for instance, when the heat at the outlet of the steam turbine of a coal-fired power plant is used as an input in a district heating system. As a result, the marginal revenues of this heat production can be subtracted from Eq. (2.2) to get the actual marginal costs of producing electricity. The typical example of this joint production of two outputs is the Combined Heat Power (CHP) plant, which are used to produce both electricity and heat. Because of this joint production, the energetic efficiency can be significantly higher than of other types of electricity generation. Such a joint production may also explain why in some circumstances such plants keep producing electricity when the electricity price is low or even negative. The economic reason for this behaviour is that the marginal revenues of heat production may be sufficient to compensate for the losses on electricity production.

Table 2.1 Efficiencies of various types of energy conversion (global estimates)[a]

Conversion output	Technology	Type of conversion	Efficiency (%)
Electricity generation	Coal-fired steam turbine	Thermal to electrical	35–46
	Gas turbine	Thermal to electrical	28–42
	CCCG plant (combined cycle of gas and steam to drive the turbine)	Thermal to electrical	60
	Gas engine (internal combustion engine)	Thermal to electrical	40–50
	Hydropower plant	Gravitational to electrical	80–90
	Nuclear	Nuclear fission to electrical	90
	Wind turbine	Kinetic to electrical	40–50
	Photovoltaic (PV)	Solar radiation to electrical	10–15
	Fuel cells	Chemical to electrical	30–50
Hydrogen production	Electrolysis	Electrical to chemical	50–70
	Steam Methane Reforming (SMR)[b]	Thermal to thermal	<90
Heat production	Combined heat-power plant (CHP) (producing both heat and power)	Thermal to thermal + electrical	60–90
	Heat pump[c] (using electricity to drive a pump that compresses heat from an external source such as air, ground water)	Electricity to thermal	300–500
	Solar water heating	Solar radiation to thermal	30–60

Sources Zweifel et al. (2017), Schavemaker and Sluis (2017), Smets et al. (2016)

[a]The estimates refer to the current situation, based on various sources. Efficiencies may improve over time due to technological developments

[b]The conversion efficiency of SMR can be defined as the ratio of the energy content of the hydrogen and the energy content of the natural gas as well as the combustion fuel used to produce the steam (Peng 2012)

[c]The efficiency of the heat pump refers to the ratio of the energy in the electricity that is used to drive the pump and the energy in the generated heat. This is not the real energetic efficiency as the energy in the external heat source is not included. Therefore, this efficiency is named the Coefficient of Performance (CoP)

2.2.2   Energy Balance

The supply, conversion and consumption of energy in an economy is statistically measured through so-called energy balances.[5] These balances measure the flows of energy from production and conversion to final consumption.

The total energy consumption in an economy is called the primary energy consumption. This refers to the use of energy in conversion processes plus the final energy consumption. The final energy consumption consists of consumption of primary energy carriers, such as natural gas that is used by residential households to heat their houses, as well as consumption of secondary energy carriers, such as households that use gasoline to drive their cars. Not all final energy consumption is directed at the use of energy, such as heating or powering electric appliances. A part of this the final consumption consists of so-called non-energy use of energy commodities, like the use of natural gas as feedstock in the chemical industry to make fertilizers or other commodities like plastics.

The total amount of energy available for final consumption (FEC) is determined by the total amount of primary energy supplied to an economy (PES) minus the use of primary energy in the conversion processes (PEC) plus the output of secondary energy carriers after the conversion processes (SE) and minus the energy used by the energy sector (EE):

$$FEC = PES - PEC + SE - EE \tag{2.3}$$

In theory, the actual final consumption should be equal to the available amount of energy, but due to statistical measurement issues there may be a small difference. Figure 2.2 shows the aggregated energy balances of the EU 28 in 1990, 2000 and 2018. This figure shows that the total primary energy available in this group of countries was about 1.6 million Mtoe in each of these years.[6] The primary energy supply is mainly used in conversion processes, as more than 90% of the primary energy is converted into a secondary energy carrier before it is used for final consumption. The overall efficiency of the conversion processes is about 75%, which can be determined as the ratio between the output of the conversion processes and the input of the conversion (i.e. ratio of bars 3 and 2 in this figure). About 90% of the energy that is available for consumption is used for energy use, while less than 10% is used for non-energy purposes.

Each of the components in the above balance results from a number of activities. For instance, the total supply of primary energy (PES) results from the domestic primary energy production (PEP) plus energy recovered from, for instance, recycled products (PER) plus imports of primary energy (PEI) minus exports of primary energy (PEO) plus changes in stocks of primary energy (PEST) plus changes in international maritime bunkers (PEM) minus primary used by international aviation (PEA):

---

[5]Eurostat, Energy Balance Guide, 31 January 2019.
[6]Mtoe stands for million tons of oil equivalents, which is the standardized heating value of one tonne of crude oil. See Appendix 1.

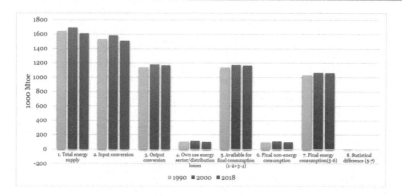

Fig. 2.2 Energy balance of EU 28 in 1990, 2010 and 2018 ($\times 1000$ Mtoe). *Source* Eurostat, EU 28 Energy Balances, 2020 edition

$$PES = PEP + PER + PEI - PEO + PEST + PEM - PEA \qquad (2.4)$$

Figure 2.3 shows that import of primary energy is a major source of the total energy supply in EU 28, as it is about twice as large as the production of primary energy in this region. The figure also shows that a major part of the import of primary energy is used for re-export, which indicates the importance of international trade in energy systems. The imports have increased strongly since 1990, to compensate for both the decreasing production within the EU 28 and the increasing exports. The net supply of energy to the EU 28 has remained fairly stable.

With the EU 28, the domestic supply of primary energy is for almost three quarter based on fossil energy (see Fig. 2.4). Oil accounts for about 1/3 of total supply of primary to the EU 28, while natural gas has a share in the supply of primary energy of about 1/4. Oil products, natural gas and electricity are the major components of the final energy consumption, while the share of heat in the final consumption is about 5% (see Fig. 2.5).

2.3   Energy Carriers

2.3.1   Coal

Coal is a solid fossil fuel which results from geological pressure on plant biomass over time. As the intensity and the time period of this pressure vary across resources, coal exists in various so-called ranks. These ranks are divided in four groups: Anthracite, Bituminous, Sub-bituminous and Lignite. Anthracite is called 'hard coal' as it has a high carbon content and, consequently, also a high heat (i.e. energy) content (Table 2.2). At the other end of the spectrum is Lignite which is a very soft coal with a relatively low heat content while the emissions are much

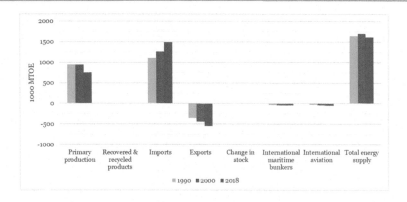

Fig. 2.3 Composition of Total Energy Supply of EU 28 in 1990, 2010 and 2018 (×1000 Mtoe). *Source* Eurostat, EU 28 Energy Balances, 2020 edition

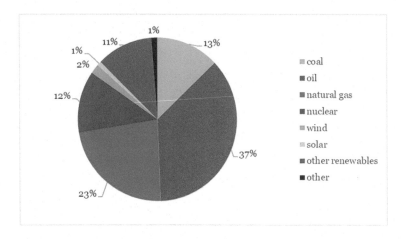

Fig. 2.4 Composition of primary energy supply by energy carrier of EU 28 in 2018. *Source* Eurostat, EU 28 Energy Balances, 2020 edition

higher than for the other types of coal. Lignite is also known as brown coal, while Bituminous is also known as black coal and Sub-bituminous as dark brown coal. Anthracite is mainly used for making coke which is used in smelting iron ore to make steel.[7] Bituminous coal can be both used for making coke and as fuel in electricity generation, while Lignite is only used for electricity generation.[8]

---

[7] When coal is used for heating purposes in the electricity or industry, it is called steaming or thermal coal, while coal that is used in the steel industry is called metallurgical or coking coal.
[8] The share of lignite in the electricity generation by steaming coal in the OECD is about 15% (IEA 2019).

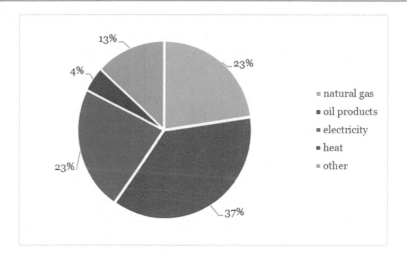

Fig. 2.5 Composition of final energy consumption by energy carrier of EU 28 in 2018. *Source* Eurostat, EU 28 Energy Balances, 2020 edition

Table 2.2 Physical characteristics of various types of coal

Physical characteristics	Ranks of coal			
	Antracite ("hard coal")	Bituminous ("black coal")	Sub-bituminous ("dark brown coal")	Lignite ("brown coal")
Carbon content (%)	85–98	45–85	35–45	25–35
Ash content (%)	10–20	3–12	≤ 10	10–50
Moisture (%)	<15	2–15	10–45	30–60
Heat content (Btu/lb)	13,000–15,000	11,000–15,000	8500–13,000	4000–8300

Source Bowen and Irwin (2008)

2.3.2 Oil

Crude oil is also a fossil natural product, which implies that its precise characteristics vary across production fields. Crude oil is a complex mixture of paraffins (long chain alkanes) and naphthene (aromatic compounds). The key characteristics that affect the quality for users of crude oil are the density and the sulphur content. The density refers to the mass per unit of volume. Based on this aspect, crude oil is characterized from light to heavy. Light oil can be more easily used to produce fuel oil, which makes that it is generally a bit higher priced than heavy oil. The sulphur content affects the smell of oil, and that's why crude oil with a low share of sulphur (<0.5%) is called sweet and oil with higher shares of sulphur is called sour. Sweet oil contains a higher proportion of elements which can be processed into gasoline and other petroleum products, which makes sweet oil a bit higher priced than sour oil. Hence, light sweet oil has the highest quality from the perspective of users.

Oil from the Middle-East region generally is more sour and heavier than oil from the North Sea region and from the USA (Fig. 2.6). The latter have about the same sulphur content (less than 0.5%), but oil from the North Sea is a bit heavier than the US WTI oil.[9] In general, the North Sea Brent and the US WTI oil belong to the group of light sweet oil.

Oil can also be distinguished in conventional and unconventional. Conventional oil is located in permeable reservoirs where they cannot leave because of a layer of impermeable rock above. Unconventional oil is not located in these permeable reservoirs but is captured in tight rocks. These tight rocks have to be fractured in a process called fracking before the oil can be mined. Hence, the difference in geological characteristics has consequences for the way of extracting the oil, the production costs as well as the emissions resulting from the production processes.

2.3.3 Natural Gas, Green Gases and Hydrogen

In the past, gas mainly referred to *natural* gas, which is a primary energy source and which is produced by depleting fossil natural resources. As natural gas is a natural product, its physical characteristics vary from field to field. Different gas fields contain different types of natural gas that have different specifications and composures. The main component is methane (CH_4) and other components are higher alkanes (so-called hydrocarbon-hydro molecules), carbon dioxide (CO_2), nitrogen (N_2), hydrogen sulphide (H_2S) and helium (He). The thermic value is determined by the content of methane. This thermic value of fuel gasses can be measured through the Wobbe index (W_i), which is the ratio between the higher heating value (V_c) and the specific gravity (G_s), which is a measure for the density of the gas:

$$W_i = \frac{V_c}{\sqrt{G_s}} \qquad (2.5)$$

As the higher heating value of natural gas is around 39 MJ/Nm^3 and the specific gravity is approximately 0.59,[10] the Wobbe index of natural gas is around 51 MJ/Nm^3 which is equal to 0.01 MWh/Nm^3.[11]

Based on this index, natural gas can be distinguished in various types of gas such as low calorific or high calorific. Gas with a low-calorific value contains a higher

---

[9]WTI stands for West Texas Intermediate and refers to oil from Texas and Southern Oklahoma. Brent refers to the oil from about 15 different fields in the North Sea.

[10]The molecular mass of natural gas is 17 gr/mol, while the molecular mass of dry air is 29 g/mol, which gives the gravity of natural gas of 17/29 = 0.59. The unit 'mol' is a constant and equal to 6,022 14 × 10^{23} and refers to the number of particles (such as molecules and atoms) in one mole of a chemical compound in grams (e.g. one mol of water is 18 gr.).

[11]Nm^3 refers to m^3 in normal (standardized) circumstances.

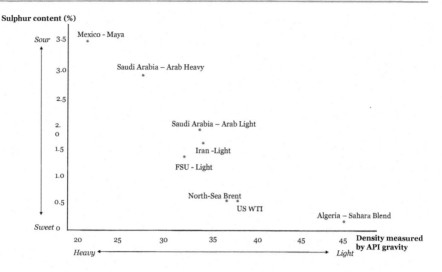

Fig. 2.6 Quality of various types of crude oil measured by density and sulphur content. *Note* FSU stands for Former Soviet Union, including the countries: Armenia, Azerbaijan, Belarus, Estonia, Georgia, Kazakhstan, Kyrgyzstan, Latvia, Lithuania, Moldova, Russia, Tajikistan, Turkmenistan, Ukraine and Uzbekistan. *Source* U.S. Energy Information Administration (see https://www.eia. gov/todayinenergy/detail.php?id=7110)

percentage nitrogen and lower percentage of methane than high-calorific gas. Therefore, the amount of thermal energy stored in a unit of low-calorific gas is lower than in the same unit of high-calorific gas.[12] The latter type of gas can be converted into low-calorific gas by either blending it with gas with a lower calorific value or by adding nitrogen to it. Through this quality conversion, it is possible to use gas from different qualities in a gas system and that users in a gas system use different gas qualities. Some countries, like the Netherlands, have formulated nation-wide quality standards for the gas that is transported and distributed (van der Wal 2003).

Besides through depletion of natural resources, gas can also be produced in a number of other ways. Through anaerobic digestion of wet organic materials (such as animal manure and food waste), biogas can be produced. Depending on the source, biogas contains various types of components. By removing these components, the biogas can be upgraded to a gas which mainly consists of CH_4. This is generally called biomethane. Another way to make biomethane is through gasification of dry organic materials (such as forest residues and residual waste). In this process, the upgrading to CH_4 is included. Biomethane can also be produced on the basis of hydrogen (H_2). If the H_2 is combined with CO_2 in a chemical methanation

---

[12]The Groningen field has a higher nitrogen (14.2%) content compared to other European gas sources, such as Russian or Norwegian gas (ca. 2%). The natural gas that is produced from the Groningen gas field is therefore qualified as L-gas, referring to low-calorific gas. The Wobbe index of this gas is about 44.

process, CH_4 can be made.[13] In Sect. 5.2, we discuss how energy consumers can be informed about the various types of gas qualities.

Hydrogen does not exist in pure form in nature, but can be produced in various ways. Currently, the most commonly used method to make hydrogen is Steam Methane Reforming (SMR). By letting steam (H_2O) under high temperature react with methane included in natural gas (CH_4), hydrogen (H_2) can be produced next to carbon monoxide (CO) and carbon dioxide (CO_2).[14] When the CO_2 is captured and stored, for instance in depleted gas fields, this hydrogen is called blue hydrogen. Another method is electrolysis. By passing electricity through water, the H_2O molecules are separated into hydrogen (H_2) and oxygen (O_2).[15] If this hydrogen is produced through electrolysis based on renewable electricity, this hydrogen is called green hydrogen. Hydrogen is an energy carrier as by combining it with oxygen again, for instance in a fuel cell, energy is released in the chemical process of making water.[16]

2.3.4 Electricity

Electricity is a secondary energy carrier that is produced by converting the energy from primary energy sources. Electricity is produced by changing a magnetic field inducing a force on a charged particle, like an electron in a conductor, or by moving such conductor in a stable magnetic field. The movement of the magnetic field, normally a rotating magnetic field, is created by the use of primary energy sources. Electricity generators, such as steam, gas, or wind turbines, produce this rotation, and the resulting mechanical energy is converted into electrical energy.[17] Other types of generators, like solar photovoltaic cells and fuel cells, convert other types of energy into electrical energy (sunlight and chemical processes, respectively).

The capacity to produce labour is called power (P), which is energy (E) per unit of time (T).[18] The capacity to produce labour by an electricity system (i.e. the

---

[13]The chemical formula is: $CO_2 + 4H_2 \rightarrow CH_4 + 2H_2O$. This process is the so-called Sabatier reaction.

[14]The chemical process of Steam Methane Reforming is: $CH_4 + H_2O \rightarrow CO + 3H_2$. Carbon dioxide ($CO_2$) is produced when the carbon monoxide (CO) reacts in an additional water–gas shift reaction: $CO + H_2O \rightarrow CO_2 + H_2$.

[15]The chemical process of electrolysis is: $2H_2O \rightarrow 2H_2 + O_2$.

[16]This chemical process is $2H_2 + O_2 \rightarrow 2H_2O +$ energy. In this way, fuel cells can be used to produce electricity.

[17]The conversion of these primary energy sources into electricity occurs in a number of steps (Schavemaker and Sluis 2017). First, the chemical energy of the primary sources are transformed into thermal energy through combustion (in the case of coal and gas) or fission (in the case of nuclear). This thermal energy is used to produce high-temperature steam under a high pressure. When this steam is released in the turbine, it is transferred into mechanical energy and when this energy passes the turbine blades the turbines is set in motion. This motion in turn is converted into electrical energy.

[18]This potential to produce labour (i.e. the power) is expressed in Watt, while energy is expressed in Joule. By definition, 1 W of power is equal to 1 J of energy per second.

power) depends on the voltage (V) and the current (I).[19] Hence, the following relationship applies:

$$P = \frac{E}{T} = V * I \qquad (2.6)$$

The voltage (which is measured in Volt) refers to the difference in electric potential between two points. Voltage can be seen as the measure for the pressure on the electrons to flow. The current (which is measured in Ampère) refers to the number of electrons which are flowing through a conductor per second. From this equation follows that the higher the difference in electric potential (i.e. higher voltage) and the higher the number of electrons that can flow per second (i.e. higher current), the higher the capacity (i.e. more power) to produce labour can be produced (i.e. more energy).[20]

The current through a conductor between two points is directly proportional to the voltage across the two points. This relationship is called Ohm's law, with R being the resistance of the conductor:

$$I = \frac{V}{R} \qquad (2.7)$$

Hence, the lower the resistance of a conductor, the more current is created by a specific difference in electric potential (i.e. voltage) and the other way around. These relationships hold for both direct current (DC) and alternating current (AC). Alternating current means that the current and voltage periodically reverse direction, in contrast to direct current which flows in one direction. As a result, power (the product of voltage and current) is also varying over time. If voltage and current are always reversing direction at the same time, the product (thus power) is continuously positive or negative. Otherwise, part of the power is also periodically reversing direction, with a zero net supply of energy. This power is called reactive power. The ratio between active power (which actually generates labour) and the apparent power, which is the sum of active and reactive power, is expressed through the power factor (or cosinus phi, cos φ):

$$\cos \varphi = \frac{active\, power}{active + reactive\, power} \qquad (2.8)$$

This power factor is a dimension less ratio measuring the efficiency of transforming power into work. This efficiency is needed to calculate the power which occurs in case of AC:

---

[19]See for more details on the physics of electricity: https://www.physicsclassroom.com/class/circuits/Lesson-1/Electric-Field-and-the-Movement-of-Charge.

[20]The ability of electricity plants to produce electricity (i.e. their power) is usually expressed in MW, which is equal to 10^6 J/s or 1 MJ/s, while the costs and prices of electricity are expressed per MWh, which is equal to 1 MJ × 3600 s, which is 3600 MJ or 3.6 GJ.

$$P = V * I * \cos \varphi \tag{2.9}$$

The power factor is 1 when there is only active power. Usual values for electroengines and turbines are in the range of 0.85–0.9, which means that the actual labour produced is about 15–10% smaller than what could be produced based on the voltage and Ampère. This also implies that more power has to be generated and transported when appliances have a lower power factor.

2.3.5 Heat

Heat is also an energy carrier which results from conversion processes. Thermodynamic laws determine the potential of these processes. The 1st law of thermodynamics is the law of conservation of energy and says that the total energy of an isolated system always remains the same. The change in the energy content (U) of a system depends on the heat added to it (Q) minus the work of the system on its environment (W):

$$dU = dQ - W \tag{2.10}$$

The 2nd law of thermodynamics says that unperturbed natural processes can only go in one direction: in the case of heat, the energy can only flow from hotter to colder objects. This law can also be formulated as follows: it is impossible to convert heat completely into work as there is always need of a reservoir of lower temperature that absorbs a certain amount of that heat (Schavemaker and Sluis 2017). Using these laws, the efficiency of a machine (η) can be defined in terms of the starting temperature (Qh) and the temperature after the process (Ql):

$$\eta = \frac{W}{Q_h} = \frac{Q_h - Q_l}{Q_h} = 1 - \frac{Q_l}{Qh} \tag{2.11}$$

In this equation, which is called the Carnot efficiency, we see that the work done by a system is equal to the difference in the heat between start and end of the process. The lower the remaining temperature, the higher the efficiency as more energy has been used to produce work.

From this follows that external work has to be added to a system when low-temperature heat has to be transformed into high-temperature heat. This law is applied in, for instance, heat pumps. In these pumps, work is added by using electricity to compress a volatile cooling liquid (such as HFO or propane) and by doing so, heat is extracted from external sources like ground water or air. The resulting higher temperature (Q_h) is the sum of the initial lower temperature (Ql) plus the work added:

$$Q_h = Q_l + W \tag{2.12}$$

The performance of heat pumps is expressed through the Coefficient of Performance (COP) which measures the ratio between the temperature that is generated and the amount of work conducted:

$$COP = \frac{Q_h}{W} \tag{2.13}$$

The COP is inversely related to the efficiency of the heat pump, as the efficiency is equal to the ratio of work and the high temperature (see Eq. 2.11). This implies that the COP of a heat pump is higher when the efficiency is smaller, and this is the case when the difference between the high and the low temperature is small. Hence, heat pumps have a low COP when they have to generate heat from a much colder source, e.g. surface water during a cold winter period. In such circumstances, they require more work, i.e. more electricity, to make heat. When the temperature of the heat source is more modest, heat pumps can perform better. This is why geothermal energy, which is transferred through heat pumps, can be an economically more efficient heat source as the COP is relatively high compared to the COP of heat pumps using air (Frauenhofer 2011).

Another option to generate heat is to make use of residual heat of other processes, such as in the chemical industry or in electricity generation. Heat production can also be combined with electricity production in so-called Combined Heat Power (CHP) plants which typically use gas as an input and are meant to produce both electricity and heat. Because of this joint production, the efficiency of this conversion process is much higher than in other types of power production (see Table 2.1).

2.4 Activities in Energy Supply Chains

2.4.1 Overview

The above description of the basic physical properties of the various energy carriers is needed to understand the economics of energy systems. The first step in the economic analysis is to analyse the supply chains. The production of energy is not an objective in itself, but it is meant to satisfy the demand of users. In order to get the energy from a primary energy source to users, various types of activities are required. This sequence of activities is called the energy supply chain. The activities in this chain can be distinguished in exploration, production, conversion, transport, storage and consumption. Exploration is particularly relevant for fossil energy (coal, oil and gas), as these energy resources have to be discovered in the earth's crust. For some non-renewable sources, however, exploration is also relevant, such as geothermal heat and the uranium needed for nuclear electricity generation.

For the various types of energy carriers separate supply chain exists, but these chains are closely connected to each other (Fig. 2.7). This connection between the

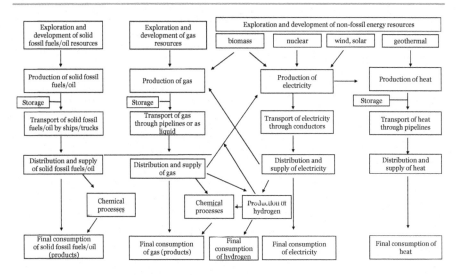

Fig. 2.7 Energy supply chains and sector coupling

supply chains is an example of sector coupling. Coal, oil and gas, for instance, are used in the electricity supply chain as input for generation, while electricity is also used to make a gas (hydrogen) through electrolysis. Moreover, when electricity is generated also heat is produced which can be used in the heat supply chain. Another example of sector coupling is provided by the supply of hydrogen, as this energy carrier needs to be produced, either on the basis of natural gas or the basis of electricity (see Sect. 2.3.3). Hydrogen can be used to generate electricity, as feedstock in the industry and as fuel by energy users, which shows that this energy carrier can play an essential role in the coupling of energy sectors. These relationships between energy sectors imply that developments in one supply chain, such as rising costs or higher demand, affect the economic conditions of other supply chains.

2.4.2 Exploration, Development and Production

In the supply chains of solid fossil fuels, oil and natural gas, the first activity is the exploration for new resources. Exploration starts with geological research, i.e. formulating and testing expectations where and how much primary energy resources may be present based on geological knowledge. Exploration activities like drilling in the underground can be seen as the empirical tests of geological hypotheses regarding the location and magnitude of prospective energy resources.

When exploration activities have provided more certainty about the contents of resources, they may be turned into reserves. The classification of resources depending on the geological certainty and economic feasibility can be presented

Table 2.3 McKelvey diagram: classification of resources on the basis of geological certainty and economic feasibility

		Geological certainty			
		High			Low
Economic feasibility	High	Proven reserves	Probable reserves	Possible reserves	Prospective resources
	Low	Conditional resources			

through a McKelvey diagram (see Table 2.3). Reserves can be distinguished in proven, probable and possible reserves. Proven reserves are resources of which the magnitude has been estimated with a high certainty (at least 90%) and which can be produced economically at current market and technological conditions. Probable reserves are reserves with a lower certainty level regarding the magnitude (50% confidence level), while for possible reserves the confidence level is 10% (Zweifel et al. 2017). When information on the magnitude of resources is fairly certain, but the resources cannot (yet) be economically produced, they are called conditional resources.

It is important to realize that the volume of geological resources is an exogenously given but not perfectly known variable, determined by the geology of the earth, but the volume of proven reserves is an endogenous variable. The level of the proven reserves results, after all, from the exploration activities of mining firms. The size of these reserves can be expressed in relation to the size of total production, which results in the so-called Reserve-to-Production (R/P) ratio.

$$R/P = \frac{Proven\,reserves}{Annual\,production} \tag{2.14}$$

The R/P ratio, however, does not say anything about the availability of resources in the long-term future, as it only refers to the proven reserves at a given moment of time. For both oil and gas, this ratio has been about 50 since many years on a global scale, while for coal the R/P ratio is about 140 (BP 2019). Across regions, however, the R/P ratio is quite different (see Fig. 2.8).

Exploration can be seen as an investment where upfront costs are made in the hope to realize revenues in (long-term) future. These future revenues depend on the expected scarcity in the market, which is related to the current level of reserves and expectations regarding future demand. Hence, exploration activities are encouraged when the R/P ratio becomes smaller.

When new resources have been discovered with a high certainty and which are economically exploitable, the resources can be developed. This means that an infrastructure has to be installed in order to be able to produce. When all installations have been installed, production can start. The maximum pace of the production depends on the capacity which has been installed. The optimal pace and, hence, optimal production capacity, depends on a number of factors: the costs of the installations, the expected future prices and the discount rate.

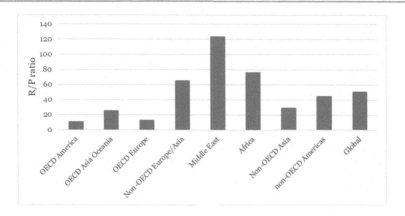

Fig. 2.8 Reserve/Production (R/P) ratio for natural gas, per region, 2017. *Source* IEA, Natural Gas Information 2019

Basically, the trade-off a firm considers is the following: postponing the production of the resource is costly because of the delay in the revenues coming from this production (measured through the discount rate), but advancing production is also costly because of the extra production and transport capacity which is required, while producing also has as opportunity costs that the produced products (oil, gas) cannot be sold in the future anymore.

These opportunity costs also depend on the expected future prices. Firms will increase their current production if they expect that the rate of increase of the energy prices will be below the discount rate, as selling the commodity and storing the financial revenues on a bank account is more profitable than waiting for higher revenues (and vice versa). This optimizing behaviour may affect the effectiveness of policies to reduce the demand for fossil energy (see Box 2.1).

This process of producers adapting the volume of their production to the expected changes in future prices makes that the changes of the prices (p) from year (t) to year become equal to the annual discount rate (r). This is the so-called Hotelling rule:

$$\frac{p_t - p_{t-1}}{p_{t-1}} = r \tag{2.15}$$

This rule only holds in case of a limited amount of resources; i.e., the total volume of resources is binding. This rule says that the price of the product will increase over time even if the marginal costs of production remains the same. The difference between the market price and the marginal costs is called resource rents, which are the extra profits the firms make because of the limited amount of available resources. In Sect. 11.3, we will discuss how societies can redistribute the resource rents through taxation.

Box 2.1 Green paradox

According to the Hotelling theory, producers of non-renewable resources like oil and natural gas optimize the timing of the depletion of these resources. As a result of this optimizing behaviour, in theory, the prices of these resources increase with the discount rate. The optimal depletion paths are, however, affected by climate policies which aim at reducing the demand for fossil energy. Because of the reduction in demand for fossil energy, the price of this energy reduces (see Fig. 2.9, left-hand side; in Sect. 4.2, we will explain how to read this graphs of demand and supply curves). When climate policies become more effective over time, this price reducing effect will become stronger as well. As a result, producers may expect that the future price path may be lower. While in the situation with climate policy, the prices increase with the discount rate, these futures prices will be lower due to the effect of lower demand for fossil energy (see Fig. 2.9, right-hand side). Because the future prices are more strongly affected than the prices in the short term, the optimal depletion path of the producers changes. As future revenues of production become less valuable, producers have incentives to bring extraction forward. This response will reduce the prices in the short term, which makes the climate policy less effective. This effect is called the green paradox, as the aim of the climate policy is to reduce the use of fossil energy, but the effect may be that the prices of fossil energy go down which stimulates its use (see, e.g., Sinn 2015).

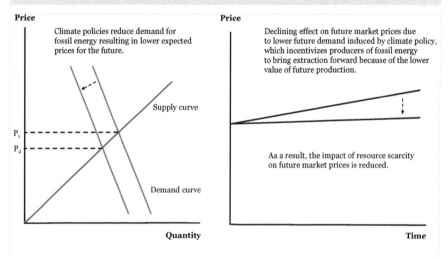

Fig. 2.9 Effects of climate policy on prices of fossil energy and responses producers

2.4.3 Conversion

As we have seen in the discussion of energy balances, a large part of the primary energy supply is converted into another energy carrier. This conversion process is done through many different technologies. In this book, we do hardly go into the chemical industry, except for the conversion of oil into gasoline and remain focussed on the energy sector. Regarding the conversion activities, we pay in particular attention to the electricity generation and to a lesser extent to the production of heat, renewables gases, like biomethane and hydrogen.

Electricity is produced by various types of plants, varying from conventional large-scale fossil-fuel based or nuclear power plants to small-scale renewable plants based on wind or solar. Power which is generated using energy from the wind or the sun differs in a number of ways from conventionally generated power.[21] Its supply is weather dependent, which implies that the maximum production does not only depend on the installed capacity, but also on external (weather) conditions. As a result, the actual production of these technologies is often significantly below the technical maximum production. This relationship is expressed through the so-called capacity factor (cf), which is defined as the ratio between the gross electricity production measured in GWh in a year (G^{prod}) divided by the net capacity measured in GW (G^{cap}) times the number of hours in a year (=365 days * 24 h/day):

$$cf = \frac{G^{prod}}{G^{cap} * 365 * 24} \qquad (2.16)$$

On average, the capacity factor of solar PV in OECD Europe in 2017 was 12% and for wind turbines the capacity factor was 25% (Fig. 2.10). For geothermal and nuclear, the capacity factor was about 80%, while for power plants using combustible fuels (as coal, gas), the average capacity factor was 42%. Although these technologies do not depend on weather circumstances, they do not produce on a constantly high level because of the volatility in demand and the resulting volatility in market prices. The dispatch of coal and in particular gas-fired power plants fluctuate with changes in demand. Note, however, that also renewable energy sources, such as wind turbines, can respond to market prices, for instance by switching off in case of negative prices In Sect. 7.5, we discuss how the volatility in electricity demand affects the investments in various types of electricity plants.

The capacity factors of wind turbines and solar PV vary strongly across countries because of differences in climatological circumstances (Fig. 2.11). Countries with relatively high capacity factors for both wind turbines and solar PV are Chile, Portugal and New Zealand, while in countries like Germany and Switzerland, the capacity factors of both renewable electricity technologies are below the average of OECD Europe. This implies that in the latter countries, investments in wind

---

[21]Renewable power is also generated from hydroelectric power, biomass, geothermal and tidal energy. The current energy transition, though, focuses on electricity generated from wind and solar energy in particular.

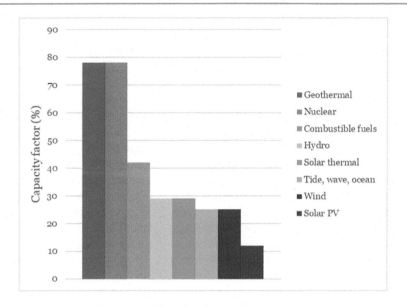

Fig. 2.10 Capacity factors of various renewable energy technologies, in OECD Europe in 2017. *Source* IEA, Electricity Information 2019

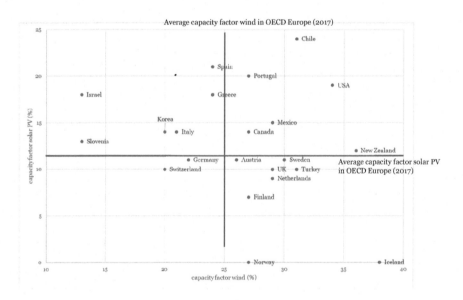

Fig. 2.11 Capacity factors of wind and solar PV in various OECD countries compared to OECD averages, 2017. *Source* IEA, Electricity Information 2019

turbines and solar PV have lower revenues. In Chile, the average production per unit of installed solar PV is about twice as high as in OECD Europe on average and about four times as high as in Finland.

Because of the weather dependency, these generation technologies are not dispatchable (i.e. the use of these technologies cannot be fully based on market circumstances), but they are controllable (i.e. the level of production can be influenced). The levels of production by wind turbines and solar panels can be influenced. In the case of negative electricity prices, for instance, they can be fully turned off. Moreover, when these technologies are not completely utilized at a particular moment of time (i.e. they are producing less than the maximum capacity given the wind speed or sun shine), the production can also be increased until to the capacity limit. This ability to adapt the production level is increasingly important in electricity markets with higher shares of renewables.

The weather dependency of wind turbines and solar PV results in more volatility in production levels. The production by wind turbines is strongly related to the wind speed. Generally, at a wind speed below 5 m/s, wind turbines do no produce. At a higher wind speed, the power of turbines increases exponentially with speed until a wind speed above 12 m/s, when they are operating on the maximum amount of power. At wind speeds higher than about 25 m/s, turbines are turned off in order to protect the turbine components (Schavemaker and Sluis 2017). The production by PV installations is highly related to the position of the sun, and, hence, the time of the day and the day within the year, which gives PV production a strong seasonal pattern (Smets et al. 2016). On top of this seasonal pattern, there are meteorological influences through the temperature and the intensity of radiation.

The volatility in production by renewable sources, however, does not imply that the production is unpredictable. Due to improved meteorological analysis methods, short-term weather forecasts are increasingly becoming more accurate (Ummels et al. 2006; Martinot 2015). Hence, more volatility in production is not equal to more uncertainty. This aspect is highly relevant for the balancing of electricity markets. In Sect. 7.4, we will further discuss the impact of increasing shares of renewable electricity generation on the functioning of electricity markets.

Another characteristic of renewable power is that its generation hardly causes marginal costs per unit of extra production. Power generation from wind and solar does, however, involve additional variable costs once the windmills and solar panels have been installed. These costs are related to operation, maintenance and management of installations (see, e.g., Blanco 2009). On top of these variable costs, the major cost component results from investment in the assets. Because of the relatively high share of these fixed costs, wind turbines and PV installations are more capital intensive than fossil-fuel power plants as these plants have also significant marginal (fuel) costs. Obviously, all the fixed capital costs of renewable power need to be covered, just as in the case of conventional power. The fact that these fixed costs have to be recovered has consequences for both the price formation in electricity markets and the incentives for investments in generation capacity. This topic is further discussed in Sect. 7.5.

Box 2.2 Comparing various generation techniques through the LCOE

Suppose a firm wants to assess the economic effects of investing in a wind farm containing a number of wind turbines in comparison to the economic effects of investing in a CCGT plant. These generation techniques differ in many aspects: investment size per MW, capacity factor, expected life time and variable costs per unit of electricity. The LCOE can be used to integrate all these variables to come to a single comparison. When information on all these aspects is collected, the LCOE of each type of plant can be calculated (see Table 2.4). A particular assumption has to be made on how fast the new plant will become operational. In this example, it is assumed that the investment is realized in the current year $(t = 0)$, while the plant becomes operational in the next year, which implies that there are no operational costs and production in the current year. Note, however, that for the full assessment of alternative technologies, the revenues which can be acquired in the market (i.e. the market-value effect) also have to be taken into account (see Sect. 7.5.3).

Table 2.4 Example of calculation the LCOE of various generation technologies

Assumptions					
Variable		Wind farm		Combined Cycle Gas Turbine (CCGT)	
Investment per MW (million euro)		1.25		1.50	
Capacity (MW)		500		700	
Expected capacity factor (%)		25		70	
Expected lifetime		20		30	
Variable costs (euro per MWh production)		20		40	
Costs of capital (%)		6		6	

Results of LCOE calculation					
			year (t)		aggragated
	0	1 .	15 .	30	numbers
discount factor $(= 1/(1+r)^\wedge t)$	1.00	0.94 .	0.42 .	0.17	
PV expenditures (million euro)					
- Wind farm	625.0	20.7 .	9.1 .	0.0	876.2
- CCGT	1050.0	162.0 .	71.6 .	29.9	3413.4
PV production (million MWh)					
- Wind farm	0.0	1.0 .	0.5 .	0.0	12.6
- CCGT	0.0	4.0 .	1.8 .	0.7	59.1
LCOE (euro/MWh)					
- Wind farm	69.8				
- CCGT	57.8				

The costs of different types of technologies can be compared with each other by using the measure Levelized Costs of Energy (LCOE). This measure shows the discounted total costs over the lifetime of a plant in relation to the discounted production volume during the total lifetime. The formula is:

$$LCOE = \frac{\sum_t^T \frac{Costs_t}{(1+r)^t}}{\sum_t^T \frac{Production_t}{(1+r)^t}} \qquad (2.17)$$

where t is the index for year, T stands for total (expected) lifetime of the plant (in years), r is the discount rate (or costs of capital), Costs stands for the investment and operational costs (in, e.g., million euro),[22] Production for the total production per year (in, e.g., million watthour (–MWh), while $\sum$ is the mathematical symbol for summation. Both Costs and Production are divided by $(1 + r)^t$ in order to express the future values of costs and production in the current value of money.[23] In order to ease the calculation the impact of the discounting can be expressed through the discount factor (df):

$$df_t = \frac{1}{(1+r)^t} \qquad (2.18)$$

The application of the LCOE method to plants that only produce one type of energy (such as electricity or heat) is fairly straightforward, but the application is more complicated when it refers to cogeneration plants, such as Combined Heat Power (CHP) plants. In order to know what the average costs per unit of electricity are, one has to control the costs for the production of heat. The allocation of costs to the production of one type of product (e.g. electricity) can, for instance, be done on the basis of the share of the product in the total output of energy.

The LCOE of renewable energy sources has gone down strongly in the past decades which is partly due to technological developments in solar and wind generation. Another factor behind the reduction in the LCOE is the decrease in the costs of capital. This decline in the discount rate in particular affects the LCOE of technologies, such as wind turbines and solar PV, with relatively low variable costs (see Fig. 2.12).

As a result of these technological developments and the reduction in the costs of capital, large-scale solar PV and onshore wind have already a lower LCOE than conventional ways of producing electricity (see Fig. 2.13). For the investment decision in these technologies, one has to compare the LCOE with the expected

---

[22]Although investments themselves are no costs, but only expenditures, during the lifetime of using the assets they are turned into costs through depreciation and costs of capital. Hence, as the LCOE is directed at the total lifetime costs, the investment sum can be treated as costs as well.
[23]In everyday language, this calculation (which is called discounting) is meant to correct for the interest rate: one Euro today has a higher value than one Euro next year because you can put the Euro now on, for instance, a savings account and receive interest. See further on this issue Sect. 6.5.

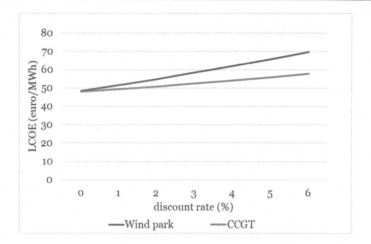

Fig. 2.12 LCOE of wind park and gas-fired power plant in relation to the discount rate (using information from Table 2.4)

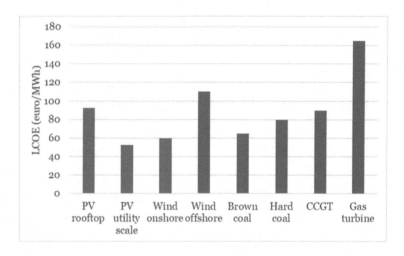

Fig. 2.13 LCOE of various electricity generation technologies, at locations in Germany in 2018 (average estimates). *Source* Frauenhofer (2018)

future electricity prices which can be captured by each of the technologies, as will be further discussed in Sect. 3.3.4.

The for the long-term expected electricity prices are only determined by the LCOE when the generation capacity is scarce during many hours in a year. In case of abundant generation capacity, the electricity prices are determined by the marginal costs of the marginal generation plants. When the average electricity prices are below the LCOE of renewable capacity, investors will not invest in that capacity, even if the LCOE of conventional capacity is higher. In such circumstances, regulators may want to give financial support, in one way or the other, to promote these investments in order to promote the energy transition. The design of these support schemes is further discussed in Sect. 8.4. In Sect. 7.5, we will further discuss the mechanism behind investments in generation capacity. In that section, we will also see that the various technologies may realize different electricity prices. Generally, the average annual price which can be obtained by wind and solar PV generation is lower than the average price a conventional power plant is able to realize. This is called the market-value effect.

2.4.4 Transport

Energy commodities can be distinguished in those which can be transported by ship or truck and those which can only be transported through a physical network. Coal, oil and biomass belong to the former category, while electricity and heat belong to the latter. Gas can be transported in both ways: in gaseous form trough pipelines as well as in liquid form as LNG (Liquefied Natural Gas). The dominant form for gas transport onshore is, however, through pipelines. Gas, electricity and heat, therefore, are also called network industries as transporting commodities via physical networks have large consequences for the organization of their markets.

In order to be able to transport gas through a pipeline, gas needs to be put under pressure. Generally, gas flows from a high-pressure network (i.e. a transport network) to a low-pressure network (i.e. a distribution network). Large users, including gas-fired power plants, are typically connected to the former part of the network, while small users, including residential consumers, are connected to the latter part of the gas network. In order to keep the gas under a certain pressure level, compressor stations need to be present at several places within the system. The process of keeping the gas under pressure requires use of energy. Hence, transport of gas causes some energy losses. The same holds for transport as LNG as both the (de)liquefication and the transport through ships or trucks require energy.

Transport by pipelines is the most efficient transport mode for not too long distances (see Fig. 2.14). This holds in particular for onshore pipelines as its transport costs per unit of gas transport are lower than the costs for LNG transport for distances below about 3000 km. Offshore pipelines are more efficient than LNG for distances up to about 1000 km. For longer distances, transport of gas in liquefied form (LNG) is more efficient, but still expensive in relation to the production costs. LNG is relatively expensive for short distances because of the distance-independence costs which have to be made for exporting terminals where gas is converted into liquid form and importing terminals where gasification of the liquids is needed in order to inject it in the pipeline system. These differences in costs explain why within Europe gas mainly is transported through a pipeline system, while global trade in gas (between USA, Asia and Europe) is done through LNG ships and terminals.

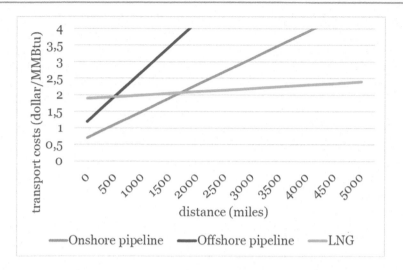

Fig. 2.14 Unit costs of gas transport in relation to distance per type of technique. *Source* Institute of Gas Technology

While electricity is also said to be transported through networks, its physical process is quite different. Electricity can be seen as the energy that is activated through electromagnetic fields within an electricity system. This system consists of various types of components, which can be distinguished in generators, transmission and distribution networks, and load (Schavemaker and Sluis 2017). Generators add energy to the electric field by converting thermal or chemical energy into electric energy (see Sect. 2.2.1), while load consists of the transformation of the electrical energy into other forms of energy, like mechanical energy, light, heat or chemical energy.

Transmission and distribution networks basically consist of overhead lines, underground cables, transformers and buses that connect generators and load to the network. Overhead lines and underground cables can be seen as power carriers. Generally, underground cables are more reliable than overhead lines, but also significantly more expensive, which is related to the material that is used. For underground cables, copper is generally used because of its relatively low resistivity which results in a low level of heat production following from so-called ohmic (i.e. resistance) losses. For overhead lines, a combination of aluminium (as conductor) and steel (as core material) is generally used.

Transmission of electricity results in ohmic losses resulting from resistance within the conductors. In a DC system, the resistance depends on the ease by which the physical material is able to let the electrons flow (i.e. the so-called resistivity (ρ),[24] the cross-sectional area (A) and the length (l) of the conductor:

---

[24]The resistivity is expressed in Ohm (ohm metre). The value of pure copper is 1.7×10^{-8}.

$$R = \rho * \frac{l}{A} \tag{2.19}$$

In an AC system, the resistance is higher than in a DC system because of the so-called skin effect, which refers to how the current uses the cross-sectional area of the conductor (Schavemaker and Sluis 2017). Because resistance in AC systems is a broader concept than in DC systems, this is called impedance.[25]

All the nodes where the power carriers connect to each other or to generators or load are called substations. When the voltage level of two connecting power carriers differ, the voltage level has to be transformed. This can be done through transformers, with less than 1% loss of energy (Schavemaker and Sluis 2017).

The higher the voltage level, the lower the current can be in order to transport the same amount of electrical energy. A lower current means lower ohmic losses. As a result, transmission lines, which connect the largest power plants to the power system, have a high voltage, while the distribution lines, which connect the load to the power system, have a lower voltage. The networks used for long-distance transmission are, therefore, also called high-voltage networks, and the networks for short-distance transmission are called medium or low-voltage transport or distribution networks.

Although power systems are designed as AC systems, which was due to the fact that in the past only AC voltage could be transformed into other voltage levels, for some transmission lines DC is used. It appears that high-voltage DC lines are more efficient in the case of long submarine lines. Figure 2.15 shows how the unit costs of transporting electricity evolve in relation to the distance for AC and DC systems. The distance-independent costs of DC line are higher because of the required equipment to convert AC into DC and v.v. on both sides of a line. The distance-related costs of a DC line are, however, lower as only two conductors are needed, while in a so-called three-phase AC system,[26] three conductors are required.

A consequence of the relationship between length of transport routes, resistance of lines and transport of power is that in a meshed network with various nodes, the transport of energy uses various paths. As we have seen above, the flow of current depends on voltage and resistance, while resistance is a function of, among others, the length of the line. Given the voltage of a network, the amount of energy that is transported through a particular line is inversely related to the length. For instance, in a network with 3 nodes where production is located at node 1 and load at node 3, and the lines in between have the same length, then 2/3 of the energy is transported through the direct connection between node 1 and node 3, while 1/3 of the energy is

---

[25]In this book, we will only use the term resistance, acknowledging that in AC systems this refers to the broader concept of impedance.

[26]In an AC system, the current is moving forward and backwards in a fixed frequency (50 Hz in Europe and 60 Hz in among others the USA). In a three-phase AC system, there are three currents having the same frequency but in a different phase. The phase difference is one-third, which means that the oscillations of each current is 120° (i.e. 1/3 of 360°) before and after the other ones. An advantage of this delay in phases is that the transfer of power is more constant.

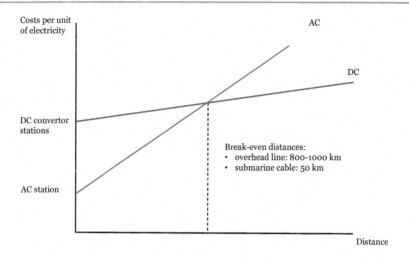

Fig. 2.15 Unit costs of electricity transport in relation to distance per type of technology. *Source* Schavemaker and Sluis (2017)

transported via the node 2 (see Fig. 2.16). This is due to the fact that the resistance in the latter is twice as high because the distance is twice as high.

This characteristic of electricity transport is of major importance for the operation of electricity markets (see, e.g., Creti and Fontini 2019). In terms of this example, if the producer in Node 1 produces electricity while the consumer in Node 3 uses this electricity, then the physical infrastructure to and from Node 2 will also be used. Because of these so-called parallel flows of electricity, the transport capacity is determined by the most constrained part of the network. For instance, if the capacity of the line from Node 2 to Node 3 is only 30 MW, while the capacity of the other two lines is 100 MW, then the flow from Node 1 to Node 3 is constrained as well. Hence, because of these physical laws, it is not possible to send more power through lines with more capacity and to use a congested line only until its capacity. When one line is congested, all transport in a network is congested, even if other lines have unused capacity.

It is important to realize that the utilization of the lines depends on what the production and loads on the various nodes will be. The actual utilization of a line may be (much) less than the aggregate of the individual production on the related nodes. For instance, if a producer in Node 2 would sell 50 MW to the consumer in Node 3, while the producer in Node 1 does the same, then the line between Nodes 1 and 2 will not be utilized as the flows cancel each other out (see Fig. 2.17). Hence, the efficiency of the utilization of the grid can be increased by taking information on the actual levels of production and load in each node into account. In Sect. 7.3, we will further discuss the consequences of these physical characteristics for the design of electricity markets.

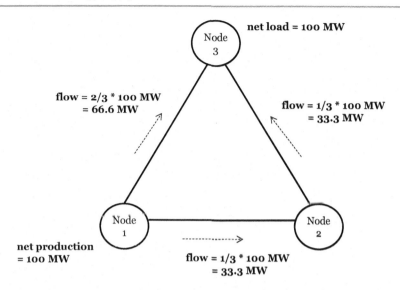

Fig. 2.16 Power flow in a 3-node network with 1 production node

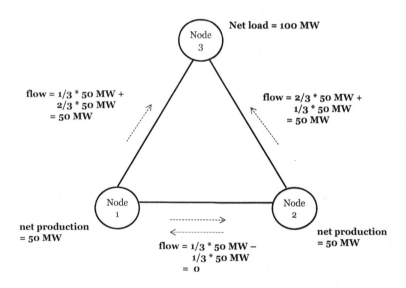

Fig. 2.17 Power flow in a 3-node network with 2 production nodes

2.4.5 Storage

Storages play a role in all energy markets, not only in oil and gas market, but also in electricity markets. In the latter markets, inputs can be stored, such as in the case of hydropower plants which store water in reservoirs, while also the outputs (i.e. electricity) can be temporarily converted into another energy carrier. From an economic perspective, storage of energy can be seen as the activity to adapt the timing of supply to the timing of consumption without adapting the timing of production, and the other way around. If markets exist, these timing differences are translated into differences between the current and the future prices of the commodities. Hence, the economics of storage is determined by the costs of storing (handling costs, investment costs) on the one hand and expected intertemporal price differences on the other. The marginal decision to use an existing storage is, hence, determined by the marginal costs of using the storage versus the expected benefits of selling the commodity at a later moment in time. The marginal costs of storage mainly consist of the costs of capital, as the opportunity costs of keeping a commodity in a storage is investing the capital in an alternative direction. These opportunity costs can be expressed through the discount rate (r). The expected benefits of selling later can be expressed through the return on the commodity prices, which is determined by the expected relative change in the commodity prices between the moment of withdrawing from the storage facility (i.e. supplying to the market) (P_{tw}) and the moment of injection into the facility (i.e. buying on the market or producing). This means that storing a commodity for a period of time (from t to t + 1) is profitable if the following condition holds:

$$r < \frac{P_{t+1} - P_t}{P_t} \qquad (2.20)$$

In other words, a trader will only buy and store a commodity, and withdraw it on a later moment of time when he expects that the capital costs of storing are below the relative change of the price of the commodity. Suppose that the price of an energy commodity at time t is 45 euro/MWh while the expected price for t + 1 is 50 euro/MWh, then the expected return in prices is (50 − 45)/45 = 0.11. If the costs of capital (r) of storing the commodity from t to t + 1 is less than this decimal interest rate, then it is profitable to buy the commodity now, store it and sell it at t + 1. After all, in such a situation, the return on capital to be achieved on the capital market is lower than the return which can be obtained by investing in the energy commodity. This is, of course, a very simplified example in which we ignored other costs of storage, such as the loss of energy during storage, and the dynamics of markets, such as the impact of storage behaviour on actual and future prices of commodities. In Chap. 3, we will discuss the latter aspect when we pay attention to the functioning of forwards markets in energy systems.

The above decision rule applies when the storage facility already exists. The costs of building such a facility don't play a role anymore in the short term; hence, these costs are viewed as sunk. The decision whether or not to invest in a new facility, however, depends on the value of the investment in a storage facility in relation to the present value of the expected revenues which can be realized by all storage operations during the lifetime of the facility. Suppose, for the sake of simplicity, that the future includes just one moment, $t + 1$, while the current moment is t. An investment in a storage facility is profitable if the discounted value of the expected price increase from t to $t + 1$ minus the marginal operational costs (mc) times the volume (Q) of the storage use exceeds the investment sum (I_s):

$$I_s \leq \frac{(P_{t+1} - P_t - mc) * Q}{1 + r} \tag{2.21}$$

In other words, an investor will only invest in a facility when the investment costs are below the present value of the expected net revenues during the lifetime of the investment.

The expected intertemporal price differences also depend on the costs of transporting energy as supply from other regions is another option to deal with unbalanced supply and consumption in a region. If the energy is imported trough pipelines with high fixed costs, this infrastructure requires, however, a high utilization ratio throughout the year. In case the demand for energy fluctuates over time (e.g. seasonal), there will be more scarcity during high-demand periods. These high prices form an incentive for investing in storage as these investments can benefit from intertemporal price differences. Hence, storage is in particular important in case of transport with high fixed costs, lack of flexible domestic producers and seasonal variation in demand. This holds in particular for most gas systems.

Storages can also be seen as a kind of insurance against shocks in the market. The value of this insurance is higher the less other options are available to deal with unexpected shocks in supply or demand. Countries that are highly dependent on import of oil or gas generally may store these energy carriers in order to be protected against sudden supply shocks (see Sect. 8.7).

2.4.6 Trade and Supply

Trade and retail supply do not have a physical role in energy systems; i.e., they do not transport or deliver the commodities to the end-users, but they take care of the coordination between production and consumption. The role of trade in supply chains is to contribute to an efficient allocation of the commodities. This means that traders search for the lowest-cost options in order to allocate them to those consumers who value the commodities the most. The role of retailers is the same but then directed at the consumption of residential and other small users.

Despite the lack of a clear physical role, traders and retailers do have a responsibility for keeping the energy systems in balance. These players have, just as other network users, a balancing responsibility, which means that they are generally obliged to act in real time according to their commercial transactions in forward markets. In addition, by providing information on these commercial transactions to the network operator, they contribute to an efficient allocation of network capacity. In Chap. 7, we further discuss the organization of these responsibilities.

Besides these functions, traders and retailers also play a role in the management of financial risks. These risks refer to the commitments of producers to produce and of consumers to pay for their consumption. If one of both parties defaults on their commitments, the other (counter) party faces the negative consequences. When retailers take over these risks, producers face less uncertainty that they will be paid for their production while consumers are less uncertain that producers will produce what they have promised to do so. Later on, in Sect. 7.4, we will see that this management of risks is a bit more complex in electricity markets because of the technical requirement that the electricity system needs to be permanently in balance.

2.4.7   Final Energy Consumption

The final and ultimate activity in supply chains is consumption. In Sect. 2.2.1, we have already seen that the consumption of energy carriers can be distinguished in use of energy and use of feedstocks.

A common characteristic of the usage of energy carriers for the purpose of energy is that the demand is not very sensitive to price changes. This sensitivity of demand (ΔQ_d) to changes in the price (ΔP) is expressed through the so-called price elasticity of demand (ε), which measures to what extent the demand changes (dQ/Q) in response to a change in price (dP/P):

$$\varepsilon = \frac{\frac{dQ}{Q}}{\frac{dP}{P}} \tag{2.22}$$

The demand for energy is generally called inelastic, which means that the (absolute value of the) price elasticity of demand is less than 1.[27] Table 2.5 shows the results of a meta-analysis on a number of empirical studies on the price elasticity for a number of energy carriers for both the short and the long term. In the short term, the price elasticity is about −0.2 and in the long term about −0.4 to −0.7. This implies that an increase in the price of natural gas, for instance, by 100%, the consumption only reduces by 18% in the short term. This low level of the price elasticity is related to the lack of substitution possibilities consumers generally have to respond to an increase in the price of gas. When the price of gas increases during a cold winter period, consumers can only respond by adapting their behaviour, such as regarding the in-house temperature. In the long term, however, consumers can also respond by (better) insulating their house, buying a more efficient gas boiler, or replacing gas by another energy carrier. The presence of more options for

---

[27]The price elasticity of a commodity is generally negative as a higher price of a commodity in most cases implies that less consumers are prepared to buy the commodity. This negative price elasticity is reflected by the downward sloping character of the demand curve. This is further explained in Sect. 4.2.

Table 2.5 Price elasticities of demand for energy in short and long term, per type of carrier

Energy carrier	Short term	Long term
Electricity	−0.126	−0.365
Natural gas	−0.180	−0.684
Gasoline	−0.293	−0.773
Diesel	−0.153	−0.443
Heating oil	−0.017	−0.185

Source Labandeira et al. (2017)

responding to price changes explains why the long-term price elasticity is always higher than the short elasticity.

The low-price sensitivity of demand for energy is also related to the fact that a number of other factors have a strong influence on demand. When energy is used for heating or cooling, the demand is strongly related to the outside weather temperature. Because of this relationship, in energy systems temperature is measured through specific energy-related metrics: heating degree days (HDD) and cooling degree days (CDD). The HDD measures how many degrees the outside temperature is below 18 degrees: the more it is below, the higher the demand for heating. The CDD measures how many degrees the outside temperature is above 18 degrees: the more it is above, the higher the demand for cooling.

Another factor that has a strong impact on energy demand is the level of income. This relationship can be expressed through the so-called income elasticity of energy demand. On the level of individual economic agents, the income elasticity of energy will be larger than one when agents buy more energy-intensive commodities when their income increases. On macroeconomic level, however, the income elasticity is generally a bit below one, which means that if the aggregated income increases by 1%, the energy demand increases by less than 1% (Burke and Csereklyei 2016). Apparently, a growth in overall income of an economy coincides with a reduction in energy intensity of economic activity. This reduction in energy intensity can be due to a shift in the structure of the economy from energy-intensive industries to less energy-intensive industries, such as services industries.

Box 2.3 Decomposition analysis of carbon emissions in Europe and East-Asia & Pacific

The development in carbon emissions in a region can be understood by looking at the underlying factors (Fig. 2.18). The EU, for instance, shows slightly declining emissions despite the increase in income per capita. The decrease in carbon emissions is realized through the decrease in carbon intensity (i.e. more renewable energy) as well as in energy intensity. In the region East-Asia & Pacific, the total carbon emissions have increased strongly, which results from a growing population, strong increase in income per capita, while also the carbon intensity of energy use increased.

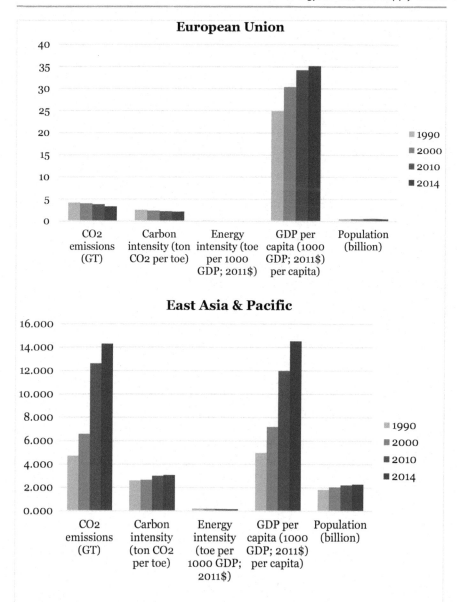

Fig. 2.18 CO_2 emissions as function of carbon intensity, energy intensity, GDP per capita and population size, EU and East Asia & Pacific, 1990, 2000, 2010 and 2014. *Source* Worldbank

The aggregated consumption of energy is, therefore, a function of energy consumption on individual and firm level and the changes within society and economy. This relationship can be expressed through the so-called Kaya identity, which states that the total use of energy (E) is equal to the product of energy use per unit of income (GDP), the income per capita and the population size (cap):

$$E = \frac{E}{GDP} * \frac{GDP}{cap} * cap \qquad (2.23)$$

This equation clearly shows that the aggregated level of energy consumption depends on several opposing factors. Reducing energy intensity of economic activities may be neutralized by increasing levels of income per capita and rising number of capita in an economy (see Box 2.3). This identity can be extended by relating energy use to the emissions of carbon (Em):

$$Em = \frac{Em}{E} * \frac{E}{GDP} * \frac{GDP}{cap} * cap \qquad (2.24)$$

This equation shows that the aggregated level of carbon emissions depends on the carbon intensity of energy use, the energy intensity of economic activity, the income per capita and the population size.

2.5 Organization of the Supply Chain

In the previous section, we have seen that a variety of activities are conducted in order to get the energy from source to end-user. The next issue is how this chain of activities can be organized. In principle it is possible that all activities are done by one firm. Such vertically integrated firms were the dominant types of firms in the energy supply chain in many countries until the liberalization of energy industries. Firms were also more horizontally integrated, meaning that they were active in various types of energy supply chains, such as gas and electricity. Since then, firms have become more specialized. The optimal way of organizing the energy supply chain depends on a number of factors, in particular economies of scope, economies of scale, productive efficiency and allocative efficiency. We first discuss these economic concepts before discussing the actual organization of supply chains in energy systems.

2.5.1 Economic Concepts

From an economic perspective, the optimal organization of a supply chain depends on the efficiency of combining activities within one firm versus conducting them in separate firms. If other firms are using more efficient technologies or using similar technologies in a more efficient way, they realize a higher productivity, which is defined as the ratio between total outputs over total costs. A higher productivity results in a higher productive efficiency, which refers to the extent the ratio of output over costs of a particular firm is close to the lowest possible costs using the most efficient technologies (see further on the concept of efficiency, Sect. 6.4.1).

Such cost advantages of firms may result from specialization in that particular type of product or from scale advantages. These latter advantages are called *economies of scale* which are defined as advantages of conducting an activity in one large plant or firm instead of conducting the same amount of activities in more plants or firms. Economies of scale exist if the averages costs (C) of a firm that produces two version of a product (x_1 and x_2) are below the sum of the average costs of two firms that produce both one version:

$$C(x_1 + x_2) < (C(x_1) + C(x_2))$$ (2.25)

These scale advantages in costs typically result from the presence of fixed costs which remain the same when the output volume increases. When economies of scale are present, it is more efficient to have larger firms. This can even result in one firm taking care of total supply. Such a situation is called a natural monopoly as a competitive situation with more suppliers would result in higher costs and is, therefore, economically not sustainable. Such natural monopolies occur in the transportation and distribution of electricity, gas and heat. This is further discussed in Chap. 6.

Costs advantages may also result from economies of scope which refers to the advantages of combining different type of activities within one firm, such as exploration, development and production of oil, or generation and transport of electricity. Economies of scope exists if the average costs of combining several types of activities (say x and y) in one firm are lower than the sum of the average costs of conducting these activities in separate firms:

$$C(x + y) < (C(x) + C(y))$$ (2.26)

Although some firms may have lower costs to produce a commodity due to economies of scale or scope, this does not necessarily imply that the customers of those firms pay lower prices as well. If the firms having low costs possess market power, they may charge high prices. As a result, for their customers it may be more efficient to produce the good themselves instead of buying it on a market. In such a case, the allocation in the market is called to be inefficient. Allocative efficiency refers to the price a firm (i.e. a consumer) has to pay for its inputs in relation to the (marginal) costs of producing these goods. The allocative efficiency is related to the functioning of markets, in particular the intensity of competition. The topic of market power and distorted market prices is further discussed in Chap. 9.

Even if market prices are below the costs of in-company production, it may still be more efficient for a firm to make a product itself instead of buying it on the market because of the presence of transaction costs. This holds in particular when complicated or highly specialized products are needed. Then, it may be costly to conclude contracts with suppliers in which all the required characteristics of the products have to be specified and monitored. Integrated production within one firm may also be efficient when it is important that investments between various types of activities are harmonized. In such situations, it is more efficient to have an integrated company where integral assessments can be made of investments in the various part of the supply chain.

2.5.2 Organization of Supply Chains in Energy Systems

Energy systems used to be characterized by vertically integrated and large-scale companies. Because of the economies of scale in the transport infrastructure for gas, electricity and heating, this part of these supply chains is generally dominated by one responsible company in a region. Consequently, these companies are operating as a monopolist, which may result in inefficient market outcomes. This topic is further discussed in Chap. 6.

When transport companies are also active in production and/or supply, they are vertically integrated. As the economics of scope in energy supply chains are, however, not that large, governments may require these companies to give other agents the possibility to make use of their transport infrastructure. This type of regulation is called third-party access (TPA), which is further discussed in Sect. 3.4.2 and Chap. 6.

In the production of energy, economies of scale are less important than in transport. The typical size of a conventional power plant is between 500 and 1000 MW, which is a large electricity plant but generally fairly small compared to the size of most markets. As a result of the energy transition, the average scale of producing electricity is likely to become even smaller (see Box 2.4).

Box 2.4 Impact of energy transition on scale of electricity generation

The average size of power plants is reducing as a result of the energy transition. Large-scale conventional power plants are increasingly replaced by small-scale renewable power plants. While building a modern coal-fired power station of about 1 GW (=1000 MW) can easily costs a few billion euros, an onshore wind turbine of about 1 MW (=1000 kW) can be built at the cost of a few million euros and solar panels of about 1 kW (=1000 W) can be installed on the roof of a house for a few thousand euros. This does not alter the fact that investments in renewable energy are also made in large-scale projects. The costs of investing in a windfarm offshore comprising, for example, 35 windmills, amount to half a billion euros.

Although the actual capital costs per installation (wind turbine, solar panel) are usually considerably lower than in common conventional installations, that does not mean, however, that renewable energy does not have scale effects. In each of the various generation technologies, economies of scale occur.

Regarding conventional technologies, also small-scale installations exist, such as gas turbines, gas engines units and micro-cogeneration, with capacities varying from approx. 15 kW up to several hundred MW (See for example: www.centraleinfo.net and www.blockheizkraftwerk.org/). These plants are used for particular purposes, such as supply during peak hours or for the combined production of heat and electricity in industries.

Regarding wind turbines, the productivity of individual turbines has increased strongly in the recent years. By increasing the length of the blades and the height of the turbines, the productivity has been raised. Although the argument of achieving economies of scale is often used in defence of a large-scale roll-out of windfarms, empirical research shows no evidence of these benefits (Dismukes and Upton 2015). The costs involved in offshore windfarms is mainly associated with the distance from the coast and the depth of the sea, not with the size of the installed capacity.

As far as solar parks are concerned, the average costs of a park with a capacity from 1 MW prove to be about 30% lower than those of installations with less capacity than 2 kW (Barbose et al. 2010). Among other things, the economic benefits of a solar farm compared with on-roof solar panels arise from the fact that the panels in a solar farm can be orientated more easily towards the sun and, besides that, are able to turn with the sun. The fact that the number of small-scale renewable energy installations has significantly increased in the recent past is most likely associated with the existence of financial support schemes (see Sect. 8.4), the limited pressure on space because solar panels can be installed on the roof of a house and people's desire to contribute directly to energy transition.

Exercises

2.1 Under which circumstances is the price of electricity directly connected to the price of gas?

2.2 Explain how the business case of hydrogen production depends on both the gas and the electricity price.

2.3 Explain how energy transition policies may stimulate the production of fossil energy.

2.4 Calculate the LCOE of an investment project in a solar park with the following characteristics:

Variable	Solar park
Investment per MW (million euro)	1.5
Capacity (MW)	100
Expected capacity factor (%)	15%
Expected lifetime	20
Variable costs per MWh production	5
Costs of capital (%)	6
Start production after. Year after start investment	1

2.5 Although electricity cannot be stored in a similar way as natural gas, storage does play a role in electricity systems. Explain.

References

Barbose, G., Darghouth, N., & Wiser, R. (2010). *Tracking the Sun III: The installed costs of photovoltaics in the U.S. from 1998–2009*. Lawrence Berkeley National Laboratory, USA, December.

Blanco, M. I. (2009). The economics of wind energy. *Renewable and Sustainable Energy Reviews, 13*(2009), 1372–1382.

Bowen, B. H., & Irwin, M. W. (2008). Coal characteristics. Indiana Centre for Coal Technology Research. CCTR Basic Facts File # 8. October.

Burke, P. J., & Csereklyei, Z. (2016). Understanding the energy-GDP elasticity: A sectoral approach. *Energy Economics, 58*, 199–210.

BP (2019). BP Statistical Review of World Energy, 69th edition.

Creti, A., & Fontini, F. (2019). *Economics of electricity: Markets, competition and rules*. Cambridge University Press.

Dismukes, D. E., & Upton, G. B. (2015). Economies of scale, learning effects and offshore wind development costs. *Renewable Energy, 83*, 61–66.

Frauenhofer, I. S. E. (2011). Heat pump efficiency: Analysis and evaluation of heat pump efficiency in real-life conditions, August.

Frauenhofer, I. S. E. (2018). *Levelized cost of electricity: Renewable energy technologies*. Freiburg: Fraunhofer Institute for Solar Energy Systems.

IEA (2019). Electricity Information 2019, International Energy Agency.

Labandeira, X., et al. (2017). A meta-analysis on the price elasticity of energy demand, *Energy Policy, 102*, 549–568.

Martinot, E. (2015). How is Germany integrating and balancing renewable energy today? Education Article, January.

Peng, X. D. (2012). Analysis of the Thermal Efficiency Limit of the Steam Methane Reforming Process. *Industrial and Engineering Chemistry Research, 51*, 16385–16392.

Schavemaker, P., & van der Sluis, L. (2017). *Electrical power system essentials* (2nd ed.). New York: Wiley.

Sinn, H.-W. (2015). The green paradox: A supply-side view of the climate problem; An introductory comment. *Review of Environmental Economics and Policy, 9*(2), 239–245.

Smets, A.H.M., K. Jäger, O. Isabella, R.A.C.M.M. van Swaaij & M. Zeman (2016). Solar energy; the physics and engineering of photovoltaic conversion, technologies and systems. UIT Cambridge, England.

Ummels, B. C., Gibescu, M., Pelgrum, E., & Kling, W. (2006). *System integration of large-scale wind power in the Netherlands*. IEEE.

van der Wall, W. (2003). The technological infrastructure of the gas chain. In M. J. Arentsen & R. F. Künneke (Eds.), *National reforms in European gas*. Elsevier, Global Energy Policy and Economic Series.

Zweifel, P., Praktiknjo, A., & Erdmann, G. (2017). *Energy economics; Theory and applications*. Springer Texts in Business and Economics.

Energy Markets and Energy Policies

3

3.1 Introduction

The most efficient way to coordinate various activities within a supply chain is to make use of markets. Markets can be defined as institutional solutions which enable economic agents to obtain inputs from other agents for their production or consumption activities or to supply their outputs to other agents. The more developed these energy markets are, the less economic agents need to realize the required inputs within their organization themselves. This chapter first discusses the various dimensions of markets (Sect. 3.2), before going into the main characteristics of the markets for coal, oil, gas, electricity and heat (Sect. 3.3). This chapter concludes by introducing the role of governments in organizing these markets, both as an active player in the supply chain and as regulator (Sect. 3.4).

3.2 Dimensions of Energy Markets

Markets are both an economic concept and something concrete in reality. The economic concept of a market refers to the interaction of supply and demand which results in a price that clears the bids of suppliers and buyers. Hence, markets are institutional devices to allocate a commodity from suppliers to buyers. All those suppliers who require a price below the clearing price will be able to supply the commodity to those consumers who are willing to pay a price that exceeds the clearing price. All other agents who want to supply the commodity at a higher price as well as those agents who want to buy the commodity at a lower price will not be successful. If the clearing process in a market operates in this way, it is called allocatively efficient. This economic concept is further discussed in Chap. 4.

Markets can be seen as opportunities for suppliers and consumers to meet each other to determine the market prices as well as how much every party can supply or buy. Such real-life opportunities are called market places (see Fig. 3.1). The most

© Springer Nature Switzerland AG 2021
M. Mulder, *Regulation of Energy Markets*, Lecture Notes in Energy 80,
https://doi.org/10.1007/978-3-030-58319-4_3

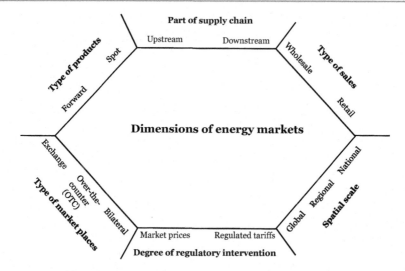

Fig. 3.1 Various dimensions of energy markets

basic kind of a market place is where economic agents can conclude bilateral contracts without any help of someone else. A more advanced market place is where economic agents are supported by brokers whose major task is to conclude contracts between individual suppliers and buyers. These markets places are called Over-the-Counter (OTC) markets. These brokers collect information from agents on both sides of the market and then try to establish the best deals for both parties. The most advanced market place is where all suppliers and buyers submit their wishes to a central coordinator who combines all this information to set the market price and to determine which market players see their wishes honoured. These market places are called exchanges. While on OTC markets, market parties can in principle conclude any type of product, this is not possible on exchanges. Exchanges offer a limited number of standardized products, because only then it is possible to ask all market parties to submit their bids. In addition, exchanges take over all the financial risk of the trading parties, while on OTC markets and also in bilateral markets, the market parties face the financial risk themselves that the counter party is not able to commit to its contract.

In all the various types of markets, different types of products are traded. An important distinction in the type of products is that between spot and forward products. Spot products are traded on markets which are close to delivery. The closest to delivery markets are real-time markets. Real-time markets are markets where the actual product is exchanged in return for a price, while in forward markets, contracts are concluded regarding the conditions under which the exchange of the commodity will occur in the future.[1]

---

[1]In electricity markets, day-ahead and intraday markets are called spot markets, but in principle these markets are also forward markets. The real-time market in electricity markets is the balancing market (see Sect. 7.4).

Forward contracts can be distinguished in physical and financial contracts. The difference between these two types of forward contracts is that the former one is directed at physical delivery of the underlying commodity, while in financial contracts the contract is settled before its maturity date (Mack 2014). Hence, physical contracts are in particular relevant for market parties who want to sell or use a commodity at a reduced price risk. These forward products are a kind of insurance products where risks are reallocated. If, for instance, a producer does not want to run the risk that the price of its outputs varies over time, it may sell the product already in the forward market where it offers to deliver the product against a predefined price (i.e. the forward price) for a period in the future.[2] If a producer is risk averse, it may be willing to pay a premium in order to have no uncertainty anymore on the future price of its output. Hence, in this case the forward price will be lower than the expected price in the real-time market (i.e. the so-called forward premium is negative). In a similar fashion, buyers may have a preference to buy the product in the forward market, which means that they prefer to have certainty about the price they have to pay for the future delivery of the product. Hence, in this case the forward price will be higher than the expected price in the real-time market (i.e. the forward premium is positive) as the consumers are prepared to pay a price (i.e. a premium) in order to have reduced their price risk in the future (see Box 3.1).

The various types of market places, where different types of products are exchanged, are implemented in various parts of the supply chain. If the markets are implemented for the trade in primary energy products, they are called upstream markets as they refer to those activities at the top of the supply chain (see Fig. 2.7). Examples of these markets are the coal and oil markets. Markets in which energy products are sold to end-users or retailers are called downstream markets. These markets can be distinguished in wholesale and retail markets. In wholesale markets, the group of buyers consist of the large industrial users and retailers. In retail markets, retailers sell the energy to residential and small-business end-users.

Box 3.1   Ex-ante and ex-post forward premiums

A forward price consists, theoretically, of two components: the expected spot price and a risk premium (Benth et al. 2014). This risk premium, which is called the forward premium, can be positive or negative, depending on which side of the market (buyers or suppliers) wants to reduce its price risk. A positive risk premium indicates that buyers want to pay more than the expected spot price, while a negative risk premium indicates that suppliers are prepared to give a discount in order to have less price risk.

---

[2]This period in the future may be shorter or longer away. It may refer to, for instance, the next day ('day-ahead contract'), next month ('month-ahead contract'), next quarter ('quarter-ahead contract') or next year ('year-ahead contract').

The forward premium can be determined in two ways: ex-ante and ex-post. The ex-ante forward premium (FP) is the difference between the current price at time t in a forward contract (F) minus the (average) expected (E) price at time of expiration (T) of the forward contract:

$$FP_{exante}(t,T) = F(t,T) - E_t|S(T)| \qquad (3.1)$$

Hence, the ex-ante premium measures which premium buyers are paying or the discount suppliers are giving based on their expectations for the future spot prices. As these expectations cannot be observed, this premium can only be estimated. The alternative method is the ex-post method, in which the forward premium is calculated as the difference between the current forward price and the (average) realized spot prices at time of expiration of the forward contract:

$$FP_{expost}(t,T) = F(t,T) - S(T)| \qquad (3.2)$$

Although the ex-post premium can be calculated on the basis of observable data, it gives a biased estimate of the premiums buyers and suppliers are paying. After all, the ex-post forward premium does not only depend on the premium buyers or suppliers are prepared to pay at time t, but also on the difference between expected future spot prices and realized spot prices.

Besides these commodity markets, there are also markets (i.e. institutional allocation devices) for access to the infrastructure, albeit that these markets are generally differently organized. The operators of transport and distribution networks sell access to their infrastructure. These markets are characterized by a natural monopoly which means that the supply side only consists of one supplier, but nevertheless it is a market where products are offered (such as various types of capacity contracts) to those who need these products (e.g. electricity producers and consumers). Because these markets are characterized by a natural monopoly, they are generally subject to regulation. For the domestic usage of these infrastructures, this regulation typically includes tariff regulation, which means that the tariffs do not result from demand and supply, but are set by the regulator or are at least subject to regulatory rules. For cross-border usage, the regulation generally makes use of market mechanisms, such as auctions, to allocate capacity. In Chap. 6, we discuss the economic principles and methods for this tariff regulation, while in Chap. 10, we also discuss how the cross-border capacity can be allocated.

Access to storage is also a product which is exchanged in energy systems, in particular in gas systems. As storage exists in electricity markets in a different form (see Sect. 2.4.5), it is not traded as a separate product. In gas storage markets, the sellers are the operators of storage facilities, while the buyers are all those who want to benefit from intertemporal price differences or who want to organize protection against such price differences. Depending on the characteristics of gas systems in a country, access to the storage is more or less subject to regulation. If, for instance, gas demand is very volatile due to weather dependency, while the number of storage facilities is limited, a regulator may conclude that the access to the facility needs to be regulated. In Chap. 6, we further discuss the need for regulation in relation to the characteristics of energy systems.

Box 3.2   Measuring liquidity of markets

A key aspect of the well-functioning of a market is its liquidity. Liquidity is related to the ability of market participants to sell or to buy products without affecting (i.e. disturbing) the market prices. In a liquid market, each participant has many choices to choose from while single bids do hardly affect the equilibrium outcomes. When in a particular market, the number of participants is low or when market participants cannot easily find each other, an individual participants may have difficulties to find a counter party that is able to buy the product at a given price. For instance, if in a regional electricity market the number of suppliers is limited due to cross-border capacity constraints, a change in the supply by one supplier may immediately have an impact on the electricity price. Such a market is called illiquid.

The liquidity of a market can be measured through various variables, such as the number of agents in relation to the size of total trade, the volatility of prices and the so-called churn rate, which measures how often a commodity changes from ownership before it is actually used. The formula for the latter measure is the ratio between the traded volume (V^T) and the actually consumed volume (V^{Ph}):

$$churn\ rate = \frac{V^T}{V^{Ph}} \qquad (3.3)$$

Markets for access to infrastructures are, by definition, regionally restricted, as the products sold on this market refer to a specific region. For instance, the so-called third-party access (TPA) to the gas market infrastructure in the EU is based on market areas which are equal to the infrastructure of a country or parts of a country. Markets for energy commodities, however, have a more international dimension. In particular, the markets for coal and oil are global markets in which all producers compete on a global scale with each other while the demand for coal or oil in one part of the world has an impact on product prices in other regions. However, as both energy resources are natural products, as we have seen in Chap. 2, there exist various qualities of both which affect the demand for these commodities. The markets for gas and in particular electricity are more restricted to a region because of the presence of transport restrictions, which implies that there may be (large) price differences between these regions. Below we will further discuss the various energy markets more in detail.

Applying these dimensions to the various energy markets, we see that the coal and oil market are more similar to each, in comparison to the other markets (see Table 3.1). All products can be exchanged on exchanges and OTC markets where both forward and spot products are traded. These markets can both be defined as upstream markets as they are closely linked to the upstream activities of exploration, developing and producing primary energy resources. In addition, these markets are global markets as the products can be fairly easily and with limited transportation costs shipped to the various global regions. Finally, both markets operate without specific regulation on prices of access to infrastructure, as they are not characterized as network industries, although some regulation exists, such as regarding strategic oil reserves (see Sect. 8.7.2).

The gas market differs from these two markets at is not only an upstream, but also a downstream market, where gas is sold to end-users, directly, but also through retailers. The market is also less global because of the relatively high costs of international transport. Related to this, it is a network industry, which implies that the access to the transport infrastructure (i.e. the pipelines) is subject to regulation, while the commodity (i.e. the natural gas) itself is traded on markets.

Compared to the natural gas market, the electricity market is even more a regional market as global transport is economically not possible, because the transport costs would significantly exceed the advantages of trade.

The heat market, finally, differs strongly from the other markets as it is highly regional (or even local) because of the high costs of transports, which results from the costs of the infrastructure and the energy losses during transport. In relation to the local character of heat markets, the exchange of heat is generally organized on a bilateral basis, with long-term contracts between producers, who are generally integrated with the transportation companies, and the users of heat.

In the remaining of this chapter, we will have a closer look at each of these markets. In the following chapters, the analysis is mainly directed at gas and electricity markets, as these markets are more regulated than the markets for coal and oil, while the markets for heat and new energy carriers like hydrogen are less developed yet.

3.3 Characteristics of Various Energy Markets

3.3.1 Coal Market

The resources of coal are fairly evenly spread around the globe, compared to the resources of oil and gas. Coal is currently produced in about 50 countries. Despite this wide global distribution in resources, the production of coal is concentrated in a few countries. The largest producer is China (45% of global production), while India and the USA are responsible for about 10% of global production and Australia, Russian Federation and Indonesia for between 5 and 10% (IEA 2019) (see

Table 3.1 Dimensions of various types of energy markets

Energy market	Type of market place	Type of products	Part of supply chain	Types of sales	Spatial scale	Degree of regulation
Coal	exchange/OTC	Forward/spot	Upstream	Wholesale	Global	Market prices
Oil						
Gas			Upstream/downstream	Wholesale/retail	Global/regional	Market prices for commodity + regulated tariffs (or prices, in case of auctions) for network access
Electricity			Downstream		Regional/national	
Heat	Bilateral	Long-term contracts		Retail	Local	More regulated

Table 3.2 Supply and use of coal in a number of countries, 2017 (in million tonnes)

Country	Production	Import	Export	Consumption per end-use sector			
				Electricity	Industry	Residential	Other
Australia	500	0	379	105	4	.	4
Germany	175	50	0	188	19	0	16
Poland	127	13	7	105	13	11	8
USA	703	7	88	607	17	.	22
Russian Federation	388	29	189	138	71	3	5
China	3397	284	8	2095	713	88	843
India	726	209	2	685	132	7	116
Indonesia	495	5	394	83	2	.	21
Total world	7596	1375	1364	5080	1197	137	1291

Source IEA, Coal Information 2019; Import and export refer to energy that has crossed national borders. When energy is transited through a country, it is included in both import and export

Table 3.2). China and India produce only for their domestic markets, while they are also large importers. Together they are responsible for about 1/3 of global imports. Australia and Indonesia produce mainly for the export: together they take care of about 1/2 of global exports.

These statistics imply that the coal market is characterized by a high degree of self-supply. China is not only by far the largest producer; it is also the largest consumer (about 50% of global use). There are only a few countries which depend on imports for their domestic needs, which is in strong contrast with the markets for oil and gas, as will become clear in the next sections.

Another difference with the oil and natural gas markets is that the marginal production costs do not strongly vary across regions. This implies that the supply curve of the global oil supply is quite flat.[3] Because coal can be very efficiently transported by ship and rail, producers compete with each other worldwide.[4] Hence, the coal market is a global market. The consequence of these characteristics is that the coal market can be seen as a competitive market where producers have to accept the market price without options to behave strategically in order to raise the market price.

Coal is extensively traded on international exchanges. Market parties can hedge their risks by trading in all kind of coal derivatives such as futures, options and indices.[5] A large part of the coal is sold through long-term contracts. To adapt their portfolio of long-term contracts to what agents actually need, coal is also purchased and sold through spot contracts, which refer to deliveries within one year (EIA, 2019). The spot prices are more volatile than the long-term prices as they depend

---

[3]In Sect. 4.2, we will explain the concept of the supply curve.

[4]The transport costs are about 5–10% of the end-user coal price paid by electricity producers (EIA 2019).

[5]See this webpage: https://www.cmegroup.com/trading/products/#pageNumber=1&sortAsc=false&sortField=oi&page=1&subGroup=9&cleared=Futures.

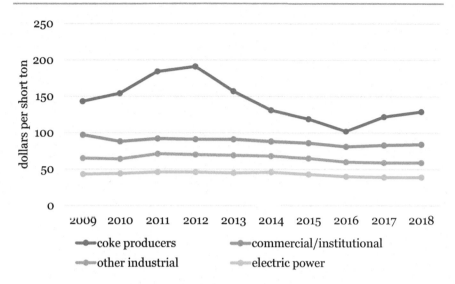

Fig. 3.2 Prices of coal for various types of users, 2009–2018. *Note* the prices are expressed per short ton, which means that both spot and contract prices are included. *Source* EIA, Annual Coal Report 2019 (retrieved from https://www.eia.gov/energyexplained/coal/prices-and-outlook.php)

more on the actual market circumstances. The long-term prices are fairly stable (see Fig. 3.2). As the quality of coal differs between the various ranks (see Sect. 2.3.1), the prices can also be quite different for the different type of users who use different types of coal. The price of coal paid by the electricity sector, for instance, is much lower than the price paid by the industry for coal which is used to make cokes, which may be related to the fact that for electricity various types of coal can be used, while for industrial purposes the coal needs to meet more restricted quality standards.

3.3.2 Oil Market

In contrast to the coal resources, the global oil resources are concentrated in a few regions: mainly in the Middle East, but also, albeit to a smaller extent, in Canada, Russia, Venezuela and Nigeria. The largest oil-producing region is the Middle East which is responsible for about 1/3 of the global oil production. This region produces about four times as much as its own consumption, while the OECD Europe, China and India, for instance, produce only 20–40% of their oil consumption (Table 3.3). These statistics indicate that the international trade is an extremely important aspect of the oil market.

Besides the large global differences in available resources per country, these countries also differ strongly in the costs of producing the oil. The oil fields in the Middle East have the lowest marginal costs implying that they are on the most left

Table 3.3 Domestic production and consumption and international trade of crude oil, in a number of countries, 2018 (million tonnes)

Country	Production	Import	Export	Consumption	Production in % of consumption (%)	Net balance[a]
Canada	265	41	192	108	245	6
USA	723	413	77	846	85	213
OECD Europe	180	600	124	667	27	−11
Russian Federation	554	1	253	157	353	145
India	40	220	0	218	18	42
China	196	419	5	573	34	37
Africa	401	32	302	195	206	−64
Middle East	1491	13	975	364	410	165
World	4511	2414	2332	4407	102	186
OPEC[b]	*1850*		*1285*			

Source IEA, Oil Information 2019; own calculations

[a]Net balance = production + import − export − consumption = change in stocks

[b]OPEC is the Organization of Petroleum Exporting Countries and has the following members: Algeria, Angola, Ecuador, Gabon, the Islamic Republic of Iran, Iraq, Kuwait, Libya, Nigeria, Qatar, Saudi Arabia, the United Arab Emirates and Venezuela. Import and export refer to energy that has crossed national borders. When energy is transited through a country, it is included in both import and export

side of the supply curve in the oil market. Offshore oil fields which are located in deep waters of Brazil and oil sands in Canada are on the rightest side of this merit order (see Fig. 3.3).

Oil is, just as coal, traded on various types of market places. On exchanges and OTC markets, different types of oil contracts are traded which enable market parties to hedge their risks. The forward and spot prices are related to each other through the presence of storages. In oil markets, crude oil is stored in among others storage tanks, while oil products are stored in among others marine and aviation bunkers. These storage facilities form an important facility to balance supply and demand. When the total supply of oil in a country, based on domestic production and imports, exceeds the total demand, based on domestic consumption and export, oil can be added to these storages. Hence, the changes in storage levels can be used as a measure for the tightness in the oil market.

The annual imbalance between supply and demand can be significant. In the USA in 2017, for instance, the total supply exceeded total demand by about 10%. In that year, the global oil market was characterized by an oversupply. In other years, the market had a negative imbalance between annual supply and demand. If the market balance is negative, the total demand exceeds total supply, which implies that oil has to be released from storages. This requires, of course, that the spot price increases as only then the operators of the storages have an incentive to supply oil

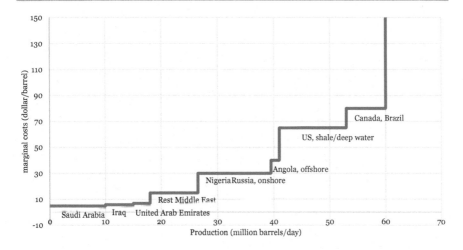

Fig. 3.3 Major part of the supply curve of the global oil market (based on the actual production of major oil-producing countries). *Note* Production refers to 2019 data; marginal costs to 2014 data. *Source* https://knoema.com/vyronoe/cost-of-oil-production-by-country

from their storages. In this situation, market parties may expect that the future price of oil will be lower because the current spot price is relatively high, and vice versa.

The relationship between inventories and the price is bidirectional as higher expected prices for the future form an incentive to market players to build up an inventory and vice versa. This relationship can be seen from Fig. 3.4 which depicts the annual change in the OECD inventories of liquid fuels and the annual change in the difference (i.e. the spread) between the futures price of oil 12 months ahead and the price for the near future.[6]

In the years 2014–2016, the inventories of liquid fuels grew which indicates that the supply in these years exceeded demand resulting in relatively low spot prices and higher futures prices. In the beginning of this period, the spread between the futures prices and the price for the near future was positive, indicating that market expects higher oil prices. Gradually the spread declined and became negative as the high inventories reduced the expected future scarcity. In 2017 and 2018, the opposite development happened, as the total demand exceeded total supply which resulted in higher spot prices and lower futures prices, which triggered the release of liquids from the inventories. This co-movement of inventory levels and futures prices show that the financial incentives play a key role to balance the oil market.

---

[6]The EIA has calculated this change as the difference in the price of the oil futures contract 12 months ahead minus the price of the next month's oil futures contract. If this difference is positive, the market expects that the oil price will increase, which is an incentive to store oil and sell it later at the higher price.

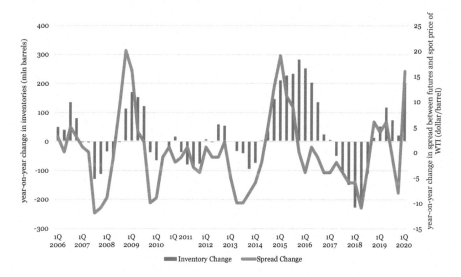

Fig. 3.4 Changes in OECD liquid fuels inventories versus annual changes in spread between futures and spot prices of oil, 2006–2018. *Source* EIA (retrieved from https://www.eia.gov/finance/markets/crudeoil/balance.php)

3.3.3 Gas Market

The resources of natural gas are less regionally concentrated than the resources of oil, but more than those of coal. The major global producers are the USA, the Russian Federation and the Middle East (Table 3.4). These countries/regions are also major consumers of natural gas, which indicates that the international trade in gas is more regionally oriented. The dependency of countries/regions on other regions is smaller than for oil. The USA, for instance, produces enough for domestic consumption, while in OECD Europe, the production is about 50% of consumption which is about twice as high as the ratio for oil. This means that the import dependence of OECD Europe for gas is 50% lower than the one for oil.

Compared to the oil market, the natural gas market is more in balance on an annual basis. The sum of production and import is more equal to the sum of consumption and export, both per country and on global level. This relatively strong annual balance is due to two facts. It is more expensive to store gas than oil and, in addition, the gas market is strongly based on transport by pipelines which requires a permanent balance between injection and withdrawal in order to function. In Chap. 7, we further discuss the issue of balancing energy networks.

When gas is transported in gaseous form, all suppliers and users of gas are physically connected to the transportation network. This network is operated by an (independent) network operator. Gas networks also facilitates the transport of gas from producers outside the region by the presence of LNG terminals where the

Table 3.4 Domestic production and consumption and international trade of natural gas, in a number of countries, 2017 (billion cubic metres)

Country	Production	Import	Export	Consumption	Production in % consumption (%)	Net balance[a]
Canada	183	21	80	114	161	10
USA	773	82	102	770	100	−17
OECD Europe	248	531	245	532	47	2
Russian Federation	695	9	228	463	150	13
China	148	89	4	238	62	−5
Africa	233	18	103	148	157	0
Middle East	627	36	155	511	123	−3
World	3793	1188	1117	3739	101	125

Source IEA, Natural Gas Information 2019; own calculations
[a]Net balance = production + import − export − consumption = change in stocks. Import and export refer to gas that has crossed national borders. When gas is transited through a country, it is included in both import and export

liquefied gas can be converted into gas and injected into the network. This facility enables the interaction with other regional markets.

The strong regional orientation of the gas market is related to the transport costs of gas. As we have seen in Sect. 2.4.4, transport by pipelines is the most efficient transport mode for not too long distances. As a consequence, gas markets for gas are mainly built on regional physical pipeline infrastructure for transport, while LNG transport connects these regional markets in case the price differences between these markets exceed the unit costs of LNG transport.

Gas markets are organized through physical or virtual hubs. In a physical hub system, gas is traded at the connection points within the network. This is the most common type of gas market in the USA, with the Henry Hub as the most liquid trading place.

In a virtual hub system, gas is traded independently of the physical location within the network (see Fig. 3.5). When gas is injected as some place within the network, it is registered by the network operator. Consequently, this gas can be traded by changing the ownership rights. These ownership rights can be transferred to someone else by sending an electronic message to the network operator stating volume, period, purchasing and selling parties. Such a message is called a trade nomination. The facility for traders to transfer and buy ownership rights is called the virtual trading point or hub. This market place acts as a virtual exit and entry points for traders. When the gas is actually used, it leaves the gas network. Note, when a trader wants to inject to or withdraw gas from a particular point of the network, it needs to have access to that network capacity and when this capacity is actually to be used, it has to inform the network operator through a so-called transport nomination.

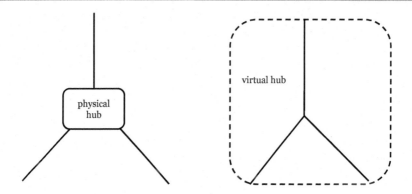

Fig. 3.5 Physical and virtual market places in gas markets

Because such a gas market is based on the physical and virtual entry and exit of ownership rights of gas, it is also called an entry-exit system. In this system, network users only need entry and exit rights and they do not need to worry about the available transport capacity between entry and exit. Taking care of transport capacity and system balance are the primary concerns of the gas network operator (see Hallack and Vazquez 2013). This form of organising gas systems is the most common form in Europe,[7] with the Title Transfer Facility (TTF) in the Netherlands, the Net Balancing Point (NBP) in the UK and the Net Connect Germany (NCG) in Germany as the most liquid hubs (OIES 2019). These hubs facilitate market places like exchanges and OTC.

3.3.4   Electricity Market

Electricity markets are, even more than natural gas markets, strongly nationally and regionally oriented because of the constraints imposed by the electricity grid. Although national networks have increasingly become interconnected, facilitating exchange of electricity on a large regional scale, further extending them to other global regions would be very costly because of the required investments in the infrastructure. More international exchange of electricity can result in more transport losses which are an increasing function of distance, although this only happens insofar the increase in commercial trade also results in larger physical flow volumes (see also Sect. 2.4.4).[8] Investments in grid infrastructure are only profitable when the costs of (further) extending a network and market are compensated by the benefits of trading electricity on a larger regional scale. These benefits depend,

---

[7]In Europe only a few physical gas hubs exist, such as Baumgarten in Vienna.
[8]The transmission losses in the current electricity systems amount to about 5% of final consumption (see Table 3.5).

among others, on the differences in generation costs and differences in load patterns between countries (see further on the benefits of international trade Chap. 8).

The relatively small regional scale of electricity markets is reflected by the low share of imports in the final consumption (see Table 3.5). On global scale this share is less than 5%, which means that almost all electricity is domestically produced. Only in OECD Europe, the share of import is above 10%, but a same percentage is exported on average, which implies that the trade in electricity is almost completely within this group of countries.

The importance of international trade varies strongly among countries. Some countries hardly import or export electricity. This holds in particular for island countries as Japan and the United Kingdom. In other countries, like the Netherlands and in particular Switzerland, import is responsible for a relatively large portion of total supply, while also export is significant. Note, that these import and export numbers may also include transit flows. The relative importance of imports and exports, however, varies from time to time, depending on the relative prices in the related electricity markets. In Chapter 10, we further discuss the factors behind international trade and how regulation may help to remove trade barriers.

Demand and supply of electricity meet each other at various types of market places (bilateral, OTC and exchanges). In the European zonal electricity markets, exchanges exists, such as EPEX and Nordpool, where standardized products are traded, while members trade anonymously and the financial risks are transferred to the exchange (see Sect. 7.3 for a discussion of zonal versus nodal electricity markets which exist in, among others, the USA). On each market place, various types of products can be sold which refer to different periods in the future. The products can be distinguished in long-term forward (years ahead), medium term forward (quarterly, monthly ahead) and short-term forward (day-ahead, intraday), while products are also further differentiated by referring to specific subperiods such as peak or off-peak hours. Compared to other energy markets, the products for the very short period ahead are more important which is related to the need to balance the grid constantly and the responsibilities market parties have to be in balance themselves. The task to permanently balance generation and consumption of electricity is extra challenging because the demand for electricity is highly volatile. In Chap. 7, we will further discuss what this means for the organization of electricity markets and how it can be realized that the decentralized supply and buy decisions match with the physical requirement of permanent network balance.

The domestic supply of electricity depends on the availability of power generation plants and their marginal generation costs. This supply to the market can be expressed through the so-called merit order. Figure 3.6 shows a typical merit order in electricity markets, with the available generation capacity on the horizontal axis and the marginal costs per unit of electricity on the vertical axis. Generally, renewable energy sources as wind turbines and solar PV and nuclear power plants have low marginal costs while gas-fired power plants have relatively high marginal costs. Consequently, the former type of plants will be more often 'in the money' than the latter. The supply by the renewable sources wind and solar may fluctuate strongly under influence of weather circumstances, and as a result, the merit order

Table 3.5 Domestic production and consumption and international trade of electricity, in a number of countries, 2017 (TWh)

	Net production (1)	Import (2)	Export (3)	Other use (4)	Supply (5) = (1) + (2) − (3) − (4)	Transmission losses (6)	Use by energy industry (7)	Final consumption (8) = (5) − (6) − (7)
USA	4086	66	9	29	4113	244	131	3738
OECD Europe	3533	408	408	51	3482	243	89	3150
Germany	619	28	80	8	558	27	12	519
Japan	1037	0	0	11	1025	41	14	971
Netherlands	113	22	19	–	117	5	6	105
Switzerland	62	37	31	4	63	5	0	58
United Kingdom	323	18	3	4	334	26	7	301
World	24,396	726	730	165	24,227	2022	796	21,410

Source IEA, Electricity Information 2019

Note Net production is gross production minus own use by power plants; other use refers to electricity use for heat pumps, boilers, and pumped storage; use by energy industry refers to electricity use for heating, tracking, and lighting purposes in conversion industry excluding electricity sector. Import and export refer to energy that has crossed national borders. When energy is transited through a country, it is included in both import and export

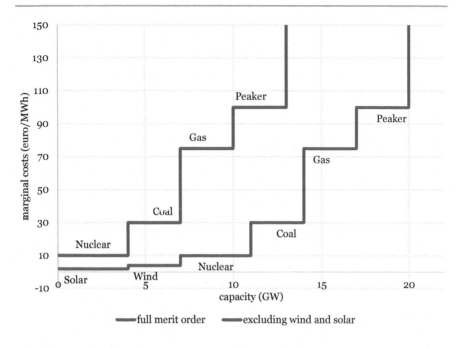

Fig. 3.6 Examples of supply curve in electricity market based on merit order, in- and excluding wind and solar power

shifts horizontally. This shift in the merit order affects the electricity price. This effect is called the merit-order effect of renewable energy (see further on this, Sect. 7.5).

In many electricity systems, gas-fired power plants (CCGT) are often so-called price setting plants, which means that a gas-fired plant is often the plant that operates with the highest marginal costs. These marginal costs are strongly related to the gas price, which connects the electricity market to the gas market (see Box 3.3). A rough estimate of the marginal costs of a gas-fired power plant (mc_g), ignoring other variable costs, is therefore given by the ratio of the price of gas (p_g) and the efficiency (η) of the power plant (see Table 2.1):

$$mc_g = \frac{p_g}{\eta} \tag{3.4}$$

The difference between the electricity price (p_e) and these marginal costs determines the margin of the plant to compensate for all other operational and fixed costs. This difference is called the spark spread (ss):

$$ss = p_e - \frac{p_g}{\eta} \tag{3.5}$$

Applying this formula to coal-fired power plants, the dark spread results. In addition, when in both cases the costs of carbon emissions are included, the outcomes are called clean spark spread and clean dark spread, respectively (see further on this issue Chap. 8).

When these spreads are negative, the electricity price is below the marginal costs of the respective plants, which means that those plants will not be used ('dispatched'). This will frequently occur for plants on the right side of the merit order, while plants on the left site, such as nuclear power plants, will still be able to profitably generate electricity. Every time, the market price is above their marginal costs, the plants will make an operational profit. One can, however, not conclude from this that it is more profitable to operate a nuclear plant instead of a gas plant, because the former type of plants has way higher fixed (investment) costs. The profits that the plants realize, based on the difference between the market price and their marginal (operating) costs, are required to recoup these fixed costs. Hence, the magnitude of these profits affects the incentives for investments in new power plants (see further on this issue Chap. 7).

The optimal composition of power plants in an electricity market is related to the volatility in demand. This volatility can be expressed through the so-called load-duration curve (see Fig. 3.7). In this curve, all hours within a year are ranked from the highest to the lowest load level. The lowest load level indicates the base load which is always present. The highest load levels refer to the peak demand which only occurs now and then (i.e. during peak hours). The shape of the

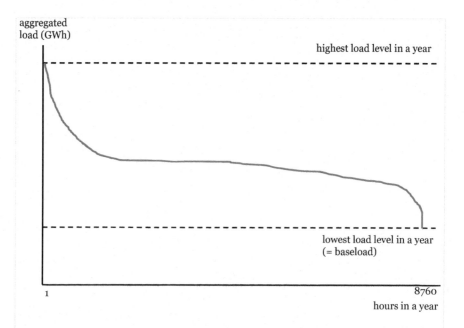

Fig. 3.7 Example of a load-duration curve in an electricity market

load-duration curve can be used to determine the optimal mix of the generation portfolio (i.e. the mix of various types of power plants). The base load can be served by power plants with high investment costs as they are able to run many hours in a year in order to realize sufficient revenues to recoup the investment costs. The plants supplying the extra demand during peak load will only operate during a limited number of hours in a year. In order to recoup their investment costs, they need to have low investments costs and/or to realize high prices during the hours of peak load. The relationship between the volatility in demand, electricity prices and the investments in generation technologies is further discussed in Sect. 7.5.

Box 3.3 Coupling between electricity and gas markets

As electricity is a secondary energy carrier, the supply curve in this market is strongly related to the markets for primary energy carriers, in particular the coal and gas market. When coal or gas-fired power plants are the marginal (price-setting) plants, the electricity price is strongly related to the price of coal or gas. In the Dutch electricity market, for instance, gas-fired power plants constitute a major part of the electricity generation portfolio and, as a result, the Dutch electricity price is strongly linked to the Dutch gas price (see Fig. 3.8). This relationship will, however, weaken when the shares of renewables increase and the price is more often determined by their marginal costs. In Sect. 7.5, we will see that these prices will be related to the long-term marginal costs (i.e. the LCOE) when the investments are only

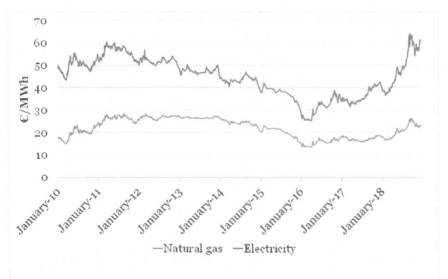

Fig. 3.8 Daily year-ahead forward prices of natural gas and electricity in the Netherlands, 2010–2018. *Source* Bloomberg

based on market prices. As, in many countries, investments in power plants (in particular renewables, but also nuclear power plants) are also based on subsidies, the market prices are more related to the short-term marginal costs, which do not reflect scarcity in the market.

3.3.5 Heat Markets

Although the share of heat in the final energy consumption is about 5% in the EU28 (see Sect. 2.2.2), markets for the allocation of heat are not well developed. This is related to the high losses of energy which occur when heat is transported. As a result, heat production and consumption are often directly connected with each other without interference of third parties for trade or transport. Heat is largely produced by combustion of fossil fuels, biofuels and waste (see Fig. 3.9).

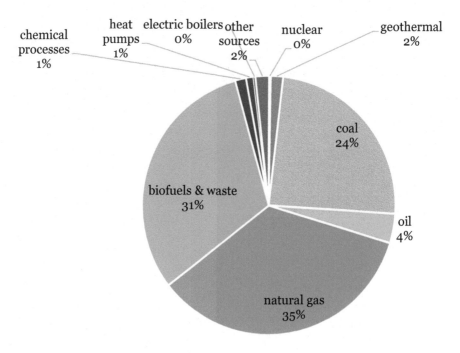

Fig. 3.9 Gross heat production in OECD Europe by source, 2017 (%). *Source* IEA, Electricity Information, 2019

When heat markets do exist, they generally operate on a local scale. The small spatial scale of these markets makes that the number of suppliers is generally limited. This low number of potential suppliers makes it difficult to realize a competitive market. A consequence of this limited number of suppliers is that heat markets are generally not liquid; i.e., there are not many traders while individual market transactions can have a strong influence on market outcomes. Generally, market outcomes (prices, quantities) are often not transparent in illiquid markets.

In line with this generally limited development of heat markets, the magnitude of these markets is fairly small compared to those of other energy carriers. Its share in the total final consumption in OECD Europe is about 2%, while the shares of both electricity and natural gas are about 20% (IEA 2018). Heat is mostly provided by Combined Heat Power (CHP) plants, which produce both electricity and heat). The major fuel source of the CHP plants is natural gas (its share is almost 50%), while the share of biomass is about 25%. Heat is mainly used by residential users, the chemical industry and commercial and public services (see Fig. 3.10).

The importance of heat markets may, however, change in the near future when governments stimulate the development of district heating systems in order to replace the use of fossil fuels (e.g. natural gas, oil) by more renewable sources like geothermal and waste heat.

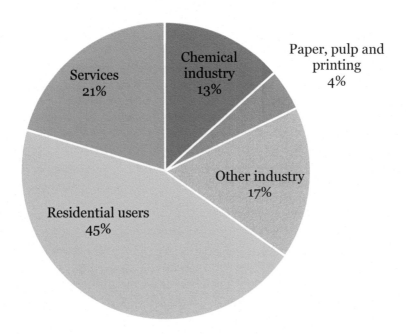

Fig. 3.10 Use of heat per sector in OECD Europe, 2016 (in %). *Source* IEA, Electricity Information 2019

3.4 Energy Policies

3.4.1 Types of Governance

Although economic agents pursue their objectives by choosing their optimal organizational structure and/or entering in market transactions with other economic agents, these processes may be subject to various types of government intervention. These interventions, first of all, refer to the general economic conditions, such as ownership rights and the legal system. In addition, government interventions may also refer to sector-specific conditions, such as the design of markets, access conditions for using networks and rules with regard to the behaviour of economic agents, such as the exploration of primary energy resources or the use of energy by residential households.

In this textbook, we do not discuss the government policies directed at the general economic conditions, but we focus on sector-specific regulation and specific measures directed at economic agents in energy systems. In the following chapters, we will extensively discuss the economics behind such measures, but first we briefly discuss the various types of measures which governments can take with regard to energy systems and how these policy interventions are subject to a legal framework. This analysis is mainly directed at gas and electricity markets, as these markets are more subject to regulation than the global coal and oil markets.

3.4.2 Designing Energy Markets

Energy markets have not always evolved automatically, as many energy systems were operated centrally in many countries until a few decades ago, while in some other countries they still are. In order to obtain markets, government intervention is often required to make this happen. This intervention can be distinguished in a number of categories: liberalization, regulation, restructuring and privatization (see Fig. 3.11). Each of these concepts refers to a different aspect of the process of designing markets.

Liberalization of energy systems means the introduction of freedom for producers and consumers to take their own decisions. In non-liberalized electricity systems, for instance, the decisions regarding the timing and type of investments in new power plants as well as the dispatch of all power plants are centrally taken by a coordinating body of the industry or the government administration. In a liberalized setting, however, this type of decisions is taken by independent firms. In a liberalized energy market, every firm is in principle free to enter the market, to decide upon all aspects of investments and how much it wants to produce at, for instance, the next day.

This freedom on decisions also holds for end-users. In a non-liberalized setting, firms that want to use energy cannot compare competing offers, but they are forced to go to one firm that has the monopoly of supplying energy. In such a setting,

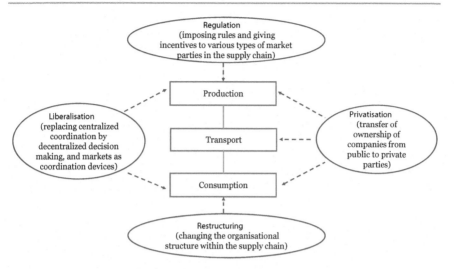

Fig. 3.11 Types of government intervention in energy markets

energy users are also not able to switch to foreign suppliers, implying that import cannot offer a competitive pressure. Generally, there is no room to negotiate the prices for energy, as the end-user prices are centrally set in close cooperation with the government administration. Because of this procedure to determine energy prices, it is more appropriate to call them tariffs. The difference between prices and tariffs is that prices result from the interplay of demand and supply, while tariffs are unilaterally set by a supplier, government or regulator. Hence, liberalization not only means the introduction of freedom for producers and consumers, but also the replacement of centrally determined tariffs by market prices which result from the interaction between supply and demand.

Although liberalization implies that economic agents are free in decision-making, this does not mean that in liberalized markets all types of behaviour of market participants are allowed. If firms obtain dominant positions in a market and use this market power to raise the price at the expense of consumers, governments may choose to intervene. For this reason, governments implement competition policy. Hence, liberalization of energy systems may still mean that governments closely monitor what happens in the market. This policy consists of two components: merger control and antitrust (see Box 3.4). The former is meant to prevent that firms obtain market power just by merging while the latter is meant to prevent that firms with market power abuse that power at the expense of consumers. Besides these general competition policy instruments, in the European Union, the behaviour of firms in energy markets is increasingly monitored in order to address insider trading and market manipulation in the so-called REMIT programme.[9] In Chap. 9, we will more closely discuss the ways regulators can detect and address market power.

Box 3.4 Competition policy versus sector-specific regulation

Competition policy and sector-specific regulation of an industry are related to each other. While regulation is meant to address specific issues in specific parts of an economy (industry, market), competition policy belongs to the general economic policy of governments.

Competition policy consists of two types of supervision: antitrust and merger control. Anti-trust focusses on detection of so-called abuse of market power through unilateral strategic behaviour or the construction of cartels to jointly behave strategically. Both types of activities are both strictly forbidden in most countries and subject to legal investigations. In these investigations, the attention is directed at the actual behaviour of firms, such as explicit agreements between firms (i.e. cartels) and supply decisions of firms with market power (in legal terms: structural dominant positions) resulting in economic or physical withholding (i.e. reducing output or charging higher prices) or predatory pricing in order obtain higher market shares. Another part of competition policy consists of merger control, which is meant to prevent that a merger of existing firms results in a new firm with a dominant position in a market.

As both anti-trust and merger control respond to activities in the economy, competition policy can be seen as a form of ex-post intervention. This is in contrast with regulation which is a form of ex-ante intervention, as the regulatory measures are set in order to influence the behaviour in the market. After all, sector-specific regulatory measures are implemented before the regulated agents act.

Generally, one can say that competition policy is applied in those industries where the structural characteristics do not hinder the possibility of a well-functioning of markets, while regulation is applied where normal competition is not possible (Motta 2004) and, as a result, abuse of market power can be expected to occur frequently while the impact of that conduct for consumers is likely substantial (Decker 2015).

Another form of government intervention is restructuring of the industry. This is the most extreme form of intervention as here the governments intervene in how the industry is organized. This type of intervention has occurred in energy systems. In the past, the energy industry in most countries was organized as a vertically integrated business where one firm was active in various layers of the energy chain

---

[9]See: https://www.acer.europa.eu/en/remit/.

(production, transport, distribution, storage, wholesale and/or retail) as was discussed in Sect. 2.5. The size of these vertically integrated firms differed strongly, varying from locally oriented firms to firms that operated on national scale. As some layers of these chains are suitable to be liberalized (production, wholesale and retail), while others cannot (transport and distribution) or only in some cases (storage), it is important to have separate companies for the different groups of activities. The reason for this is that firms that operate on a market, such as the gas or electricity wholesale market, and have to compete with other firms in that market, should not have a competitive advantage by also being the operator of an essential facility, such as the gas and electricity transmission grids. An independent operator of such facilities is viewed to be crucial in order to realize a level playing field for access to the networks for all participants (see, e.g., Baldwin et al. 2012). This kind of restructuring is called vertical unbundling. Vertical unbundling can be done in various intensities: administrative, legal and ownership unbundling (see Table 3.6). For the gas and electricity sector, European legislation requires all Member States to unbundle network operations from other, commercial activities, at least in a legal sense (see Opolska 2017).

Another type of restructuring is horizontal splitting. Here, the activities of a firm in one layer are allocated to a number of firms. The reason for this form of restructuring is to enforce competition by creating a number of competing firms.

Although liberalization of energy systems is often accompanied by privatization, this is not a necessary condition. Privatization means that the ownership of firms is transferred from the state or another public authority to private firms or even citizens. In the past, most energy firms were fully owned by national, regional or local authorities. The reason for this so-called public ownership was that the supply of energy was viewed to be of crucial importance for society and, hence, that energy systems should, therefore, be managed by public authorities. Later on, however, it appeared that public management did not guarantee an adequate supply of energy at reasonable costs. Many publicly owned energy firms turned out to be inefficient and not really innovative, while also the security and reliability of supply were often

Table 3.6 Types of vertical unbundling in energy sector

Type of unbundling	Description
Administrative	Network activity and commercial activities are conducted in the same legal entity but for both types of activities separate financial administrations are applied
Management	In addition to administrative unbundling, separate business units are implemented
Legal	Network activity and commercial activities belong to the same holding but are conducted in legally separate units
Ownership	Network operators are not permitted to be part of a group which is active in the generation or trade in energy and that their full focus must be on managing the networks well

below acceptable levels. In addition, it became clear that publicly owned companies are not necessarily directed at the public interests, as governments generally have multiple (political) objectives which also vary from time to time (Decker 2015).

Starting in countries such as the UK, it became more a general belief in many countries (but not all) that privately owned parties are better equipped to manage various activities within energy systems (as well as other parts of the economy), provided that they operate in well-functioning markets or that they are subject to appropriate government regulation. A factor that contributes to this higher efficiency of companies that are owned by private investors or that are listed on a stock exchange compared to companies in public hands is that the former face stronger incentives from the capital market. If they are less efficient than what investors expect or require, they will experience more difficulties to attract capital, which means that they have to pay a higher return to the investors.

The last type of public intervention in industries is sector-specific regulation. By regulation of a market, we mean that governments intervene in the decisions that producers, consumers or other agents, such as network operators, are making.[10] In a well-functioning market, microeconomic theory states that there is no need to regulate as all players are taking the optimal decisions for themselves and, as a result, the system moves to the optimal situation.

This theoretical concept, which is further discussed in Chap. 4, is challenged in many circumstances, as often economic agents cannot make the best decisions, for instance because market prices are distorted as not all costs or benefits are included in the prices or when no market exists. The latter holds, for instance, for the network parts of energy systems. Here, it is not possible to have competing firms as that would be too expensive for society. As a result, operators of networks generally have a natural monopoly which would give them the possibility to abuse that power by raising prices for consumers or to save on costs by reducing the quality of supply. In order to prevent this to happen, many governments have implemented sector-specific regulation for energy networks. Consequently, network operators are not free to decide upon the tariffs for using their grids or to choose the quality levels of the service, but they have to meet the standards and rules set by the government, which is generally represented by a governmental body which is called regulator. The design of this type of regulation of natural monopolies is the topic of Chap. 6.

The regulatory intervention in markets can be of various intensities (see Fig. 3.12). When the regulation only consists of the threat of regulatory measures in case the regulated parties do not behave in line with the regulatory objectives, this is called light-handed regulation. The opposite of this type of regulation is heavy-handed regulation, when the regulator heavily intervenes in the management decisions of the regulated party. Both have their pros and cons. Light-handed regulation requires less regulatory effort, but has a higher chance not to be effective,

---

[10]Regulation of an industry can be defined in various ways. Baldwin et al. (2012) distinguish (a) regulation as a set of rules, (b) regulation as all actions by the state to influence business or social behaviour and (c) regulation as all forms of influence (by state or market participants) on behaviour of agents. In this book, we follow the second definition.

Fig. 3.12 Various intensities of regulatory intervention

while heavy-handed regulation causes large regulatory transaction costs, is likely more effective, but also coincides with the risk that the regulator prescribes sub-optimal measures. Most regulators, therefore, choose for an intermediate form in which actual regulatory interventions are made, but these are more directed at the outcome of the process, instead of the activities of the regulated parties (see Decker 2015).

Besides this sector-specific regulation, governments may also implement a number of other regulations in order to steer the decisions of producers and consumers in what is seen as the right direction from societal perspective. This type of regulation includes, among others, environmental regulation. In Chap. 8, we will further discuss the contents of this type of regulation.

3.4.3 Legal Framework for Sector-Specific Regulation

The policies of governments regarding markets are subject to legal procedures. The main principle behind the governance of the government policies is the separation of power within the state. This separation is inferred from the classical *trias publica*, which means the separation of three different public responsibilities: law making, execution of laws and administration of law. The first responsibility is given to the parliament representing the society, and the second responsibility is given to governments, while the third responsibility is given to the courts.

The separation of power is also a fundamental aspect in the regulation of energy markets. A particular aspect of energy regulation in countries (i.e. Member States) of the European Union (EU) is that the regulation is imbedded within the EU governance which basically prescribes that the supranational (i.e. EU) regulations form the legal framework for the regulations of national and regional authorities. The regulatory measures taken by the EU consist of several types of measures. The most important are the Regulations and the Directives. Regulations are regulatory rules which directly hold in all EU Member States, while Directives contain principles and policy goals which have to be implemented in the national legislation by the Member States. As an example, Box 3.5 gives an impression of the contents

Table 3.7 Network codes and guidelines regarding the EU electricity market

Type of codes	Name of code (guideline (GL)/network code (NC))	Published on
Market codes	The capacity allocation and congestion management guideline (CACM GL)	25.07.2015
	The forward capacity allocation guidelines (FCA GL)	27.09.2016
	The electricity balancing guideline (EB GL)	23.11.2017
Connection codes	The network code on requirements for grid connection of generators (RfG NC)	14.04.2016
	The demand connection network code (DC NC)	18.08.2016
	The requirements for grid connection of high voltage direct current systems and direct current-connected power park modules network code (HVDC NC)	8.09.2016
Operation codes	The electricity transmission system operation guideline (SO GL)	25.08.2017
	The electricity emergency and restoration network code (ER NC)	24.11.2017

Source Schittekatte et al. (2019)

of the EU Regulation of the internal electricity market. An example of an EU Directive is the Renewable Energy Directive which sets the rules for the EU how to achieve its renewable energy targets.[11]

The more detailed implementation of the EU Regulations occurs through network codes and guidelines. Generally, network codes are more detailed than guidelines. For the electricity market, eight detailed regulations have been published, divided in three groups (see Table 3.7). Both network codes and guidelines are regulations, but the fundamental difference between the two is that guidelines require that the Transmission System Operators and the so-called Nominated Electricity Market Operators (NEMOs) must develop methodologies. In most cases, also the Network of Transmission System Operators (ENTSO-E), the European energy regulatory (ACER) and the European Commission participate.

The above regulations are in particular directed at promoting the internal EU electricity market. Comparable regulations exist for the internal EU gas market. A part of these regulations have to be translated into national legislation. This legislation consists of primary and secondary legislation. The former is formulated in national energy laws, while the latter consists of the decisions of the national regulatory authority (NRA) (i.e. the regulator). This regulation consists, among others, of the regulation of the maximum tariffs the operators of the electricity and gas networks are allowed to charge (see Fig. 3.13).

---

[11]See https://ec.europa.eu/energy/topics/renewable-energy/renewable-energy-directive_en.

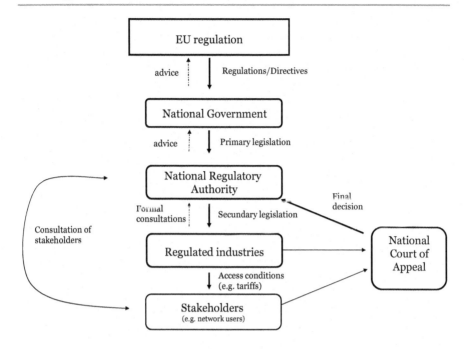

Fig. 3.13   The legal process of tariff regulation of energy networks within an EU Member State

Box 3.5   EU Regulation of the internal electricity market

EU regulation (EU) 2019/943 provides the rules for the functioning of the internal market for electricity. The following quotations give an impression of the content:

General principles: (Art. 3): 'Member States, regulatory authorities, transmission system operators, distribution system operators, market operators and delegated operators shall ensure that electricity markets are operated in accordance with the following principles: (a) prices shall be formed on the basis of demand and supply; (b) market rules shall encourage free price formation and shall avoid actions which prevent price formation on the basis of demand and supply; (c) market rules shall facilitate the development of more flexible generation, sustainable low carbon generation, and more flexible demand'.

Balance responsibility (Art. 5.1): 'All market participants shall be responsible for the imbalances they cause in the system ('balance responsibility').

To that end, market participants shall either be balance responsible parties or shall contractually delegate their responsibility to a balance responsible party of their choice. Each balance responsible party shall be financially responsible for its imbalances and shall strive to be balanced or shall help the electricity system to be balanced'.

Day-ahead and intraday markets (Art. 7.1): Transmission system operators and NEMOs shall jointly organise the management of the integrated day-ahead and intraday markets in accordance with Regulation (EU) 2015/1222."

Forward markets (Art. 9.1): '(…) transmission system operators shall issue long-term transmission rights or have equivalent measures in place to allow for market participants, including owners of power-generating facilities using renewable energy sources, to hedge price risks across bidding zone borders, unless an assessment of the forward market on the bidding zone borders (…) shows that there are sufficient hedging opportunities in the concerned bidding zones'.

Dispatching (Art. 12.1): 'The dispatching of power-generating facilities and demand response shall be non-discriminatory, transparent and, unless otherwise provided under paragraphs 2–6, market based'.

Network congestion (Art. 16.1): 'Network congestion problems shall be addressed with non-discriminatory market-based solutions which give efficient economic signals to the market participants and transmission system operators involved'.

Network charges (Art. 18.1): 'Charges applied by network operators for access to networks, including charges for connection to the networks, charges for use of networks, and, where applicable, charges for related network reinforcements, shall be cost-reflective, transparent, take into account the need for network security and flexibility and reflect actual costs incurred insofar as they correspond to those of an efficient and structurally comparable network operator and are applied in a non-discriminatory manner. Those charges shall not include unrelated costs supporting unrelated policy objectives'.

Resource adequacy (Art. 20.1): 'Member States shall monitor resource adequacy within their territory on the basis of the European resource adequacy assessment'.

Capacity mechanisms (art. 21.1): 'To eliminate residual resource adequacy concerns, Member States may (…) introduce capacity mechanisms'.

Network codes and guidelines (Art. 58): 'The Commission may (…) adopt implementing or delegated acts. Such acts may either be adopted as network codes (…) or as guidelines'.

Source Regulation (EU) 2019/943 of the European Parliament and of the Council of 5 June 2019 on the internal market for electricity (Text with EEA relevance.) (Source: https://data.europa.eu/eli/reg/2019/943/oj).

NRAs generally have some discretion how to implement the national primary energy legislation into secondary legislation. For instance, the primary legislation may prescribe that the tariffs for using the electricity grid should be related to the costs of an efficiently operating operator, but the NRA has to determine what is precisely understood by efficient operation. In Sect. 6.5, for instance, we will see that there are many methodological choices to make in order to be able to calculate the level of efficient costs.

The primary national legislation, based on the EU directives, does not only form the framework for the content of the secondary legislation by the NRAs, it also prescribes the process of defining the regulation. In many countries, it is prescribed that regulators should closely involve the stakeholders in the process of defining the tariff regulation. During the process of drafting regulatory decisions, stakeholders need to have the opportunity to participate in the discussion. Moreover, draft regulatory decisions have to be published in order to enable all stakeholders to submit comments.[12] After having considered and explicitly discussed all comments, the NRA decides upon the final regulatory decision. If stakeholders do not agree with (elements of) this decision, they can go the court of appeal. Hence, the final decision on the implementation of regulation is made by this court.

In this book, we will use the term 'regulator' in order to refer to the group of governmental agencies together that decide upon how to regulate the energy markets. This also implies that we do not go into the legal and administrative process of implementing the regulation, but that we focus on the economic argumentation to find the optimal design of regulation.

3.5 Exercises

3.1 What is the difference between liberalization and privatization?

3.2 What makes that firms still make use of bilateral or OTC markets while exchanges are well developed?

3.3 In which situation is the forward premium negative?

3.4 What is the essential characteristic of a liquid market?

3.5 Which information can be inferred from the imbalance in the oil market?

3.6 To what extent do participants in the gas markets have to worry about the transport capacity within a gas network?

3.7 Explain what the merit-order effect of renewable energy is.

---

[12]Note that the relevant stakeholders do not only include the regulated firms, but also those economic agents that are directly affected by the regulatory decisions, such as the users of networks who have to pay the regulated tariffs.

3.8 What is the load-duration curve?

3.9 What is meant by vertical unbundling?

3.10 What is meant by secondary legislation?

References

Baldwin, R., Cave, M., & Lodge, M. (2012). *Understanding regulation; Theory, strategy and practice* (2nd ed.). Oxford University Press.

Benth, F. E., Kholodny, V. A., & Laurence, P. (Eds.) (2014). *Quantitative energy finance; Modelling, pricing and hedging in energy and commodity markets.* Berlin: Springer.

Decker, C. (2015). *Modern economic regulation; An introduction to theory and practice.* Cambridge University Press.

EIA, Annual Coal Report (2019).

Hallack, M., & Vazquez, M. (2013). European Union regulation of gas transmission services: Challenges in the allocation of network resources through entry/exit schemes. *Utilities Policy, 25*(2013), 23–32.

Mack, I. M. (2014). *Energy: Trading and risk management; A practical approach to hedging, trading and portfolio diversification.* Singapore: Wiley.

Motta, M. (2004). *Competition policy; Theory and practice.* Cambridge University Press.

Oxford Institute for Energy Studies. (2019). European traded gas hubs: A decade of change. Energy Insight 55, July.

Opolska, I. (2017). The efficacy of liberalization and privatization in introducing competition into European natural gas markets. *Utilities Policy, 48*(2017), 12–21.

Schittekatte, T., Reif, V., & Meeus, L. (2019). The EU electricity network codes. European University Institute/Florence School of Regulation, February.

Microeconomic Perspective on Regulating Energy Markets

4

4.1 Introduction

The analytical framework for the economic analysis of regulating energy markets is derived from microeconomics. Section 4.2 discusses the basic microeconomic concepts and the theoretical benchmark of perfectly functioning markets. It shows that in such markets optimal welfare is achieved, which means that the allocation of goods is done in the best possible way. This chapter also explains the conditions for having perfectly functioning markets and that, in practice, almost never all conditions are satisfied. Although energy markets have become fairly mature in most regions, many of them suffer from market failures. Besides discussing the various types of market failures, attention is paid to the possibility of regulatory failures. Section 4.3 discusses how costs of and benefits of regulatory measures can be assessed.

4.2 Microeconomic Concepts[1]

4.2.1 Consumers Maximizing Utility

In energy economics, the focus is on how people, firms, organisations, etc. (hereafter: economic agents) use energy commodities, how resources are used to produce energy commodities and how the revenues from these activities are distributed among a group of economic agents. These topics can be analysed from various

---

[1]This chapter only describes the basic microeconomic concepts. For further reading, see e.g. Varian (2003).

© Springer Nature Switzerland AG 2021
M. Mulder, *Regulation of Energy Markets*, Lecture Notes in Energy 80,
https://doi.org/10.1007/978-3-030-58319-4_4

economic perspectives (Bhattacharyya 2019). Within business economics, the focus is on how firms organize their business, i.e. how they allocate business resources to activities regarding energy, such as depletion of oil fields or generation of electricity. Within macroeconomics, the above topics are analysed from the perspective of society, aggregating all activities of individual economic agents and analysing relationships between these aggregated activities, such as the relationship between total consumption of energy and GDP. The perspective of this book is microeconomics, as this field of economics is directed at the functioning of markets, i.e. at how the allocation process functions.

The key concept in any economic analysis is scarcity of commodities. By scarce commodities are meant all types of things (physical commodities, virtual commodities or services) which can be used in alternative ways but only in one specific way at a specific moment of time. The value consumers, which form a specific group of economic agents, attach to the use of a good is called utility. Using scarce commodities implies that these economic agents have to make choices among alternative potential usages with a different impacts on their utility. Suppose an agent can choose to consume a bundle of two commodities: products 1 and 2. If she consumes a number of products 1, then she gets a certain utility. When the number of consumed products 1 is reduced, she needs to consume more of product 2 in order to remain on the same level of utility. All bundles of the products which give the same level of aggregated utility can be depicted by so-called indifference curves (Fig. 4.1). These curves are generally convex to the origin because of the diminishing marginal utility when more of a product is consumed. This means that, when an agent consumes a lot of product 1 and hardly anything of product 2, she is prepared to reduce relatively strongly the consumption of product 1 in order to be able to consume more of product 2, and the other way around.

These indifference curves can be constructed for many different bundles. Assuming that a consumer prefers to have more of product 1 if this does not have consequences for the number of products 2, the indifference curves more to the top right of the graph are preferred above indifference curves closer to the origin since the former result in a higher utility than the latter. Hence, many different indifferences curves exist which are parallel to each other (such as IC1, IC2 and IC3 in the graph). As consumers generally face a budget constraint, they cannot afford to buy too many of the products. This is depicted by the so-called budget line. The budget line shows the amount of the various products which can be bought given the available budget and the prices of these products. For instance, a budget can be completely spent on product 1 or on product 2 or to buy a combination of both. The slope of the budget line gives the relative prices of these products. Combining the indifference curves with the budget constraint gives the optimal choice which is, in Fig. 4.1, on the tangent of indifference curve I2 and the budget line. With this combination of products, the consumer reaches, given her budget, the highest possible level of utility.

The indifference curves in combination with the presence of a budget constraint show that consumers have to assess trade-offs: using a commodity for a particular purpose means that it has consequences for the ability to use it for other purposes.

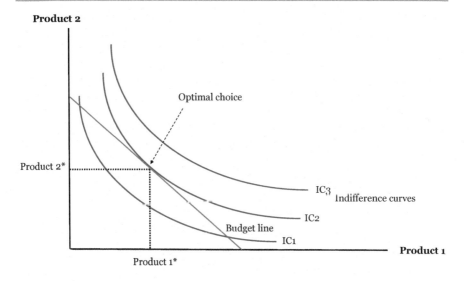

Fig. 4.1 Indifference curves and budget line

The missed benefits of these alternative usages are called the opportunity costs of a particular choice. If a consumer has to decide how to spend a particular budget, it may choose from a number of options, such as buying a new car or going on holiday. A choice for one option has consequences for the financial leeway to choose for the other. The opportunity costs of buying a new car, in this example, consists of the net benefits of going on holiday, and v.v.

When making such choices, economics agents are assumed to behave rationally. This assumption regarding rational behaviour must not be interpreted as that agents are assumed to make explicit calculations of all pros and cons of alternative choices. Rationality in economics means that agents are assumed to behave according their preferences. If a consumer prefers green energy above grey energy, but she does not want to spend time on searching for the greenest energy product, it can be fully rational for that consumer to stay with a default energy contract based on grey energy.

The assumption that economic agents are always behaving rationally implies that agents are assumed to maximize their utility when making a choice how to use a specific commodity. The utility cannot be measured itself, as this concept refers to the value someone derives from enjoying a specific commodity in relation to the value it attaches to the foregone benefits of alternatives. It is also a basic assumption in economics that the utilities of agents cannot be compared with each other, as human beings cannot see, let alone assess, what is in the mind of others.[2]

The utility that an economic agent derives from using a commodity in a specific way can be expressed through the so-called willingness-to-pay (WTP). This key

---

[2]This economic principle is called the impossibility of interpersonal comparison of utility.

concept in microeconomics measures the maximum price an agent is willing to pay in order to be able to have (e.g. to consume) that commodity. This maximum price is the value agents attach to a commodity which is equal to the sacrifice an agent is willing to make to enjoy the consumption of that commodity. This maximum price can also be seen as a measure of the preference of an economic agent. This preference (i.e. WTP) can be expressed in any type of unit. Generally, the WTP is expressed in monetary terms (i.e. money is used as a numeraire), but theoretically, it can be expressed in any other commodity. In the above example, the WTP of product 1 can be expressed in terms of the quantity of product 2. For instance, the WTP for product 1 is equal to x times the quantity of product 2.

The WTP for a specific commodity may vary from consumer to consumer as they may have different preferences. This heterogeneity of consumers is also an important element in the analysis of economic systems. In order to maximize welfare of society, economic theory prescribes that consumers with the highest WTP for a commodity should have a priority treatment above those consumers with a lower WTP, provided that the welfare of individuals is valued equally. If a consumer with a low WTP for particular commodity, say electricity generated by a wind turbine, would get that electricity while someone else with a higher WTP would not be able to consume that electricity, the allocation of the commodity to the first user is called inefficient. Therefore, one of the conditions for an efficient allocation is that a commodity is consumed by those agents with the highest preference (WTP) for it. Note, however, that a society may define another social-welfare function, which means that the welfare (WTP) of individuals is not treated equally. In such cases, governments may implement policies to redistribute income or to adapt the allocation of goods by the market (see further Chap. 11 on distribution and equity concerns).

4.2.2 Producers Minimizing Costs

Another condition for an efficient allocation is that for the supply the lowest-costs options are used. Generally, commodities can be produced through various technologies operated by different economic agents, which also holds for energy commodities as we have seen in the previous chapters. It would be allocatively inefficient if firms with relatively high costs would produce the commodity, while firms with relatively lower costs would not be able to do so. Hence, the heterogeneity of producers is also an important element in the analysis of economic systems.

Just as consumers are assumed to maximize their utility by choosing the optimal bundle of commodities given their budget restriction, producers are assumed to maximize their profit by choosing the optimal bundle of inputs to produce a commodity. At least, this holds in the case of perfect competition. In such markets, the only option firms have to maximize their profits is to reduce their costs as they cannot influence the price of a product in competitive markets. Hence, profit maximization is done by choosing the optimal mix of inputs. This optimal bundle of inputs depends, theoretically, on the technical relationships to produce an output from a number of inputs and the relative prices of these inputs.

These technical relationships can be depicted through so-called isoquants (Fig. 4.2). These curves show by which bundles of inputs a specific number of products can be produced. In principle, a firm wants to reduce all inputs as far as possible in order to minimize costs and maximize profits, but this is restricted by the technological options. The isoquant can also be seen as the technological frontier: all possible combinations of lowest inputs to produce a specific number of outputs. In Sect. 6.4, we will further discuss the use of this frontier concept in the process of benchmarking in tariff regulation.

The other aspect affecting the optimal bundle of inputs are the relative prices of the inputs. This is depicted by the isocost line in Fig. 4.2. At each point of this line, the total costs of using the inputs is the same. In order to minimize costs, a firm wants to be on an isocost curve which is closest to the origin. All points on the isocost line I1 have lower costs than the points on I2 and I3, but they are technically not possible as the line I1 is below the isoquant. The points at the isocost line I3 which are on or above the isoquant are technically possible, but they are not the most efficient ones, while other points on I3 are technically not possible. The lowest possible combination of inputs is there where an isocost curve is tangent with the isoquant. At this point, the production is on both the technological frontier and on the lowest possible cost line. This point reflects the optimal choice of a producer regarding the mix of input to be used for producing a product.

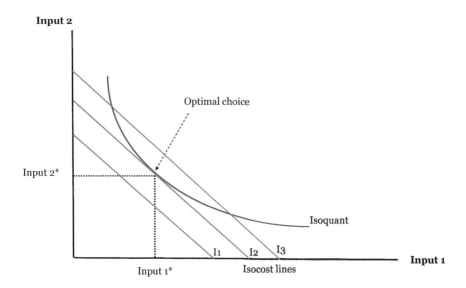

Fig. 4.2 Isoquant curve and isocost lines

4.2.3 Allocation of Commodities Between Producers and Consumers

The allocation of a commodity from a subset of a group of potential producers to a subset of a group of potential consumers can be expressed through the concepts of supply curve, demand curve, and market equilibrium. The supply curve shows the ranking of all offers by suppliers from the lowest to the highest minimum price required by the suppliers in order to recoup their cost for supplying the commodity. In a competitive market, this minimum required price is related to the optimal production choices of suppliers. As firms may use different technologies and operate on different scales, they may also have different required minimum prices.

In Fig. 4.3, supplier S1 requires the lowest price, while supplier S5 requires the highest price. These minimum required prices only refer to the marginal costs, which are the extra costs of supplying one extra unit.[3] The steepness of this curve reflects the differences in production costs among the group of producers. The steeper the supply curve, the more producers have different marginal costs for supplying one extra unit of the commodity.

The demand curve shows the ranking of all bids by consumers from the highest to the lowest WTP. In this figure, consumer D1 is prepared to pay the highest price, while consumer D6 has the lowest willingness-to-pay. The steepness of the demand curve reflects the differences in preferences among the group of consumers. The steeper the demand curve, the more consumers have a different WTP for a commodity. The larger the differences in WTP in the group of consumers, the less the total demand is affected by changes in the price, which is measured through the price elasticity of demand (see Sect. 2.4.7).

The supply and demand curves form the basis for the allocation of a commodity. In the theoretically ideal situation, the first unit will be supplied by the producer having the lowest costs (S1) to the consumer with the highest willingness-to-pay (D1). This will go on until a unit of the commodity is supplied by a firm which requires a price which is equal to or at least not higher than the highest remaining maximum price of a consumer (S4 and D4, respectively). More units will not be supplied as for the other units, the minimum required price exceeds the maximum price that the remaining consumers are prepared to pay (i.e. that part of the supply curve is above the demand curve). In terms of Fig. 4.3, supplier S4 is only able to supply a part of its potential supply to consumer D4. For the remaining of its potential, the willingness-to-pay of consumer D5 is below the price S4 requires. Hence, a market equilibrium is reached where all the demands of consumers D1 to D4 are met by the supply of the producers S1 to S4. In this equilibrium, the market price is about equal to the marginal costs of the last producer S4.

---

[3]The marginal costs do not need to be equal to the short-term operational costs. This is only the case when the investments in the fixed assets already have been made. Then, the investment costs can be viewed as sunk, as they do not change anymore when a firm supplies a commodity. When a firm considers to supply a commodity for a long period of time, it may also consider to invest in fixed assets. In that case, investment costs belong also to the marginal costs affecting the minimum required price.

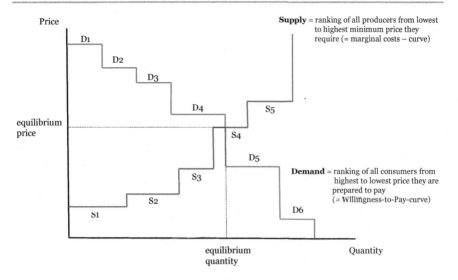

Fig. 4.3 Supply and demand curves constructed from individual bids

In this figure, we assumed a limited number of agents on both demand and supply sides. When there are many agents, the differences in WTP and marginal costs of consecutive bids in the demand and supply curves are smaller. Therefore, demand and supply curves are generally depicted as continuous curves instead of the step-wise curves as shown in the above figure. In the remaining of this book, we will generally assume such continuous curves to simplify the analysis (see Fig. 4.4).[4] The equilibrium outcome in such a stylized situation is also more straightforward: the equilibrium price is equal to the marginal costs of the last unit allocated and equals the willingness-to-pay of the last consumer whose bid is satisfied. This equilibrium is described by two elements: the equilibrium price (i.e. the market price) and the equilibrium quantity (i.e. the realised volume of market transactions).

Suppose the supply ($S(p)$) and demand ($D(p)$) curves can be described by the following functions of price (p):

$$S(p) = \frac{1}{2} * p \tag{4.1}$$

$$D(p) = 90 - p \tag{4.2}$$

This specification of the supply function means that the supply increases by 50% of the increase in the price. The economic interpretation behind this is that the marginal costs of supplying the commodity increases twice as much as the increase

---

[4]The exception to this rule is when we discuss merit orders.

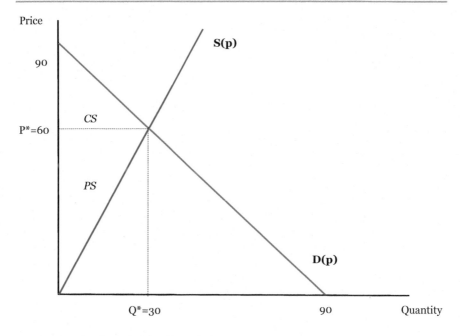

Fig. 4.4 Supply, demand and welfare effect of allocation

in quantity. The supply curve can also be written in a reverse way, by relating the minimum required price to the amount of supply: $P = 2 * q$. This reverse formulation is used to draw the graphs, where the quantity is on the horizontal axis and the price on the vertical axis.

The specification of the demand curve means that at a price of 90 euro/unit, no one is prepared to buy the product, while at a price of 0 the demand reaches is maximum level of 90 units. This demand curve can also be expressed in the inverse way in which the maximum price consumers are prepared to pay is a function of the quantity, just as is done in Fig. 4.4. In this case, the inverse demand function is $p = 90 - q$.

The equilibrium solutions can be found by determining at which price levels the quantities of demand and supply are equal:

$$90 - p = \frac{1}{2}p \tag{4.3}$$

which gives the equilibrium price of 60 (euro/unit) and the equilibrium output of 30 (units).

If a commodity is allocated in this way, the welfare is maximized. The welfare consists of two components: consumer surplus (CS) and producer surplus (PS). CS is the difference between what consumers are prepared to pay (i.e. their WTP) and what they actually have to pay (i.e. the market price). PS is the difference between the minimum price producers require and the market price. The PS and CS can be calculated be applying the so-called rule of half (when D(p) and S(p) are linear):

$$PS = \frac{1}{2} * (60 - 0) * 30 = 900\,euro \tag{4.4}$$

$$CS = \frac{1}{2} * (90 - 60) * 30 = 450\,euro \tag{4.5}$$

Given the preferences of consumers and the costs of producers, any other allocation of goods from producers to consumers would result in a lower welfare. It is important to recognize that this welfare concept refers to the market as a whole. The concept of allocation, however, should not be confused with the concept of distribution. The latter concept refers to how the returns of an allocative process (such as the profits) are distributed among members of a specific group. This topic is further discussed in Chap. 11.

4.2.4 Perfect Markets as Benchmark

In the microeconomic theory, the paradigm of a well-functioning market forms the benchmark for analysing real markets. This is a market characterized by perfect competition where all costs and benefits are taken into account by market participants. In such a market, goods are produced and allocated to users in the most efficient way, as described above. This implies that perfect competition between firms results in the highest possible social welfare, which means that only in such a situation the sum of producer and consumer surplus is maximized. Competition between multiple producers for providing one particular type of a (homogeneous) good or service may result in lower costs for producers as well as lower prices for consumers. Because suppliers need to take the price as exogenous and consumers can compare all offers of the various suppliers, firms are forced to reduce their costs. Only the firms with the lowest costs will survive, while high-cost producers are forced to leave the market. Because of this cost reduction, market prices go down. This mechanism is one explanation why well-functioning markets increase productive efficiency.

If prices go down, consumers realize a higher welfare (consumer surplus), as there is a larger difference between what they would be prepared to pay for the product (WTP) and what they actually have to pay. Lower prices also imply lower profits (producer surplus) for the firm (when costs remain the same). Hence, a waterbed effect exists between consumer surplus and producer surplus: lowering the price shifts welfare from producers to consumers. In most countries, competition authorities and regulators are focused on increasing consumer surplus, even though this may go at the expense of producer surplus.

In case of a perfectly functioning market, a number of conditions have to be fulfilled (Table 4.1). One of these conditions is that no market player is able to

Table 4.1 Conditions for perfectly functioning markets

Condition	Description
No transaction costs	Agents can switch to other suppliers or consumers without any cost
Transparency	All agents have full information on the offers made by other agents
Homogeneous goods	Agents can only compare offers by others if the commodities have the same characteristics
No possibility to behave strategically	No agent can influence the market prices and quantities
No externalities	All effects of supplying or consuming a commodity are taken into account by the agents

strategically influence market outcomes, for instance, by asking successfully a minimum required price which exceeds the actual marginal costs (this is called economic withholding) or by not offering all products to the market (this is called physical withholding). When no market player is able to act strategically then the market price is fully exogenous to all suppliers and consumers. As a result, in a perfectly competitive market the only option firms have to make more profit is to either reduce their costs per unit (this is called productive efficiency) or to improve the quality of their products and to sell the products to consumers having a higher willingness-to-pay for such products (this is called dynamic efficiency).

Generally, one can say that the higher the number of producers in a market, the less firms are able to act strategically. This relation between market structure (i.e. degree of concentration) and intensity of competition is, however, not that straightforward as even in a market with only two firms competition may be intense.[5] Firms may also have market power when they have a strategic advantage over others, for instance, by having access to a specific, essential, infrastructure which cannot be used by others. In other words, in order to obtain a market with perfect competition, all firms should operate in the same circumstances, i.e. there should be a level playing field.

Another key condition that has to be realized in order to get perfect competition is the presence of full transparency. Producers and consumers need to know the relevant product characteristics and what the price and other conditions are for a market transaction. Related to this is the requirement that products should be homogeneous because only in that case the various offers of suppliers can be compared by consumers. In addition, agents should not face transaction costs when they want to sell or buy, as that hinders the options for agents to make the best choice. Finally, economic agents should take into account all effects of their decisions, which means that there should be no externalities.

---

[5]This is so-called Bertrand competition, where firms compete on prices without any quantity restrictions. As a result of this competition, prices will become equal to marginal costs.

4.2.5 Market Failures

The above theoretical notion of well-functioning markets is not meant to describe actual market behaviour, but to be used as a benchmark for assessing real markets. In practice, many markets suffer from fundamental shortcomings which prevent that the market results in an efficient allocation of goods and maximization of welfare. Although energy markets have become fairly mature in many regions, many of them suffer from a number of fundamental shortcomings. These shortcomings are called market failures. In theory, a number of market failures can be distinguished (Table 4.2).

Economies of scale and economies of scope are in themselves not market failures as they refer to causes behind cost advantages, but they may result in firms having market power. Firms may have a cost advantage because they operate on a larger scale or they combine different types of activities. Cost advantages due to scale are

Table 4.2 Types of market failures

Market failure	Description	Example
Natural monopoly due to economies of scale	Average costs per unit decline if the quantity of production increases	Electricity and gas networks
Market power due to economies of scope	Average costs per unit decline if the number of types of products increases	Supply of gas and electricity to residential users by retailers
Market power due to pivotal position in supply	Supplier is indispensable to supply all the commodities demanded	Electricity producer during hours with high demand, while other producers are operating on their capacity constraints
Negative externalities	Economic agents do not take into account all costs of their activity	Carbon emissions resulting from use of fossil fuels
Positive externalities	Economic agents do not take into account all benefits of their activity	Knowledge spillover resulting from R&D; operation of gas and electricity grids (semi-public goods)
Network externalities	Benefit of an activity depends on the number of other agents also conducting that activity	Development of local energy system where value depends on the number of participants
Information asymmetry	Economic agents who want to enter in a transaction don't have the same information	Consumers cannot see how electricity is produced
Hold-up	The outcome of a decision (such as an investment) by an agent depends on what other agents do once this decision has been made which gives the latter a strong ex-post negation position and which makes the former uncertain ex-ante the decision	Investors who have invested in an asset (such as a grid) that can only be used by a limited group of users without the possibility to liquidate the investment without a loss

related to the presence of fixed costs which do not depend on the volume of the production. The larger the production, the more these fixed costs can be shared over more products, reducing the average costs per unit. An example of such costs is investments in an electricity grid. Cost advantages due to scope are related to the presence of common costs which have to be made anyway, independent of the number of types of products produced. An example of such costs is costs of customer relationships, which do usually not strongly increase if the variety in products increase.

If such costs advantages exist, it is more efficient to operate on larger scale or a broader scope. The presence of these costs advantages may distort the functioning of markets if the result is that only a few firms or even one firm is able to supply a good. In such a situation, the market is characterized by a structural position of dominance of these firm(s), which enables them to behave strategically and to set prices above the competitive level (which is the theoretical benchmark) in order to maximize profits. So, the actual market failure resulting from economies of scale and economies of scope is the presence of dominant players who can influence market outcomes (see Fig. 4.5). This market failure may also occur when some producers are indispensable for supplying the required commodities, for instance, during hours of peak demand. The regulatory measures to address market power in wholesale and retail markets are discussed in Chaps. 9 and 10.

In particular, economies of scale cause a market failure in energy markets. The transport of natural gas, electricity and heat is characterized by the presence of natural monopolies. This market failure means that it is not possible to introduce competition because it is more efficient to have one firm providing the service to the whole market (i.e. being responsible for the transport) than to have more firms doing the same task. This natural monopoly follows from the high investment costs required to set up a network. Once a network infrastructure has been built, the marginal costs of one extra user are negligible. Because competition is not possible, regulation is required to prevent the adverse effects of having a monopolistic firm responsible for transport. The main objective of this regulation is to guarantee all gas and electricity users access to the network infrastructure against reasonable tariffs. This is further and more extensively discussed in Chap. 5.

Another, large group of market failures consists of externalities, which are effects of decisions taken by economic agents that are not taken into account by them because of lacking markets. There are three types of externalities: negative, positive and network.

Negative externalities occur when economic agents do not take into account all costs of their activities. An example of this market failure is carbon emissions resulting from the use of fossil energy. Without any regulation, users of fossil energy do not (directly) face any costs of their emissions, and, hence, they will not be inclined to take them into account. This market failure results in a too high level of activities from a social point of view. The welfare loss here is the too high level of production (difference between actual and optimal quantity) times the difference between the WTP and the actual social costs of production. This welfare loss is called the deadweight loss of a negative externality. In order to solve this market

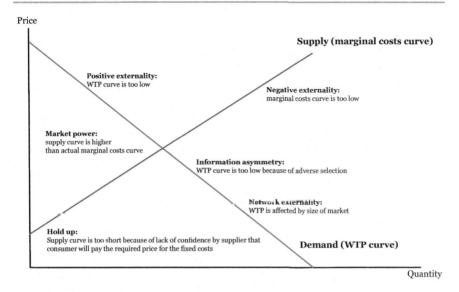

Fig. 4.5 Impact of various types of market failures on supply and demand

failure, agents should be made subject to regulation which let them consider these costs when making a decision regarding production or consumption. This regulation can consist of standards, obligations, subsidies, taxes of emissions trading schemes. In Chap. 8, we will further discuss the design of various regulatory measures to address negative environmental externalities.

Positive externalities occur when economic agents do not take into account all benefits of their activities. This may result in a suboptimal level of activities as agents cannot capture all benefits of their activities. This may occur, for instance, if the benefits of innovation cannot be protected by innovative firms. If everyone can freely benefit from a new innovative product, then no one is prepared to make costs for this innovation, unless the private benefits exceed the private costs. If the private benefits are below the private costs, firms will not innovate. In Chap. 8, we will also further discuss regulatory measures to address positive externalities.

Some commodities with positive externalities are called *public goods*, which are commodities characterized by two aspects: non-excludability and non-rivalry. The first aspect means that no agent can be excluded from enjoying the commodity, which has as a result that no one has an incentive to pay for it because it is more efficient for everyone to freeride on someone else who supplies the commodity. Non-rivalry means that the consumption of a commodity by one consumer does not affect the consumption of others. The classical example of public goods is dikes to protect against sea water. Once a dike has been built, everyone behind the dike is protected while consuming this protection by one agent does not affect the availability of it for others. As a result, dikes will only be built when there is collective arrangement how to share the costs.

A special group of public goods are the so-called semi-public goods, in which case only one of the two conditions hold. An example of a semi-public good are the operation of gas and electricity networks. If the network operator maintains the quality of the network (i.e. the network balance), all network users benefit and no one can be excluded. Consequently, maintaining system balance requires regulation to get this activity done. Another example of a positive externality is the efficient utilization of energy networks which facilitates trade. This is further discussed in Chap. 7.

Network externalities occur when the benefit of consuming a commodity depends on the number of other agents also consuming that commodity. An example of such a market failure is the market for telecommunication products like smart phones. The value an agent attaches to a smart phone increases when more other agents have also one because this raises the usability. Another example is the development of a local energy system: the value for an individual participant increases when there are more participants as that raises the options for local exchange of energy. The occurrence of such a market failure may result in a limited number of suppliers capturing the full market and, as a result, other firms being unable to enter the market. If network externalities exist in a market, market parties should coordinate how they want to organize the market or a regulator should impose regulations on market design, such as the obligation on network operators to connect also users of other networks.

Another type of market failure refers to the process of making transactions. Both ends of a potential transaction need to be well-informed in order to find the best solution. *Information asymmetry* may occur if economic agents cannot fully assess the quality of a commodity and, as a result, they may not be inclined to pay the full price. The presence of this market failure results in so-called adverse selection, which is that high-quality products with relatively high prices are driven out of the market as consumers are not prepared to pay the higher price if they cannot observe the higher quality sufficiently well. As an example, consumers are not prepared to pay their maximum price for green energy if they are uncertain about the true characteristics (quality) of a product. Because gas, electricity and heat are transported through networks, consumers cannot see how these energy carriers have been produced. Without any regulation, consumers may not trust statements of suppliers saying that they produced the energy in a renewable way. As a result, markets for renewable energy will not be possible or can only exist to a lesser extent unless the information asymmetry problem is solved. If this market failure occurs, coordination or regulation is required, for instance, by organizing a trustworthy certification scheme. The regulation of this market failure is further discussed in Chap. 5.

The final type of market failure refers to the confidence agents have about the future conditions. This is in particular relevant when agents have to make decisions with long-term commitments, while their negotiation position is weakened once the decision has been made. For instance, when an investment in a gas grid has been realized, the investment costs are so-called sunk and do not play a role anymore in the negotiation about access conditions with potential users. This market failure is called *hold-up* which occurs when, for instance, in the case of long-term investments, investors do not want make as investment as long as they are uncertain about the future usage and returns on the investment. This hold-up problem may, for instance, occur in the development of a new district heating or hydrogen system, where a new network will only be built when the investors have sufficient certainty about the usage by producers and consumers. The ex-ante uncertainty (before the investment is made) is strongly related to the irreversibility of an investment, i.e. the limited options to undo an investment. This irreversibility weakens the ex-post negotiation position of an investor. Hence, the uncertainty about the future position is related to the presence or lack of liquid markets or the (in)ability to conclude long-term contracts with customers. The occurrence of this market failure may result in a too low level of investments. If this market failure exists, coordination or regulation is needed to give investors more certainty about the future market circumstances. The regulation of this market failures is further discussed in Sect. 6.7.

Besides the above market failures, regulators may have other concerns which may legitimate intervention from an economic perspective. These other concerns refer to the distribution of welfare. In the analysis of efficiency of market, the initial distribution of resources is taken as given. When, however, another distribution of resources is assumed, a different efficient outcome follows. Therefore, it is important also to address the distributional effects of welfare, which is done in Chap. 11.

4.2.6 Regulatory Failures

If markets fail, regulatory intervention may be effective to overcome this failure. However, regulators face a fundamental challenge which is related to the information asymmetry between regulators and regulated agents. Generally, it is impossible or highly expensive for a regulator to acquire the same level of information and knowledge as regulated firms have about their characteristics and potential options to implement measures, such as regarding energy efficiency. Because of this lack of information, governments run the risk that they take inappropriate regulatory measures, such as support levels which are higher or lower than needed to incentivize economic agents to take specific measures (see Table 4.3). In the first case, the support results in windfall profits for the recipients of the support, while in the latter case the support is ineffective. See further on this type of regulatory failure, Sect. 8.4.

Another aspect of the information asymmetry is that a regulator is generally not perfectly able to monitor the behaviour of the economic agents (see e.g. Veljanovski 2012). Because of this, these agents may have options to deviate from what they should do according to regulatory measures. This information asymmetry results, hence, in moral hazard, which is the behavioural response that an agent takes more risk or make higher costs that what it would do without the regulation. In order to overcome these information problems, regulators may try to invest in extensive monitoring activities, but this, of course, increases the costs of regulatory intervention. See further on this issue Sect. 6.3.

Table 4.3 Examples of welfare effects of various types of inefficient regulatory measures

Examples of inefficient measures	Welfare consequences
Too high subsidy levels	Windfall profits and extra deadweight loss due to taxation to finance the subsidies
Too low taxes on fossil energy	Remaining deadweight loss of too less reduction in fossil-energy consumption
Too high emission standards	Agents are making too high costs to achieve an objective
Subsidies for production without presence of market failures	Loss of productive efficiency as a product with higher costs replaces a product that could have been produced with lower costs

Another source of ineffective or inefficient regulatory policies is the presence of regulatory capture by interest groups (see Sect. 1.4.2). Such a capture may result in regulatory measures which may be good for specific groups of economic agents, but not beneficial for society as a whole (see Decker 2015). Examples of such measures are support schemes which in particular favour particular groups or regulatory standards on product quality which protect incumbent suppliers.

Governments may also formulate policy measures to overcome some market barriers for specific technologies or specific groups, without addressing a market failure. In this respect, it is important to realize that market barriers always exist, even in perfectly functioning markets. Market barriers can be defined as factors which make that specific types of commodities cannot enter a market. For instance, the magnitude of demand (i.e. WTP and number of consumers) may be too small for a particular product because of high investment costs. If governments give financial support to the supply than the barrier may be effectively solved, but at the expense of a lower welfare.

Hence, even in the presence of market failures, it may be complicated to formulate effective and efficient regulatory measures. Therefore, it is important to assess the costs and benefits of regulatory measures upfront. An economic method to do this is the so-called cost–benefit analysis based on welfare economics, in which total costs and benefits from societal perspective are assessed (see Sect. 4.3). Another method to evaluate regulatory policies is cost-effectiveness analysis by which the costs for the realisation of regulatory objectives are determined (see Sect. 4.4).

4.3   Cost–Benefit Analysis

4.3.1   Welfare Economics

The theoretical basis for cost–benefit analysis (CBA) is welfare economics that analyses how a change in the allocation of commodities in an economy affects total welfare. Welfare (W) is defined as the sum of producer surplus (PS) and consumer surplus (CS), which implies that a change (Δ) in welfare is defined as a change in this sum:

$$\Delta W = \Delta PS + \Delta CS \qquad (4.6)$$

The change in the sum of producer and consumer surplus due to a policy measure is called the efficiency effect of that measure. Next to this effect, distributional effects can be distinguished. Although this distributional effect of welfare can be formulated in terms of producer and consumer surplus, these economic concepts are not really useful for policy discussions in practice. These concepts are theoretical concepts, not saying much about which groups in society are receiving which part of the improvement in welfare. Consumer surplus refers to the surplus realized by all users of a commodity, which means in the case of energy that the 'consumers' mainly consist of large energy-intensive industries, possibly partly owned by a group of international investors. For producer surplus holds more or less the same: the 'producers' of energy are all those agents who produce electricity, including residential prosumers and small-scale wind parks. In Chap. 11, we further discuss the distributional effects. In this chapter, we focus on efficiency effects.

An allocation of commodities can change due to many different factors, including regulatory measures. Suppose a regulator introduces a standard for the quality of a commodity which raises the costs of the supplying companies. For instance, the regulatory standard may refer to the level of environmental emissions of power generation. Suppose that the initial supply curve is S(p) = 1/2 p and this curve changes into S'(p) = 1/3 p, which means that the marginal costs have increased by 50%. Assume that the demand curve remains the same (D = 90 − P). The new supply curve results in a new equilibrium price of 67.5 euro/unit and quantity (22.5 units) (see Fig. 4.6, panel A).

Hence, due to the higher marginal costs, the market price becomes higher, while less commodities are exchanged. This lower level of production and consumption and higher costs reduce the welfare realized in this market. The resulting change in welfare can be calculated by determining the difference in PS and CS in the new situation versus the initial situation (900 and 450, respectively; see Sect. 4.2.3):

$$\Delta PS = \left(\frac{1}{2} * (67.5 - 0) * 22.5\right) - 900 = -140 \, euro \qquad (4.7)$$

$$\Delta CS = \left(\frac{1}{2} * (90 - 67.5) * 22.5\right) - 450 = -197 \, euro \qquad (4.8)$$

$$\Delta W = -140 + (-197) = -337 \, euro \qquad (4.9)$$

Only looking at the increasing costs due to the new regulatory measure, this measure has a negative effect on overall welfare (−337). If this would be the only

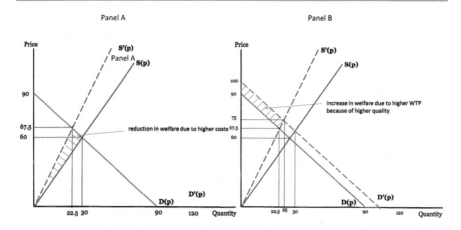

Fig. 4.6 Welfare effects of regulatory measures which raises costs as well as quality

effect, it is not efficient to take such a measure. The regulator, however, will take the measure for a reason. Hence, taking the measure will have some benefits as well. In order to have a positive overall welfare effect, the benefits of the measure should exceed the above negative welfare effect. A benefit of the regulatory measure can be that consumers appreciate the commodity more than they did before. For instance, if the standards refer to product quality, consumers may have a higher willingness-to-pay for the commodity because of the higher quality. Suppose that the new demand curve has become $D'(p) = 100 - p$, which states that the WTP of all consumers have linearly increased (see Fig. 4.6; panel B). Using this new demand curve, the new equilibrium price and quantity become 75 euro/unit and 25 units, respectively. The change in welfare of this new situation with both higher costs and higher quality can be calculated in a similar way:

$$\Delta PS = \left(\frac{1}{2} * (75 - 0) * 25\right) - 900 = +37 \, euro \tag{4.10}$$

$$\Delta CS = \left(\frac{1}{2} * (100 - 75) * 25\right) - 450 = -137.5 \, euro \tag{4.11}$$

$$\Delta W = +37 + (-137.5) = -100.5 \, euro \tag{4.12}$$

Hence, although the quality of the product has increased resulting in a higher WTP of consumers, the overall welfare effect remains negative. This means that this regulatory measure is inefficient.

4.3.2 Policy Variants and Scenarios

The general objective of cost–benefit analysis is to estimate the welfare effects of specific policy measures, paying attention to both efficiency effects and distribution effects. The first step in a CBA is determining the precise characteristics of a (proposed) policy measure. Examples of such measures in the field of energy policy are the implementation of a tax on the use of fossil energy, imposing the requirement on vertically integrated energy companies to unbundle network management from production and retail activities, implementing environmental standards on electricity generation processes, and investments in (strategic) energy storages.

The policy measure that is analysed in a CBA is the so-called project alternative as the implementation of the policy measure is the alternative for the business-as-usual (BAU) where not any new measure is taken (see Table 4.4). The description of the BAU forms the counterfactual in the CBA. Similar to the above example, the welfare effects of a policy measure are determined by comparing the consumer surplus and producer surplus in the project alternative with the surpluses in the counterfactual. The project alternatives can often be defined in various variants, as regulators may have several choices to pursue a particular objective.

Since a regulatory measure is generally meant for a longer period of time, it is essential to conduct the CBA over that longer period as well. In that future period, many other factors may affect the outcome of markets, which means that the impact of these factors has to be included in the analysis of the BAU. This can be done by formulating scenarios regarding the possible future development of these factors. Scenarios can be seen as internally consistent story lines about the future. These stories need to be internally consistent because the external factors are related to each other. For instance, assumptions on future prices of oil, gas and electricity should not be independent from each other as these prices are linked, as we have seen in Chap. 3. Scenarios are, therefore, internally consistent stories about how various relevant external factors jointly affect the variables of interest.

As the future is generally uncertain, it is useful to define a number of alternative scenarios to explore the range of uncertainties in exogenous factors. Such scenarios are typically defined on the basis of two key external factors that determine many other conditions. Such factors can, for instance, be the intensity of global climate policy and the role of public agencies and market parties in an economy. If these factors are plotted in a diagram, four quadrants follow where each one reflects one particular scenario.[6] Figure 4.7 shows an example of scenarios constructed in this way. In the scenario in the right-top quadrant in this figure, a strong climate policy is combined with liberalized markets, which means that climate policy is implemented through economic measures like energy taxes and emissions-trading schemes. In the quadrant in the left-top quadrant, a strong climate policy is

---

[6]In scenario analysis, it is common to have an even number of scenarios in order to prevent that users (e.g. policy makers) view one of the scenarios as the most likely. When scenarios are well defined, every scenario is even likely and conceivable, which makes them real alternative views on the future.

Table 4.4 Difference between policy variants and scenarios illustrated through two dimensions of a table

Policy variants (options for regulatory measures)	Scenarios (internally consistent stories about future external circumstances)		
	Scenario A	Scenario B	Scenario C
Null-alternative (Business-as-Usual)			
Project alternative 1			
Project alternative 2			

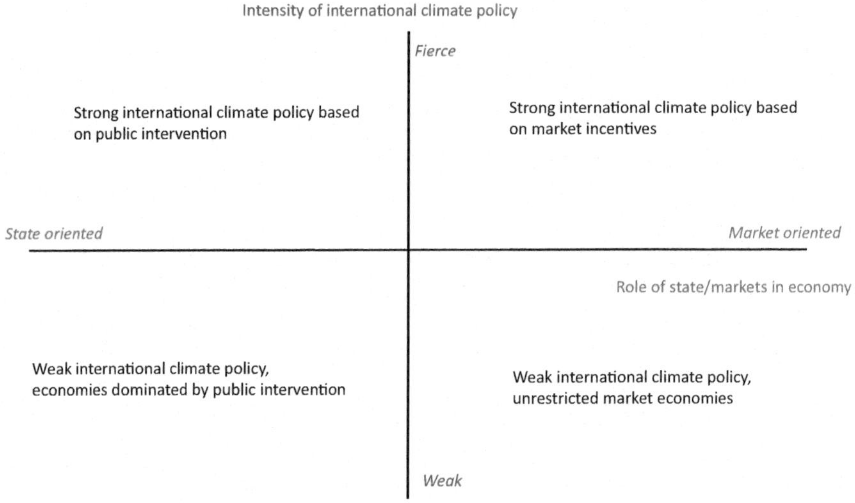

Intensity of international climate policy

Fig. 4.7 Example of definition of scenarios on the basis of two external factors

implemented through intensive regulatory interventions by the state, for instance, through public ownership and command-and-control measures. Hardly any climate policy is implemented in the scenarios in the two quadrants at the bottom. In the left-down quadrant, economies are based on a high degree of state intervention, while in the right-bottom quadrant, economies are characterized as unconstrained markets without much policy intervention in any domain.

For each scenario, the welfare effects of alternative policy variants can be analysed. These effects can be distinguished in various types.

4.3.3 Various Types of Welfare Effects

In a CBA, it is of crucial importance to clearly distinguish various types of welfare effects in order to prevent double counting and incorrectly define effects as cost or benefit.

The first obvious distinction is between costs and benefits on the one hand and expenditures and revenues on the other. On firm level, a subsidy is a revenue but from society perspective subsidies are transfers within society. Hence, subsidies do not affect the social costs or benefits of a project, they only affect the distribution of welfare. The same holds for taxes which are in principle only a transfer within the group of economic agents. Collecting taxes and distributing subsidies, of course, require efforts and, therefore, they result in costs, but these are called the transaction costs of taxation and subsidisation. In addition, financing of subsidies by raising taxes may create extra real costs through what is called the deadweight loss of taxation.

Another crucial distinction is between direct and indirect effects. The former type of effects occurs in the markets where the intervention is implemented, while the latter type of effects occurs in related markets. For instance, the direct effects of the implementation of an energy tax occur in the energy market, such as the effect on electricity use, while indirect effects occur in other markets, such as in the market for new power plants. Generally, the direct effects are the most important ones to analyse. If all markets operate efficiently, then one can assume that the indirect effects are negligible. For instance, if the price of new power plants decreases due to a lower demand for electricity, then this effect is included in the analysis of the direct effects as the lower price of plants is a benefit for power producers in the electricity market. If the market for power plants, however, is distorted, then the change in price of the power plants is not directly related to a change in market circumstances. The price of power plants may go down, but this is not necessarily related to an increase in efficiency as it can also result in lower profits for plant producers, which is a distribution effect, not an efficiency effect.

Box 4.1 Cost–benefit analysis (CBA) of unbundling of the electricity industry

One of the conditions for well-functioning electricity markets is that all producers and consumers of electricity have access to the grid against reasonable conditions. Before the liberalisation of these markets, electricity companies used to be vertically integrated, meaning that they were active in transport and distribution of electricity on the one hand, and in production and supply to end-users on the other. This position gave the vertically integrated incumbents a powerful market position, enabling them to obstruct competitors. In order to realize a level playing field among all players in the wholesale and retail markets, regulators in many countries have imposed unbundling, as was discussed in Sect. 3.4.2. The decision to unbundle the network operation from the other, commercial activities, may also have some negative effects. Unbundling may result in a loss of economies of scope as the coordination

Table 4.5 Elements in a CBA of the regulatory measure to impose ownership unbundling in an electricity industry

Costs	Benefits
One-off transaction costs (e.g. administrative efforts needed)	Better access of new producers to the network, affecting competition in the wholesale market (lower generation costs, lower margins)
Loss of economies of scope (e.g. more costs to be made to adapt timing of investments in network to developments in generation and load)	Better access of retailers to the network, affecting competition in the retail market (lower retail costs, more innovation, lower margins)
More risk for investments in generation capacity as the (regulated) network revenues cannot be used to attract capital	Side benefits, for instance: ownership unbundling may enable the shareholders to sell parts of the formerly vertically integrated company

Source Mulder et al. (2007)

within the supply chain and, in particular, between the network operator and electricity producers may become more complicated. In order to make the optimal decision for society, a CBA can be helpful as it is a systematic analysis of all effects of a specific intervention. Table 4.5 gives an overview of elements which could be included in a CBA of ownership unbundling.

Another crucial distinction in CBA is between internal and external effects. Internal effects are all effects which occur in markets, i.e. they are expressed in changes in prices. As an example, the benefit of the regulatory measure may be reflected in a higher willingness-to-pay for the commodity because the measure has resulted in a higher product quality. For such internal effects, one can make use of actual (or expected) market prices to determine their magnitude. In some cases, however, the benefits of a regulatory measure are not reflected in the demand for a commodity because of lacking markets for those benefits. As an example, a regulator may impose a restriction on electricity producer regarding the emissions of small particles or nitrogen in order to protect nature in the surroundings of the power plant. The value of this benefit cannot be derived from market prices or actual consumer behaviour, as there is no market for it. In order to estimate such external effects, they have to be monetarized in one or the other way, typically on the basis of non-market valuation techniques.

The theoretical basis for monetarizing the value of these external, so-called unpriced effects is given by the utility theory. In order to determine the value of a negative externality, such as emissions of electricity generation, one has to estimate the WTP of economic agents for a reduction of this externality. In other words, the marginal value of an externality can be determined by analysing which sacrifice agents are prepared to make in order to lower it, or the other way around, which compensation the agents require in order to accept a deterioration of the quality of the environment.

Market commodities (Y)

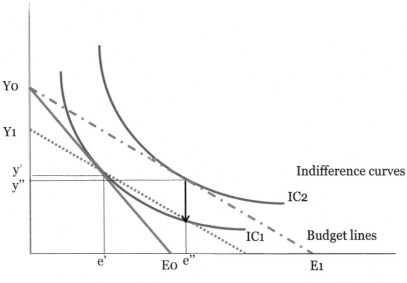

Environmental quality (E)

Fig. 4.8 Determination of WTP for environmental commodities

Using the concept of indifference curves, Fig. 4.8 depicts the analytical framework for this method. Suppose there are two types of commodities: a market commodity Y and an environmental commodity E. Let assume further that economic agents derive utility from consuming both types. More Y and a same amount of E means a higher utility, and vice versa. If Y increases and E decreases the utility can remain the same. The set of combinations of Y and E in which the agent enjoys the same utility is expressed through the indifference curve. In the figure, two indifference curves are depicted by IC1 and IC2. An important aspect in any economic analysis is that economic agents are restricted: we cannot have everything; hence, choices have to be made. The necessity to make a choice is measured here by the so-called budget line. The initial budget line is shown by the steep bold curve Y_0E_0. This curve shows that an agent can either have Y_0 of the market commodities or E_0 of the environmental commodity or any combination on or below the line Y_0E_0. Given the shape of the indifference curves and the budget line, the utility of the agent is maximized by choosing y' market commodities and e' environmental commodities.

Suppose that there is an exogenous increase in the quality of the environmental commodity. As a result, economic agents can have a higher environmental quality for the same amount of market commodities. This implies that the budget line shifts from Y_0E_0 to Y_0E_1. This changed budget line enables the agent to reach the indifference curve IC_2 which is above the curve IC_1, implying that a higher utility can be reached. The new optimal composition of the goods becomes $y''e''$, which means that new optimal choice consists of a higher environmental quality and less market commodities. This makes sense, because the environment has a higher quality which is equal to saying that the price per unit of quality has decreased.

The question now is: what is the monetary value the economic agent attaches to this increase in quality of the environmental commodity? Theoretically, this value is equal to the sacrifice in market commodities the economic agent is willing to make in order to return to the initial level of utility (IC_1). In the figure, this value is depicted by the length of the arrow. This value is called the Hicksian compensation variation.This theoretical framework for estimating the monetary value of unpriced effects is straightforward, but applying this framework in specific cases may be more challenging. As discussed above, the basic idea is to estimate the sacrifice agents are prepared to make to realise an improvement in quality or to determine the minimum compensation they want to have in order to accept a deterioration of the quality. The former estimate is called the WTP of a good, and the latter the WTA (willingness-to-accept). Several methods exist to estimate these values, such as the Contingent Valuation Methods (CVM) and Choice Experiments (see Box 4.2).

Box 4.2 Estimating the WTP for CO_2 reductions through a Choice Experiment

In the absence of markets, it is not possible to use market prices to infer the value consumers attach to a commodity. This holds, for instance, for environmental effects, such as a reduction in the emissions of CO_2. For governments, it can be valuable to know which premium citizens are prepared to pay for a commodity with a lower amount of emissions. After all, when this WTP is known, governments can try to develop a market that enables consumers to actually pay that premium. Examples of such markets are green-certificate systems (see Sect. 5.2.3). If consumers pay a higher price for products with lower emissions, less government subsidies are required to compensate for the extra costs of products with less negative environmental effects.

A technique to estimate this WTP is the Choice Experiment. In this experiment, a sample of citizens (consumers) is offered a series of bilateral choices for products: each time they have to choose the product of their preference. The products are described on the basis of a number of attributes, including a financial attribute and the environmental attribute. On the basis of their choices, the WTP for the environmental attribute can be expressed in terms of the financial attribute.

An example of such a study is the analysis by Hulshof and Mulder (2020) who estimate to what extent consumers are prepared to pay extra for a passenger car with lower emissions. The attributes that describe the characteristics of the products the respondents could choose from are the purchase price of a car, the fuel type, the fuel costs and the CO_2 emission per kilometre. The authors find that the average consumer is prepared to pay about 200 euro per ton of CO_2 emission reduction. Among the group of respondents, however, a large variety in preferences exists. It appeared that in particular females, people aged 65+ and in individuals with higher education have a higher WTP for CO_2 reduction than others.

A caveat of this kind of so-called stated-preference methods is, of course, that the analysis is based on an artificial situation, which may not perfectly reflect the actual situation consumers have to make choices. When in reality, however, markets no not exists, it is more difficult to use revealed-preference methods.

4.3.4 Discounting

As the effects of regulatory measures generally extend to the future period, one has to control for the timing of costs and benefits. If costs only occur in the short term, while benefits occur in the more distant future, it is crucial to take this difference in timing into account. This can be done by calculating the net present value (NPV) using the following formula, in which r refers to the discount rate, t to years and T to the expected lifetime of a project in years:

$$NPV = \sum_{t}^{T} \frac{(Benefits - Costs)_t}{(1+r)^t} \tag{4.13}$$

The NPV of a regulatory measure should be at least positive; otherwise, the measure results in a loss of welfare. The crucial parameter in this analysis is the discount rate. The key factors behind the discount rate are the time value of money, the required premium to compensate for the risks and the inflation. The first factor is related to the fact that economic agents generally value the present more than the future, the second factor is related to the uncertainty and the risks regarding the future benefits, while the third factor is related to the fact that the value of money as numeraire (generally) decreases over time.

The value of the discount rate also depends on the perspective that is taken to do the analysis. When the perspective is the one from a private investor who is considering the costs and benefits of a particular investment project, one has also to take into account project-specific risks. For instance, when a project refers to building an offshore wind park, investors typically also want compensation for risks related to the construction and operation of the wind park. These risks are less relevant when the analysis is done from a society point of view, as the project-specific risks are neutralized by specific risks of other investment projects. Moreover, from a society point of view the future benefits may be more relevant than for a private investor (see e.g. Zweifel et al. 2017). After all, a society can be seen as constituted of not only the current generations living, but also future generations. As a result, the effects of a project for them may be given an equal weight, resulting in a much lower discount rate. On the other hand, future generations may enjoy a higher income, which may be a reason to give a higher weight to the short-term economic effects compared to those for the long term. In general, however, the social discount rate is typically significantly lower than the discount rate of a private investor (Nas 2013).

4.4   Cost-Effectiveness Analysis

Another method to assess economic effects of alternative policy measures is cost-effectiveness analysis (CEA). The key difference between CEA and CBA is that in the former analysis the magnitude and monetary value of the benefits are not estimated, but used as denominator for the costs. If the objective of a policy measure, for instance, is to reduce the emissions of carbon or to increase the amount of renewable energy, then the costs can be expressed per unit of carbon reduction or per unit of installed renewable energy capacity. This type of metric enables regulators to assess by which policy measure the policy objective can be realized at the lowest costs. The costs can be determined from both a social perspective or from the perspective of a particular actor, such as the regulator.

When the costs are determined from a society perspective, one has to keep in mind the distinction between costs on the one hand and expenditures on the other hand. Subsidies given to, for instance, residential households for carbon reduction are not necessarily equal to the costs these households make to reduce the emissions. As we will see in Chap. 8, support schemes are often vulnerable to over-subsidization resulting in windfall profits realized by the recipients of the support.

When the costs are determined from the perspective of one particular agent, then all expenditures made by this agent to realize the objective are relevant. Typically, this type of analysis is done to analyse how much government support is needed to realize a particular objective. If governments have a preference for lower government spending, than they may want to look for those policy measures which realize an objective at the lowest public expenditures. This is not necessarily equal to the policy measures having the lowest cost for society.

Exercises

4.1 Explain why a positive externality is seen as a market failure and result in a suboptimal welfare.

4.2 Suppose a market is characterized by the social supply curve $S(p) = 3q$ and the demand curve $D(p) = 50 - 2q$.

(a) Calculate the equilibrium price and quantity

(b) Suppose the firm does not include all costs in its private supply curve because of negative externalities. The supply curve the firm takes into account is given by $S(p) = 2q$. What will be the equilibrium price and quantity in this situation?

(c) Which conclusion can be drawn on the impact of a negative externality on market outcomes?

4.3 What are sources of market power in energy markets?

4.4 Why may the development of a district heating system be hampered by a hold-up problem?

4.5 What is the difference between a variant and a scenario in a cost–benefit analysis?

References

Bhattacharyya, S. C. (2019). *Energy economics; Concepts, issues, markets and governance* (2nd ed.). Berlin: Springer.

Decker, C. (2015). *Modern economic regulation; An introduction to theory and practice.* Cambridge University Press.

Hulshof, D., & Mulder, M. (2020). Willingness to pay for CO_2 emission reductions in passenger car transport. *Environmental and Resource Economics, 75*(4), 899–929.

Mulder, M., Shestalova, V., & Zwart, G. (2007). Vertical separation of the Dutch energy distribution industry: An economic assessment of the political debate. *Intereconomics, 42*(6), 292–310.

Nas, T. F. (2013). *Cost-benefit analysis: Theory and application.* Lexington Books.

Varian, H. R. (2003). Intermediate microeconomics; A modern approach (6th ed.). New York/London: W.W. Norton & Company.

Veljanovski, C. (2012). Economic approaches to regulation. In R. Baldwin, M. Cave, & M. Lodge (Eds.), *The Oxford handbook of regulation.* Oxford University Press.

Zweifel, P., Praktiknjo, A., & Erdmann, G. (2017). *Energy economics; Theory and applications.* Springer Texts in Business and Economics.

Information Asymmetry in Retail Energy Markets

5.1 Introduction

Economic agents need to have full information on characteristics of commodities in order to successfully conclude contracts with each other. In real markets, however, consumers and producers often don't have the same information which results in so-called information asymmetry. Regulatory options to address this information asymmetry are discussed in Sect. 5.2. For residential energy consumers a specific information problem occurs, which is that they sometimes cannot oversee all the products offered by retailers. The regulatory options to address this are discussed in Sect. 5.3. Finally, residential consumers may not be able to oversee the ability of energy retailers to commit to supply contracts which may affect the reliability of supply. Section 5.4 discusses the regulatory options to protect residential consumers against this risk.

5.2 Information on Product Characteristics

5.2.1 Market Failure

In well-functioning markets, consumers are able to obtain the necessary information on the qualities of a commodity, while producers have the incentive to be fully transparent (Varian 2003). The qualities of commodities may refer to their physical characteristics (e.g. energy content in MWh per m^3), but also to financial (e.g. price) and other contractual characteristics (e.g. duration of a contract). In a competitive market, suppliers have an incentive to distinguish themselves by offering, for instance, different types of products and different types of contracts as this may enable them to realize higher prices or higher market shares. In addition, in the ideal situation, suppliers also have the incentive to be transparent about the product

© Springer Nature Switzerland AG 2021
M. Mulder, *Regulation of Energy Markets*, Lecture Notes in Energy 80,
https://doi.org/10.1007/978-3-030-58319-4_5

characteristics. After all, only by being transparent firms can distinguish themselves and realize higher revenues and profits.

Despite these incentives, sometimes it may be impossible or very costly for consumers to be fully certain about the information received from suppliers. This may be the case, for instance, when they cannot experience themselves how a product is produced. Such an inability to check the true characteristics of a commodity may give the suppliers the ability to sell a low-quality product as a product of higher quality and, hence, realize a higher price.

When consumers are aware that the supplier may not tell the true story and it is costly or impossible to collect the relevant information on the product characteristics, they have to make an estimate of the product quality themselves. If consumers make an estimate of the product quality, they can be wrong in two directions. If consumers underestimate the quality of a commodity, their willingness-to-pay for that commodity is below the price they would be prepared to pay when they are perfectly informed. Hence, the actual demand curve is positioned below the demand curve in the situation where consumers are well informed as for each commodity the marginal willingness-to-pay is lower (Fig. 5.1). If consumers would know the actual quality, the equilibrium level of consumption would be higher (Q1 instead of Q2) and the same holds for the equilibrium price (P1 instead of P2). This underestimation of the quality of a commodity results in a deadweight loss, indicated by the blue triangle in the figure. Besides this loss of welfare, the

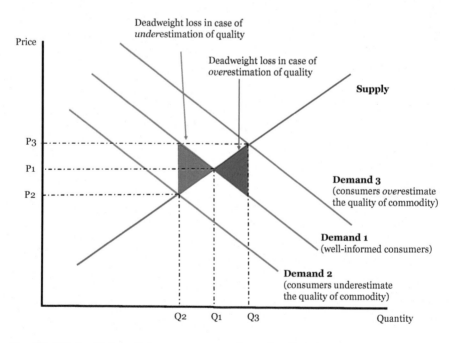

Fig. 5.1 Effects of information asymmetry on market and welfare

lower market price implies that the inframarginal suppliers (i.e. the suppliers left from the equilibrium outcome) realize lower revenues which affect their willingness to invest in equipment to supply the commodity.

The impact of an underestimation of the true quality is also called the problem of hidden information or adverse selection. This means that products of higher quality and higher production costs are not bought by consumers when they are not sure that the higher price they have to pay is related to a higher quality. This may the case when, for instance, it is too costly for them to acquire all the necessary information. As a result, firms cannot profitably offer the high-quality and high-cost product to the market. Hence, such products leave (or do not enter) the market.

The opposite situation occurs when consumers overestimate the quality of a commodity. This over estimation may be due to incorrect information provided by suppliers, while consumers are unable to control that information. In such a situation of overestimation, the actual demand curve is positioned above the demand curve of well-informed consumers. For each unit of the commodity, consumers are prepared to pay more than what they would do if they were aware of the true (lower) quality. This results in a deadweight loss indicated by the red triangle in the figure. This loss of welfare means that the actual consumption (Q3) is above the social optimal level (Q1), which implies that for the extra consumption (i.e. Q3-Q1), their true WTP is below the marginal costs of supplying these commodities. Note, however, that an overestimation of the true qualities of a commodity may be good for welfare in the case of market power. After all, in such a situation, the deadweight loss resulting from market power (i.e. consumers consume too less) can be neutralized by the deadweight loss resulting from the overestimation of the quality (i.e. consumers consume too much) (see further on the welfare effects of market power Chap. 9).

5.2.2 Variety in Energy Products

Because energy products are produced in many different ways and from different sources, the quality of energy products can vary strongly (see Sect. 2.2). There are, for instance, many different types of oil with various qualities (such as North Sea Brent, US WTI and Arab Light). The same holds for coal with four different groups of coal qualities (hard, black, dark brown and brown coal) and also for natural gas with various thermic values based on the content of methane. Electricity is neither a homogenous product as its quality is determined by the electric voltage, frequency (in case of alternating current) and the capacity of conductors determining the amount of current. Heat can also be offered in different variants determined by the temperature.

This large variety in the various types of energy imply that energy markets can only function properly if buyers can be well informed about the specific quality of a particular energy product. Producers have an incentive to provide this information as consumers are, in principle, able to check whether the provided information is correct. They can measure the relevant physical characteristics (such as sulphur content of oil, methane content of gas, voltage of electricity and temperature of heat) and, as a result, they are able to determine the maximum price they want to pay based on the true characteristics of a commodity and their own preferences. Hence, for product characteristics which can be measured by consumers themselves, markets are able to take care of sufficient information provision.

For some product characteristics, however, consumers are unable to measure the value themselves. In energy markets, this holds in particular for the way energy is produced. Electricity with a specific set of physical characteristics (power, voltage and current) can be generated in various ways: through a nuclear power plant, wind turbine or coal-fired power plant (see Sect. 2.3.4). The different sets of generation technologies produce the same type of end-product, but users may be not indifferent regarding the way electricity is generated. The same holds for gas (i.e. methane) which can be produced by depleting natural-gas fields or through gasification or anaerobic digestion of biomass and upgrading to the same physical qualities as natural gas (see Sect. 2.3.3).

This lack of information gives rise to the hidden information problem: consumers do not have the same information as producers. If consumers are uncertain about how the energy is actually produced, producers of particular types of energy (like renewable energy) may not be able to obtain a premium for the specific way their energy is produced. Without such a premium, more costly energy production techniques may be unable to obtain a (higher) market share unless other kinds of support are given (see Sect. 8.4). The outcome of the hidden information problem may also be that consumers overestimate the true qualities of the energy product, such as regarding the carbon emissions. If consumers believe that an energy commodity is generated without any carbon emissions, while this is actually not the case, they may be prepared to pay a price for the energy which exceeds their WTP when they would be well informed. In both circumstances, the market for renewable energy is hampered due to a lack of transparency regarding the product qualities.

5.2.3 Regulation

In principle, the problem of hidden information can be addressed by market parties themselves. After all, suppliers of high-quality products have an interest to solve this problem as that helps them to realize higher prices and higher market shares. This response by suppliers is called signalling (Varian 2003). The challenge of this response is to develop a system of information provision that is credible to consumers (Viscusi et al. 2005).

Methods for signalling are systems of warranties or certificates. These certificates disclose information on the production process and can be seen as proofs that something (a commodity or a process) meets certain criteria. These systems have been introduced in energy markets to address the problem of hidden information on how energy is produced. In the case of renewable energy, certificates provide evidence that the energy is produced in a renewable way. The certificates for renewable energy may also give information on the carbon emissions per unit of energy.

Voluntary systems of green-energy certificates have been introduced in the gas market. These systems are organized in collaboration between producers, traders and consumers (see Box 5.1). Regulated systems of green-energy certificates exist in the electricity market (see Box 5.2). Because of the presence of certificate systems, producers of green energy can realize extra revenues on top of the price of the energy commodity. Hence, green-energy producers sell two products: energy as commodity and a certificate that tells that the energy is produced in a renewable way. Consequently, the marginal revenues (ΔR) of a green-energy producer depend on the energy price realized in the energy market (p_e) as well as the price received for the certificates (p_c):

$$\Delta R = p_e + p_r \tag{5.1}$$

The better the market for the green-energy certificates functions, the more the producers will be able to receive extra revenues from consumers having a positive willingness-to-pay for the renewable character. These extra revenues may help these producers to compete with the suppliers of conventional energy.

Depending on the regulation in place, green-energy certificates may also be needed to receive subsidies. In that case, the certificates have to be handed in to the public agency responsible for the support scheme. It is sometimes also possible that green certificates can be converted into renewable fuel certificates in another industry, such as transport.

When green-energy producers operate in the same network system as conventional producers, the commercial exchange of the energy and the certificate is done next to the actual injection of energy into the network (see Fig. 5.2). Hence, in systems with green-energy certificates, producers of green energy do in principle three things: (a) injecting energy into the network (which meets the physical requirements determined by the network operator), (b) selling energy in the wholesale market and (c) transferring the green-energy certificates to, for instance, retailers or end-users, or to an agency responsible for green-energy support schemes.

The issuance of the green-energy certificates is generally done by a certification body that is closely connected to the network operator in order to guarantee that the issuance is directly related to the injection of energy into the grid. A green-energy producer only receives certificates for units of energy that are actually injected. This mechanism prevents that certificates are given for energy that is produced, but not really injected and used. The presence of this mechanism indicates that the key condition for the well-functioning of the certificates system is the ability to convince end-users that the certificates can be trusted to give the correct information on the characteristics of the product they want to buy.

To assure that the issuance and use of certificates is directly linked to the actual flows of energy into the grid, the issuing bodies may use the so-called mass-balancing approach. Under this approach, the transaction flows of the certificates are connected to the physical flows of the underlying energy commodity. This approach is used in the gas market because of the European Renewable Energy Directive and the Fuel Quality Directive, which only recognize international trade

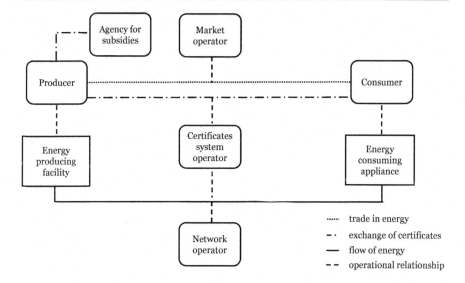

Fig. 5.2 Energy certificates system in relation to energy market and network

in certified liquid and gaseous biofuels when the physical transfer is coupled to the trade in certificates. Consequently, international trade of renewable gas certificates can only be done between countries for which a physical connection exists. In addition, when green-gas certificates are used to make transport fuels green, the certificates should also be linked to the mass-balancing approach.

The mass-balancing approach is meant to prevent double counting of the renewable value of the gas that is produced in a renewable way, but a caveat is that it restricts international trade as certificates can only be traded if there is a connection with physical exchange of gas in the same network region (see CertifHy 2015). For domestic gas systems, however, it is possible to use an alternative system, which is the so-called book-and-claim approach.

Box 5.1 Organization of a voluntary system of renewable-gas certificates

In European gas markets, voluntary renewable-gas certificates have been introduced in several countries (Moraga et al. 2019). The main ingredients of renewable-gas certification systems are depicted in Fig. 5.3. Renewable gas as biomethane is made on the basis of biomass (see Sect. 2.3.3 for the physical characteristics of biomethane). Therefore, the starting point for renewable-gas certificates consists of information on the inputs (i.e. the biomass) that is used to make biomethane. As not all types of biomass may be called renewable or sustainable, standards are required for the types of biomass which may be used to make biomethane. These standards can be developed by the industry itself through the definition of common industry standards (such as NEN standards).

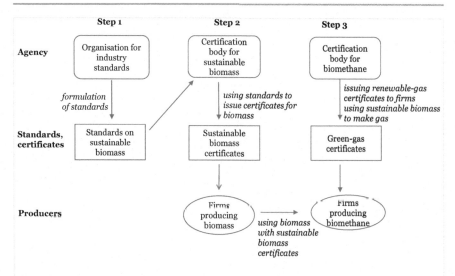

Fig. 5.3 Ingredients of the organizations of a system of green-gas certificates (based on Moraga et al. 2019)

In the next step, a certification body can issue certificates for sustainable biomass to producers of biomass that meet the criteria based on these industry standards. When this biomass is used to produce biomethane, i.e. gas that has the same characteristics (such as methane content) as natural gas, it can be injected into the gas grid. In order to enable the producer of this renewable gas to receive a premium for the greenness of the gas, a certification body for renewable gas may issue renewable-gas certificates to these producers that use sustainable biomass as input.

In the book-and-claim method, there is no relationship between the physical flows of the underlying commodity and the trade in the certificates. This approach is used in the European electricity market. This means that the international trade in green-electricity certificates is not constrained by restrictions within the European electricity network as they don't play any role. As a result, for instance, Norway may export amounts of green-electricity certificates to Germany and Netherlands that significantly exceed the capacity of the respective cross-border interconnections; likewise, Iceland may export certificates to the continent in the absence of any power interconnection.

Box 5.2 European system of green-electricity certificates

The green-certificates systems for electricity in Europe are based on a European certification scheme for renewable energy [the so-called Guarantees

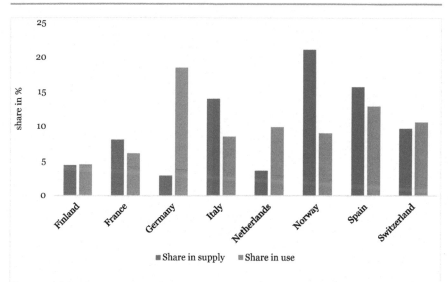

Fig. 5.4 Major European countries issuing and cancelling green-electricity certificates. *Source* AEB (2018)

of Origin of Electricity Produced from Renewable Sources, based on EU Directive (2009/28/EC)]. In all EU Member States, producers of renewable energy receive an EU-certificate for each MWh that they produce. These certificates may be traded, both nationally and internationally. The issuance, trade and cancelation of certificates are standardized through the European Energy Certificate System (EECS), which is organized by the Association of Issuing Bodies (AIB).

The major supplying countries of certificates in Europe are Norway, Spain, Italy and Switzerland, together responsible for 61% of total supply (AEB 2018). The major cancelling countries (i.e. countries where the certificates are used to make electricity consumption green) are Germany, Spain, Switzerland and the Netherlands (see Fig. 5.4). The large differences between issuance and cancelation on country level indicate the magnitude of international trade in these certificates.

Although Germany is the main producer of renewable electricity in Europe, its share in the supply of certificates is small because a large part of this production is funded under the Feed-in-Tariff scheme (see Sect. 8.4), which implies that the renewable electricity is not offered on the market but transferred to the network operator in return for the subsidy. Therefore, in order to be able to sell green electricity to consumers, German retailers have to import green certificates on the international market.

Although the implementation of green-electricity certificates is based on European regulation, EU Member States have some freedom to determine the precise design of the scheme. It appears that the countries have made different choices, in particular regarding the market design and ownership of the certifier. In markets where the certificates are based on international standards, the liquidity of the market appears to be significantly higher than in markets where the certificates are based on national standards (Hulshof et al. 2019). Apparently, international standards facilitate international trade and, hence, the options for traders to buy and sell certificates. In addition, public ownership seems to affect the trust market participants have in the certificates system, as in markets where the certifier is publicly owned the liquidity of the market is higher.

Despite the many years of experience, green certificates system in the electricity market cannot be viewed as mature (Hulshof et al. 2019). The liquidity of the market is still fairly low, reflected in the fairly low churn rates, which measures how often a commodity is traded before it is actually used (in this case, cancelled) (see Sect. 3.2). Moreover, this market is still not very transparent, as full information on the volumes and prices of transactions is lacking. Despite this low liquidity, international trade in green-electricity certificates has increased. The import of these certificates is playing an increasing important role in European retail markets.

An alternative for the green certificates market would be a system in which the information on production characteristics is directly revealed during the trade in a commodity. In such a system, instead of selling just electricity or gas in terms of energy content (MWh), producers sell electricity or gas not only in energy content but also differentiated according to the production characteristics. For a specific hour, there would be several electricity prices instead of one if there are several types of products offered to the market. Such a design of the electricity market, which currently does not exist, requires intensive use of information technology, making this market even more complex than the market already is, while this also may negatively affect the liquidity of these energy markets.

5.3 Information on Supply

5.3.1 Market Failures

In competitive markets, suppliers can only make profit by operating more efficiently than their competitors or by distinguishing themselves from their competitors by offering different types of products. When suppliers are successful in this latter strategy, they create a kind of niche markets where they can charge higher prices. Such niche markets may even occur in markets where the products are homogenous in physical sense, such as in retail electricity and gas markets. These products are homogenous in physical sense in the way there are transported and supplied to residential consumers, but these products can be differentiated in many ways. The differentiation can be implemented through the use of certificates, as discussed in the previous section, enabling suppliers to supply various kinds of electricity (like grey, green, European wind, or domestic wind). In addition to this differentiation based on certification, retailers can differentiate their products in terms of contract attributes, such as length of contract period, structure of price, level of price(s), extra discounts, or gifts.

A consequence of these differentiation activities by retailers is that the total supply of energy products to consumers may consist of a large number of different products (see Box 5.3). This increase in product variety can be seen as a benefit of competition, which has stimulated retailers to innovate and develop new products which may be appreciated by consumers (Littlechild 2000).

Despite the benefits for consumers to have so many options to choose from, they may experience problems in overseeing the total supply. This holds in particular for residential consumers and less for business consumers, which is due to the limited ability of the former type of consumers to collect and assess information in combination with the presence of psychological biases in decision making (Kahneman 2011). Even if the characteristics of separate products are completely clear to consumers, it may still be difficult for them to make the best choice for themselves (Creti and Fontini 2019).

Box 5.3 Variety in retail energy products in EU

Since the liberalization of retail energy markets in the EU, the variety of products offered to consumers has increased strongly (see Fig. 5.5). In many EU countries, consumers can choose, for instance, between contracts with fixed prices during a period of time, contracts with prices which can change at any time and contracts with prices which are directly linked to the actual development in the wholesale spot market. They can also choose between grey and various types of green energy. In some countries, consumers can also choose contracts that give a discount for having only online contact with the retailer or contracts with extra services, such as advice on energy efficiency measures.

Because of the limited information processing abilities of residential consumers, it may be economically rational for them not to explore the full market and to search for the best deal, but to stay with the incumbent supplier and to keep the contract details unchanged. This behaviour may be efficient for these consumers when they do not want to make the effort to search and assess new information. Because of these search and information costs, consumers may be willing to pay more for the existing contract with the incumbent supplier instead of searching for a cheaper, more attractive contract, provided of course that the price difference with the alternative suppliers is not too big. If the price difference between the current contract and new offers in the markets, however, becomes large, also these consumers may want to switch. From this follows that the difference in prices charged

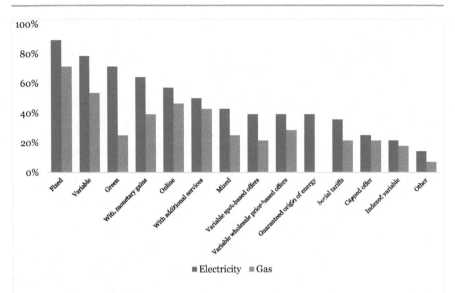

Fig. 5.5 Types of products offered in retail markets, in % of EU Member States, 2017. *Note* 'The offers are defined as follows: • Variable pricing: Price paid per unit of electricity used can change at any time • Fixed pricing: An offer that guarantees that the price paid per unit of electricity used will not change for a given period of time • Mixed pricing: i.e. based on both fixed and variable components • Variable spot-based offers: Variable price based on the wholesale market spot price • Variable wholesale price-based offer: Settled against monthly/weekly average wholesale price • Capped offer: The offer guarantees that the price paid per kWh for electricity will not rise beyond a set level for a given period of time, but may go down—usually for this certainty customers pay a small premium • Indexed variable: Similar to spot-based, which is linked to wholesale, but linked for example to standard incumbent offer with guaranteed discount of x% or to RPI • Green offers: Based on renewable generation resources like hydro, solar, wind, biomass etc. • Online offer: With savings/discount for managing accounts online, online billing • Social tariffs e.g. offers for vulnerable consumers • Guaranteed origin of energy (any energy source other than green or country-specific) • With monetary gains (e.g. discount, supermarket vouchers, etc.) • With additional services (e.g. energy efficiency, boiler maintenance etc.) • Other: special offers for electrical vehicles or dual fuels'. *Source* CEER (2018)

by the various retailers in a market can be used as an estimate for the search and information costs consumers are willing to make to switch from retailer and/or contract. After all, if the price difference would be larger, consumers would switch. Hence, these search and switching costs (ss) can be inferred from the actual difference between the maximum (p^{max}) and the minimum (p^{min}) price in the retail market:

$$ss < p^{max} - p^{min} \qquad (5.2)$$

This equation says that if the price difference is less than what consumers perceive as search and switching costs, they will not switch to another contract. Because of this response of consumers to price differences, suppliers will set their maximum price below the sum of the minimum price of competitors and the costs consumers have to make to search for alternatives and to switch. Note that that the price difference among products can only exist in the case of heterogeneity among consumers. After all, if consumers would be homogeneous and experience the same level of search and switching costs, then no retailer would have an incentive to offer a lower price. Just because some consumers experience lower costs, a retailer may make more profit by offering a lower price to these consumers. This heterogeneity will be further discussed below.

The search and switching costs do not only refer to (perceived) financial costs consumers have to make, but also to the unpriced effort which they believe is required. This explanation of consumer behaviour is based on the common microeconomic analysis of markets. Sometimes, however, consumer behaviour cannot be completely understood from this standard economic approach. In such cases, this behaviour is not rational from an economic perspective. In this perspective, by irrational behaviour is meant consumer behaviour that is not in the interest of the consumers themselves (see also Sect. 4.2.1).

To understand such irrational behaviour, the insights from behavioural economics and psychology are useful (see Table 5.1). These approaches can give valuable explanations for consumer behaviour which cannot be called economically rational, for instance, when consumers do not switch to another retailer, while even when controlling for search and information costs such a decision would be beneficial for them. It appears, for instance, that consumers can suffer from loss aversion, which means that the risk of potential loss is valued stronger the risk of a potential benefit. When loss aversion is present, consumers tend to be risk averse which means regarding their behaviour in retail markets that they are not inclined to switch from their default retailer, because the potential losses of such a switch are valued more strongly than the potential benefits.

Another psychological factor affecting consumer behaviour is the status-quo bias, which means that consumers tend to attribute a higher value to the current situation compared to any other situation. Consumers also may suffer from time inconsistency, which means that they are making inconsistent decisions when they

Table 5.1 Behavioural-economic explanations for consumer behaviour in retail energy markets in the UK

Observed behaviour of consumers	Explanation from behavioural economics
70% finds it difficult to oversee the market	Limited capabilities to deal with a lot of information
60% has never switched from supplier; 77% is satisfied with current supplier	Preference for the status quo
58% believes switching from supplier will create many problems	Loss aversion
Many consumers are sensitive to advertising and, as a result, make decisions which are not wise from a long-term perspective	Time inconsistency

Source OFGEM (2011)

have to value effects in the short- and long-term future. As a consequence of time inconsistency, a consumer may underestimate the benefits of having a long-term fixed retail price and overestimate the benefits of having a discount in the short term.

Because of the search and information costs as well as these behavioural factors, many consumers do not to behave like active consumers in energy retail markets. Some appear to be passive and a bit naïve, meaning that they do not have a good picture of the market circumstances and the methods suppliers use to sell their products. Because of this naivety, some consumers are also vulnerable to exploitation by suppliers. Besides these so-called non-savvy consumers, there is also a group of consumers who are active players in the retail market and who fully understand how the game of selling and buying is played. This group can be called savvy consumers (Armstrong 2014).

Between these groups of consumers, two types of externalities exist (see Fig. 5.6). The group of non-savvy consumers may pay a higher price for its products because retailers can charge higher prices without the risk that these consumers will switch to a competitor. The resulting higher profits realised in the sales to these consumers enable the retailers to behave more competitively towards the savvy consumers. Hence, the latter may benefit from the lack of activity of the previous group. This is called the rip-off externality by Armstrong (2014). This is, of course, only possible, when retailers can distinguish between the different types of consumers. Insofar they are not completely able to do this, there exists another externality by which the non-savvy consumers benefit from the active searching behaviour of the other group of consumers. Because the active behaviour of the latter group of consumers incentivizes retailers to make their offers more competitive, the less active consumers may benefit from this as their prices may also go down. This is called the search externality.

5.3.2 Regulation

In order to help consumers to get a better overview of the various products they can choose from and to protect them against exploitation by suppliers, regulators can take a number of measures. The general objective of this regulation is to make information on supply better available to consumers (Baldwin et al. 2012). These measures refer to information on product characteristics, the process of information provision and retail behaviour regarding consumers (Table 5.2).

The process of searching and assessing information on product characteristics becomes easier when this information on the product characteristics is structured in a standardized way. Regulators can impose rules for the standardization of information which describe product characteristics. An example of such a standard is the rule that retailers have to inform consumers about the total costs per unit of energy, based on their average consumption during a year. When all retailers use this metric to describe the product, then consumers are better able to compare the various offers.

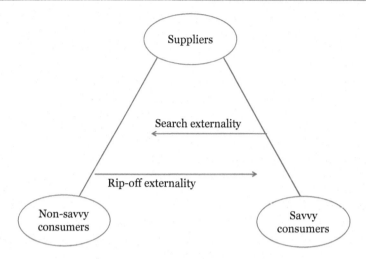

Fig. 5.6 Search and rip-off externalities between consumers in retail markets. *Source* Definition of externalities is derived from Armstrong (2014)

Table 5.2 Regulatory measures to support consumers in making decisions in retail energy market

Objectives of regulation	Examples of regulatory measures
Product characteristics	• Imposing standards for information content and structure of product offers and contracts • Protecting consumers against unreasonable prices
Process of information provision	• Imposing standards for price-comparison websites
Retail behaviour	• Giving consumers the option to cancel decisions within a short period after the decision is made

In addition to this regulation on the type of information provided on product characteristics, regulators may also want to implement more intervening types of measures. An example of such an intervention is a regulation that explicitly aims to protect consumers against too high prices (see also Schotter 1996). This regulation is in particular meant to protect the less active consumers who may be exploited by retailers. This regulatory supervision on the energy prices charged by retailers requires that the regulator is able to determine what (un)reasonable prices are. In general, this can be defined on the basis of the wholesale prices of energy and the required margin for retailers. Determining the relevant wholesale price is, however, not straightforward as retailers have many options to determine their portfolio of wholesale contracts to meet their commitments in the retail market, as will be discussed in Sect. 7.4. Another caveat of this regulatory measure is that restricting the retail prices may negatively affect the competitive pressure in the retail market (Mulder and Willems 2019).

Regarding the process of information provision, it appears to be important that all information is available on a limited number of places and that the information is provided in a transparent and independent way. In order to realize such a situation, regulators can impose rules on the information provision on websites which consumers can use to compare various types of contracts. An example of such a rule is that the website has to explicitly mention which company is operating the platform and how that company is related to retail companies. This rule makes it possible for consumers to assess to what extent the price-comparison websites are operated in an independent way.

In addition to these regulations, regulators may also want to impose rules on how retailers should behave when they contact potential customers. Because we know from behavioural economics that human beings tend to make irrational decisions when they are forced to make these decisions in a short period of time,[1] regulators can protect residential consumers against this risk by giving them the ability to cancel purchase decisions within say two weeks of time. Such a rule protects consumers against the errors occurring due to forced fast decision making.

5.4 Information on Firm Characteristics

5.4.1 Theory

Besides being well informed about all product characteristics and the differences among the various products offered by suppliers, consumers also need information on the ability of firms to supply a commodity in future. This holds in particular if consumers are about to enter into a longer-term contract with suppliers. In that case, they need to know to what extent a firm is really able to supply the commodity also in future against the conditions which are agreed in a contract. If consumers are uncertain about the real ability of the firm to commit for a longer period of time and if the consumers are not able to conclude an extra contract to reassure this commitment, they may not want to buy from a specific supplier. This market failure is called the problem of hidden action or 'moral hazard', which is that one side of the market cannot observe the actions of the other side (Varian 2003).

In energy markets, this kind of information asymmetry is relevant in the relation between retailers and residential consumers. The latter cannot easily oversee the financial strength of retailers and their ability to commit to contracts concluded in the retail market. Because retailers enter into financial contracts, which basically consists of linking the contracts on the supply side (i.e. the wholesale market) with their contracts with customers (regarding duration and prices), they run a large financial risk (see Box 5.4). If a retailer defaults on its financial obligations, its customers face the risk of interrupted supply of energy.

---

[1]See Kahneman (2011) for the difference between fast and slow thinking.

Box 5.4 Dealing with financial risks by energy retailers

Retailers do not produce, transport or distribute energy, but they are intermediate agents between producers and consumers of energy. Residential consumers cannot go to the wholesale market themselves in order to directly buy from producers because of the transactions costs and the financial risks of taking care of the balancing responsibility (see Sect. 7.4). It is way more efficient to have intermediate parties that pool all the demanded energy of individual consumers and to find counterparties in the wholesale market that want to supply the required energy. In many countries, therefore, residential consumers are not allowed to buy their energy directly on the wholesale market. As many consumers turn into producers themselves (so-called prosumers), however, this may change in future when prosumers can directly exchange energy with each other through so-called peer-to-peer trading.

Despite this development, retailers still play a key role in energy markets by providing commercial contracts between producers and consumers. The value added of a competitive retail market is that it incentivizes retailers to develop and offer those contracts which are most favourable for their customers (Littlechild 2000).

The activity of retailers is, therefore, to a large extent financial as retailers have to deal with a number of financial risks:

- counter-party risk, which is the risk that the counter party is not able to supply (but on an exchange this risk is taken over by the exchange);
- price risk, which is the risk that the input prices (i.e. the prices to be paid in the wholesale market) deviate from the output prices (i.e. the prices to be received in the retail market);
- volume risk, which is the risk that the actual volumes of consumption deviate from what has been bought in the wholesale market;
- imbalance risk, which is the risk that the changes in volumes result in an imbalance and resulting in imbalance costs.

Retailers have a number of options to deal with these risks:

- choosing a balanced portfolio of different types of forward contracts (varying from year-ahead up to day-ahead), matching the duration of contracts with the expected variation in the load profile of the customer base (see also on this topic Box 7.5);
- spreading the price risk regarding wholesale prices by diversifying the purchases over time, which is in particular relevant for fixed-price contracts for consumers. In those cases, retailers face the full risk of changes in wholesale prices.

- buying a contract for differences in which the retailer is compensated for the difference between actual wholesale price (P_{actual}) and wholesale price in a contract ($P_{contract}$) and for which a premium has to be paid (P_{CfD}). Consequently, the wholesale price for the retailer is equal to: $P_{actual} - (P_{actual} - P_{contract}) + P_{CfD}$. The counter party in such a contract can be a wholesaler or a financial party.

Retailers may have the incentive to inform consumers adequately about their financial position and how they handle their financial risks, but consumers may be uncertain about this. Because of this risk, consumers may decide not to switch to another retailer, but to stay with the incumbent retailer and the existing contract. If this happens, competition in retail markets is weakened, giving more power to retailers to charge higher prices. Hence, the adverse effect of this problem of hidden action is that consumers pay more for their energy than what they would do otherwise, while retailers may also have less incentives to innovate due to the lower intensity of competition.

5.4.2 Regulation

As residential consumers cannot adequately observe the financial behaviour and risks of retailers, regulators can do this on their behalf. Regulators may introduce licensing for new retailers by which they check the financial position of a retailer before giving it permission to enter the retail market. Such a licensing policy is generally used by regulators when consumers cannot assess the behaviour of firms (Schotter 1996). Next to providing a license to operate, regulators may also want to regularly monitor whether a firm still meets the criteria used for the license. In the case of retail energy markets, as the financial position of retailers may change over time, this position can be frequently monitored to control for adverse developments in the ability of a retailer to deal with financial shocks. The most important criteria to be used refer to the financial strength to commit to the long-term contracts with suppliers and customers.

In addition, regulators may also introduce a policy for last-resort supply in case of bankruptcy of retailers (ACER 2019). After all, it may happen that a retailer turns into bankruptcy despite a close monitoring of the financial position by the regulator. If a retailer is not able to meet its financial obligations regarding the purchase of energy, its suppliers will stop supplying which makes it impossible for the retailer to continue supplying its customers. Hence, these customers face the risk of interrupted supply. A regulatory measure to mitigate this risk is to give the network operator the responsibility to temporarily take over the supply commitments of a retailer in case the latter defaults on its financial obligations.

Exercises

5.1 What is meant by signalling?

5.2 Why can the presence of adverse selection be a problem for renewable energy?

5.3 What is the difference between book-and-claim approach and mass-balancing approach?

5.4 What is the downside of a large number of product offers in the retail market?

5.5 What is the rip-off externality?

5.6 What is the price risk for a retailer?

5.7 Explain why a low level of switching may indicate a lack of competition between retailers, but that it may also indicate a high degree of competition.

References

ACER. (2019, October 30). Market Monitoring Report 2018—Consumer Empowerment Volume.

Armstrong. (2014, July). *Search and rip-off externalities*. University of Oxford.

Association for Issuing Bodies (AEB). (2018). Annual Report 2018, Brussels.

Baldwin, R., Cave, M., & Lodge, M. (2012). *Understanding regulation; Theory, strategy and practice* (2nd ed.). Oxford University Press.

CEER. (2018, December). Monitoring report on the performance of European retail markets in 2017. Council of European Energy Regulators. Brussels.

CertifHy. (2015). A review of past and existing GoO systems; Deliverable No. 3.1.

Creti, A., & Fontini, F. (2019). *Economics of electricity: Markets, competition and rules*. Cambridge University Press.

Hulshof, D., Mulder, M., & Jepma, C. (2019). Performance of markets for European energy certificates. *Energy Policy, 128,* 697–710.

Kahneman. (2011). *Thinking, fast and slow*. Penguin Books.

Littlechild. (2000). *Why we need electricity retailers: A reply to Joskow on wholesale spot price pass-through*. University of Cambridge.

Moraga, J. L., Mulder, M., & Perey, P. (2019). *Future markets for renewable gases and hydrogen; What would be the optimal regulatory provisions?* Brussels: CERRE.

Mulder, M., & Willems, B. (2019). The Dutch retail electricity market. *Energy Policy, 127,* 228–239.

Ofgem. (2011). What can behavioural economics say about GB energy consumers? London.

Schotter, A. (1996). *Microeconomics; A modern approach.* Addison-Wesley Educational Publishers.

Varian, H. R. (2003). *Intermediate microeconomics; A modern approach* (6th ed.). New York/London: W.W. Norton & Company.

Viscusi, W. K., Harrington, J. E., & Vernon, J. M. (2005). *Economics of regulation and antitrust.* The MIT Press.

Natural Monopoly in Transport and Distribution

6

6.1 Introduction

Most energy is transported and distributed via networks to end-users, which holds in particular for electricity, gas and heat. These networks are often characterized by a natural monopoly. This chapter first discusses the economic definition of natural monopolies and what the consequences are if a firm operating a natural monopoly is not regulated (Sect. 6.2). Section 6.3 discusses the principles of tariff regulation and how to deal with the information asymmetry between regulator and regulated firm. Next, attention is paid to benchmarking methods (Sect. 6.4), the WACC (Sect. 6.5), the design of tariff structures (Sect. 6.6), relationship with investments (Sect. 6.7) and, finally, how regulators can evaluate the effectiveness of tariff regulation (Sect. 6.8).

6.2 Natural Monopolies

6.2.1 Monopolies and Competition Policy

Before going into the peculiarities of natural monopolies, we first discuss the differences between a market characterized by perfect competition and a market characterized by the presence of a monopoly.[1]

The key difference between a competitive market and a monopolistic market is that a monopolist is able to choose the price of its products, while firms in a competitive market have to accept the prices as given. Therefore, firms in a com-

---

[1]The name 'monopoly' is derived from the Latin *monopolium* which means single seller.

© Springer Nature Switzerland AG 2021
M. Mulder, *Regulation of Energy Markets*, Lecture Notes in Energy 80,
https://doi.org/10.1007/978-3-030-58319-4_6

petitive market are called 'price takers'; whereas, a monopolist is a 'price setter'. In theory, a monopolist is able to choose a price which maximizes its profits. This is the so-called monopoly price. The profits are maximized when the marginal revenues of selling one extra product equal the marginal costs of supplying that product (Fig. 6.1). The marginal revenues are defined as the difference between the total revenues at a certain level of output and the total revenues at a one unit lower or higher output level. As long as these marginal revenues exceed the marginal costs, extending production is profitable for a monopolist.

As the total revenues (R) are equal to the product of price (p) and quantity (q) (i.e. $R = p*q$), the marginal revenues are equal to

$$\frac{dR}{dq} = p * \frac{dq}{dq} + q * \frac{dp}{dq} = p + q * \frac{dp}{dq} \tag{6.1}$$

In a competitive market, the additional (marginal) supply of an agent does not have any influence on the market price, which means that the marginal revenues in that case are equal to the market price (i.e. $dp/dq = 0$ which results in $dR/dq = p$). In a monopoly, however, the supply to a market depends fully on the monopolist's supply. In a monopolistic situation, the marginal revenues result from two effects of volume changes on revenues: on the one hand, a higher volume means more revenues as more products are sold (=output effect), but on the other hand, a higher volume implies a lower equilibrium price for all products sold (=price effect). The latter effect depends on the price elasticity of demand (i.e. the steepness of the demand curve). If demand is highly inelastic, a small decrease in volume supplied to the market results in a strong price increase (i.e. a strong price effect). In such cases, the monopoly quantity lays significantly below the competitive quantity, while the monopoly price is significantly above the competitive price. This also implies that the incentive of a monopolist to charge higher prices is higher, the less consumers are having an alternative choice option, i.e. when they are less able to respond to price changes.

Suppose the demand function is $D(p) = 90 - p$ and, hence, the inverse demand function is $P = 90 - q$, then the marginal revenues for a monopolist are equal to:

$$\frac{dR}{dq} = \frac{d((90 - q) * q)}{dq} = (90 - q) + q * \frac{d(90 - q)}{dq} = 90 - 2q \tag{6.2}$$

Suppose further that the costs (C) of the firm are given by $0.25\,q^2$, which means that the marginal costs are equal to:

$$\frac{dC}{dq} = \frac{d(0.25q^2)}{dq} = 0.5q \tag{6.3}$$

The profit maximizing quantity for the monopolist is there where the marginal revenues equal the marginal costs:

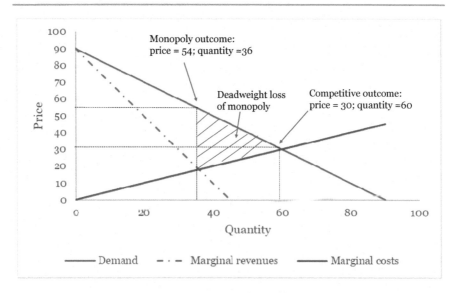

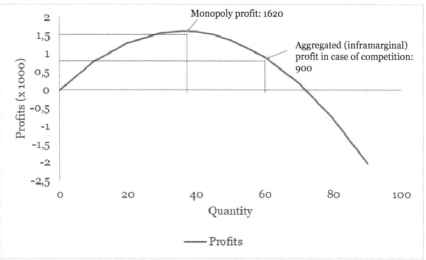

Fig. 6.1 Market outcomes in cases of competition and monopoly

$$\frac{dR}{dq} = \frac{dC}{dq} \overset{yields}{\rightarrow} q = 36; p = 54 \tag{6.4}$$

If a monopolist would sell more than 36 in this example, then the extra revenues would be below the extra costs, so it is not rational for the monopolist to sell more than that quantity.

In case of competition, a supplier doesn't have any influence on the market price, which means that the market price remains the same no matter what the size of the supply of a supplier is. As a result, a competitive supplier produces that volume where its marginal costs equal the price that consumers are prepared to pay:

$$0.5q = 90 - q \overset{yields}{\rightarrow} q = 60; p = 30 \tag{6.5}$$

Although the competitive players are making a positive aggregated profit, for the marginal unit the costs are equal to the revenues (see Fig. 6.1). This positive inframarginal profit results from the increasing shape of the marginal costs curve and the fact (which is actually an assumption) that the market price is only driven by the marginal costs of the last unit needed. Note that this is precisely the way the price is set in electricity markets. In these markets, the supply curve is based on a merit order (i.e. a ranking of all supply bids from the lowest to the highest marginal costs) while the equilibrium price is based on the marginal costs of the price-setting plant (i.e. the last plant needed to clear the market). As a result, the so-called inframarginal suppliers are making a profit. This way of price setting, which is called an energy-only market, plays a key role in incentives for investments in new power plants (see further Sect. 7.5).

As a consequence of the fact that the monopoly output is below the output level in case of competition, the welfare of market transactions in a monopolistic situation is also lower than in a competitive situation. This loss of welfare is equal to the area between the price consumers would be prepared to pay (i.e. their WTP) and the marginal costs for the volume of output that is not consumed. This loss of welfare is called the deadweight loss of a monopoly, indicated by the striped area within the figure.

A monopolist may not only charge a higher price, but may also perform worse in terms of productive efficiency and quality of its products. A monopolist will reduce its costs if this results in a higher profit, but cost reductions are not required for a monopolist to maintain or raise its market share. In a competitive market, however, increasing efficiency is necessary for a firm to gain sufficient profits or even to survive, for instance by reducing overhead expenses or by applying innovative technologies that generate the same output with less inputs. A firm operating in a competitive market is forced to continuously improve the productive efficiency, because its competitors are doing same, and as a result the market price is determined by the marginal costs of the most efficient producer.

In a similar fashion, a monopolist does not have any reason to invest in quality improvements because clients cannot switch to another supplier. As a result, profits realized on a new (innovative) product only replace profits on the existing products. This effect is, therefore, called the replacement or Arrow effect of a monopolist (named after the economist Kenneth Arrow). A firm in a competitive market, however, has to improve quality to keep its customers satisfied and to prevent them from switching to a competitor as the competitors will also continuously look for quality improvements.

Because of these drawbacks of monopolies, most societies have implemented competition policy to prevent that monopolies are created. Competition policy consists of two types of policy interventions: merger control and antitrust policies (see Box 3.4). Here, it is important to recognize that competition policies generally do not forbid firms to have market power, but it does not allow that these firms abuse market power by, for instance, charging above competitive prices or hindering competitors. Antitrust policies are also directed at preventing unfair competitive behaviour, such as the construction of cartels of suppliers who make agreements on their joint production in order to influence market prices.

Despite the potentially negative effects of markets with firms having market power, one should also keep in mind that market power can also have positive effects on innovation. This may occur when firms need to make high investments for innovation, which can only be recouped when the firms are able to supply to the full market and to charge prices above marginal costs. Hence, in order to foster innovation, it may sometimes be welfare improving to allow a lower level of competition (see, e.g. Decker 2015). In such cases, investments in innovation activities may be protected by patent law; see further on this topic Sect. 8.2.

6.2.2 Definition of Natural Monopoly

Although competition is generally preferred above monopolistic situations, sometimes monopolies cannot be prevented. This is the case when the monopoly can be characterized as a 'natural monopoly'. In such a market, competition cannot exist. An industry has a natural monopoly when the average costs (C) of producing all commodities (Q) by one firm are less than the sum of the average costs of producing those commodities (q1 and q2 which sum op to Q) in separate firms. Hence, a natural monopoly exists when the average costs are subadditive over the whole range of market demand:

$$C(Q) < (C(q1) + C(q2)) \qquad (6.6)$$

According to this definition, a natural monopoly exists when one firm is able to supply the commodities at lower (average) costs to the market than when several firms would together do the same. In general, a natural monopoly exists if the average costs decline when the volume of production increases over the whole range of the market. The declining average costs result from large initial investments (say in infrastructure) and low variable costs for each unit of production.

Suppose that the initial investment is 1000 euro, while the variable costs are 5 euro per unit until an output level of 60 units, while they increase afterwards. Then the resulting average costs curve is as shown by Fig. 6.2. At a production volume of for instance 10, the average costs are (1000 + 10 * 5)/10 = 105, while at a volume of 50, the average costs have decreased to (1000 + 50 * 5)/50 = 25 euro per unit, which indicates a strong reduction in average costs the larger the production. If the inverse demand curve is given by D(p) = 90 − q, then this market is clearly characterized by a natural monopoly as the average costs form a declining curve over the full market demand.

If the market demand increases strongly, then the natural monopoly vanishes when at higher production levels (in this case, above an output volume of 60) the

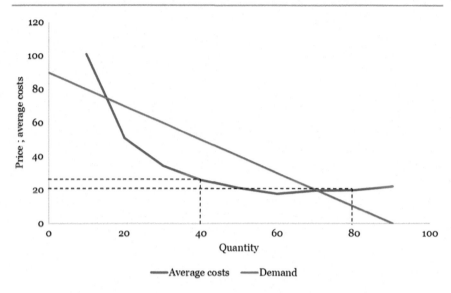

Fig. 6.2 Definition of natural monopoly based on declining average costs

average costs increase. This occurs if at some point the infrastructure is fully utilized and new investments have to be done to expand capacity, while also the operational costs may become larger at a high utilization rate. When the variable costs exceed the average costs, the latter will rise again.

It is important to realize that an increase in average costs does not necessarily imply that there is no natural monopoly anymore. In Fig. 6.2, after a production quantity of 60, the average costs increase again because of, for instance, extra retrofit investments. The firm may still have a natural monopoly as average costs of one firm are still below the average costs of more than one firm producing the same amount (i.e. the costs can still be subadditive even in case of diseconomies of scale). For instance, at a market size of 80, the average costs of one firm are below the average costs of two firms of the same size (both 40). However, the natural monopoly of a firm is only sustainable until the output level of 60, because if the firm produces more, another firm could enter the market, produce only 60 and, hence, charge a lower price. Consequently, one can say that a sustainable natural monopoly is characterized by decreasing average costs.

The above natural monopoly is due to economies of scale, but a natural monopoly can also result from economies of scope, which means that the average costs of producing a number of products (X_a ... X_n) by one firm are lower than the sum of the average costs of producing these products by more than one firm (see Sect. 2.5.1).

Examples of natural monopolies are the infrastructures for transporting and distributing electricity, gas and heat. Because of the high investment costs to build these infrastructures combined with the (very) low marginal costs per unit of using the infrastructure, the average costs decline when the utilization of the infrastructure increases (see Box 6.1).

Box 6.1 Sources of natural monopoly in gas pipeline infrastructure

A gas transport infrastructure is characterized by strong economies of scale because the capacity of a gas pipeline is an increasing function of the diameter of the pipes and the pressure put on the gas (see Fig. 6.3). For instance, a pipe with a diameter of 0.60 m and a pressure of 80 bar has a transport capacity which is 4000 MW, while a pipe with a diameter of 0.30 m and the same pressure has a transport capacity of about 1000 MW. Hence, by doubling the diameter size, the transport capacity quadruples. Although a pipe with a larger diameter requires more material, the total costs increase much less than the increase in capacity. More importantly, in order to have the same capacity of a 0.6 m pipe, 4 pipes of 0.3 m are required. This not only results in higher capital costs, but also in higher administrative costs, such as for permitting procedures. Hence, it is way more efficient to have one large gas pipeline instead of a few smaller ones.

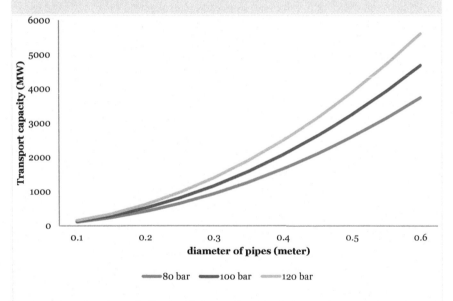

Fig. 6.3 Capacity of gas pipeline in relation to diameter of pipes and pressure. *Source* Mulder et al. (2019)

6.2.3 Policy Responses to Natural Monopolies

In case of a natural monopoly, competition is not sustainable because one firm will have lower costs than more firms producing the same amount of output. Although this results in the lowest average costs for society, this does not mean that the customers of a natural monopolist pay the lowest price possible as the monopolistic firm has both the possibility and the incentive to charge monopoly prices. To prevent this to happen, governments may consider to intervene in one or the other way (Table 6.1). In the past, many countries chose for the option to set the management of the firm under direct control of the public authorities, i.e. by making it part of the public administration subject to direct political influence. This is a highly intensive (i.e. a heavy-handed) form of intervention; see Sect. 3.4.2. Public authorities may also choose not to intervene at all, but just to rely on general competition law. This policy is, however, only effective when at some point competition is possible, which means that the natural monopoly characteristic should vanish in the future.

In between these two extremes, there are basically two other types of policy interventions. The one closest to the latter one is the organization of a competitive tender, which means that firms have to compete to be allowed to operate the natural monopoly. This is called competition *for* the market, as compared to competition *in* the market in the case of normal competition. Such a competition for the market can only work and result in efficient outcomes, when a sufficient number of firms are able to participate in the tender. After all, without a sufficient number of bidders, the bidder(s) can behave strategically and raise the price of the permission to operate the natural monopoly. In addition, competitive tendering does not guarantee that the winner of the tender will charge efficient prices and that it will be directed at the

Table 6.1 Possible regulatory responses to the presence of a natural monopoly

Response	Condition	Risks
General competition policy (Antitrust)	Competition should be possible in the (near) future	• Ineffective when competition will not emerge • Difficult to prove that firm has abused market power
Competitive tendering (=competition *for* the market)	Sufficient number of firms that can participate	The winner of the tender obtains a strong negotiation power afterwards which may result in holding up the investment (i.e. postponing it) in order to renegotiate the conditions
Sector-specific regulation	Independent regulator	Regulatory capture
Public agency subject to direct political control (e.g. under the direction supervision of the government)	Sufficient public resources (as capital market is not an option for financing investments)	Captured by short-term political interests

interest of the customers (Decker 2015). In general, competitive tendering is more likely to be successful when it refers to large and distinct projects, in which the precise conditions for the tender can be well formulated without too much transaction costs (Baldwin et al. 2012; Decker 2015).

When such a competition for the market is not possible, the only remaining option is to regulate the firm which has a natural monopoly. In general, the objective of regulating a natural monopoly is (a) to prevent monopoly pricing, (b) to foster efficient behaviour and (c) to guarantee a minimum level of service quality. This type of intervention is called sector-specific regulation.

In this chapter, we focus on sector-specific regulation to address the presence of a natural monopoly. We discuss the principles of tariff regulation (Sect. 6.3), methods for benchmarking (Sect. 6.4) and for determining the costs of capital (Sect. 6.5), the relationship between tariff regulation and the quality of performance (Sect. 6.7) and finally, the evaluation of the effects of tariff regulation on profits and financeability of the regulated company (Sect. 6.8).

6.3 Principles of Tariff Regulation

6.3.1 Introduction

An essential element of sector-specific regulation of a natural monopoly consists of tariff regulation. This regulation refers to the tariffs the operator of a natural monopoly is allowed to charge from its customers. Generally, regulators want that tariffs are reasonable, but regulators can give different answers to the question what a reasonable level is. A reasonable level can be defined as tariff levels that result in revenues which are not higher than the actual costs of the firm itself, which means that the regulated firm is not making excess profits. The profits can also be capped at the level of the costs of a similar firm that operates efficiently, which means that the consumers should not pay higher prices than those which would occur in competitive circumstances. After all, in a competitive market, the most efficient firm sets the prices. The way a regulator determines the tariffs affects the incentives for the regulated company to improve its operation as well as its ability to make a profit. The structure of tariffs also affects the allocative efficiency of network usage.

Hence, in order to set tariffs, the regulator has to make three types of choices. These choices refer to (a) the incentives for the regulated company to operate efficiently, (b) the tightness of the allowed revenues in relation to the costs of the company and (c) the design of the structure of the tariffs. In each of these choices, the regulator has to deal with a fundamental problem in his relationship with the regulated firm and that is the problem of information asymmetry, which will be discussed first.

6.3.2 Information Asymmetry and Regulatory Solutions

The origin of the problem in the relationship between regulator and regulated firm is that the regulator pursues its own objectives, but is completely dependent on the behaviour of the regulated firm for the realization of these objectives. The regulator may have as objective that the network operator works as efficiently as possible, and that an efficiently operating firm should be able to generate a 'normal profit'. The latter can be defined as a profit which is sufficient to give the investors a market-based return on their investment (Varian 2003).[2] This normal profit can be seen as the financeability or break-even constraint of tariff regulation, which means that the allowed tariffs leave the firm with sufficient returns to recoup the fixed costs of the investments and to give an appropriate reward to investors.

In order to design tariffs such that they give both incentives for efficiency improvement and meet the financeability constraint, the regulator faces two types of information asymmetry problems: hidden information and moral hazard (see Fig. 6.4). Hidden information refers to a lack of information on the precise characteristics of the regulated firm. Because of this lack of information, the regulator may make wrong assumptions on the actual costs of the regulated firms and, as a result, set the allowed tariffs on a too high or low level. Moral hazard refers to the behavioural risks that arise from a lack of information on the actual behaviour of this firm. This risk may refer, for instance, to a situation in which a firm executes less effort to improve productive efficiency as the costs can be passed on to its customers or to tax payers.

One option to solve the problem of hidden information is to use information from the regulated firm on its actual costs and to set the allowed revenues equal to the actual costs of the firm. This type of tariff regulation is called cost-plus regulation.[3] In such a type of regulation, all actual costs of the regulated firm are reimbursed through the network tariffs, no matter what the level of these costs are.[4] However, in this type of tariff regulation, the regulated firm does not have any incentive to improve its efficiency. In other words, cost-plus regulation may solve the problem of hidden information, but it creates the problem of moral hazard.

The problem of moral hazard can be solved by giving the firm an incentive to realize a higher level of efficiency. This can be done by determining the allowed level of revenues on the basis of exogenous information, i.e. information which is not at all related to the actual costs of the regulated firm. An example of this kind of tariff regulation is price-cap regulation where the tariffs are capped independently of

---

[2]The profit that exceeds this normal profit level is called 'supra-normal' or 'economic' profit. See further on this issue Sect. 6.8.1.

[3]The 'plus' refers to the compensation for the costs of capital, which implies that cost-plus regulation realizes a compensation for all (operational and capital costs) of the firm. This compensation can be calculated on the basis of the accounting profit (i.e. the difference between revenues and operational costs) and then adding a margin (the 'plus') as a compensation for the required rate-of-return.

[4]Because of the information asymmetry between regulator and regulated firm, regulators usually require cost reports by independent accountants, which reduces the risk that a regulated firm will try to misinform the regulator by exaggerating its actual costs.

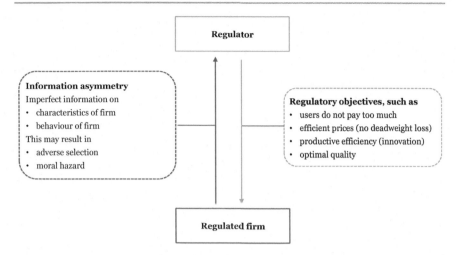

Fig. 6.4 Information asymmetry between regulator and regulated firm

what the actual costs of the regulated firm are. This cap only holds for a number of years, the so-called regulatory period. During that period, the regulated firm can be seen as a price taker, just as firms operating in competitive markets.[5] Price-cap regulation, therefore, gives strong incentives to improve efficiency. Therefore, price-cap regulation solves the problem of moral hazard since the regulated firm faces all consequences of changes in its costs. The downside of this type of regulation, however, is that the allowed level of revenues is not related to the actual costs of the regulated firm, which may have as a consequence that the revenues are way too high (resulting in supra-normal profits) or too low (resulting in finance-ability problems).

The characteristics of cost-plus and price-cap regulation can be shown by using the following formula for tariff regulation, where R stands for revenues, C for costs of the regulated firm, B for the costs based on external (benchmark) information while i is the incentive power parameter:

$$R = (1 - i)C + iB \tag{6.7}$$

If i is set equal to 1, the revenues are fully based on external information from a benchmark, such as in price-cap regulation. If i set equals to 0, the revenues are fully based on the own costs of the regulated firm, which is cost-plus regulation. Because tariff regulation affects the revenues of the regulated firm, it also affects its profits (Profit) which is a function of the incentive power parameter, the actual costs of the firm and benchmark information:

---

[5]At the end of a regulatory period, a regulator may reconsider the cap on tariffs, taking into account information on the actual costs of a regulated firm. In this resetting of the cap, the regulator may make use of benchmarking (see further Sect. 6.4).

$$Profit = R - C = (1 - i)C + iB - C \qquad (6.8)$$

The incentive power measures to what extent a change in the costs of the regulated firm affects the profits of the firm. Hence, the incentive power can be calculated as the effect of changes in costs (dC) on profits (dProfits).

$$\frac{dProfit}{dC} = (1 - i) - 1 = -i \qquad (6.9)$$

When the firm sees the full consequences of a change in its costs on its profits, then the tariff regulation is said to be high-powered, while when the regulated firm does not experience any change in its profits when its own cost changes, then the tariff regulation is said to be low-powered. An example of the first type of tariff regulation is price-cap regulation (where $i = 1$) and an example of the second one is cost-plus regulation (where $i = 0$) (see Fig. 6.5). In case of $i = 1$, a change in costs is fully translated into a change (in the opposite direction) in the profits, while in the case of $i = 0$, the profits remain the same no matter what happens with the costs.

An intermediate form of tariff regulation, between cost-plus and price-cap regulation, is rate-of-return regulation (sometimes abbreviated as 'ROR'). It allows a regulated firm to cover its own costs, but the compensation for the costs of capital is based on a fixed compensation for the capital invested (i.e. rate-of-return).[6] An advantage of rate-of-return regulation is that a network firm has a guarantee that it will always make a minimum rate-of-return, but this rate-of-return may not be sufficient to compensate for its actual costs of capital. With the passage of time, a regulated firm will typically argue that the rate-of-return is too low, while the regulator has the task to check whether such claims are correct or that this rate is just right or perhaps even too high. As long as the allowed rate-of-return stays the same, there is an incentive for managers to attract capital at the lowest possible costs. In this respect, rate-of-return regulation somewhat resembles price-cap regulation as the 'regulatory lag' (i.e. the time period between the review of the allowed revenues or the rate-of-return, which is equal to the length of the regulatory period) gives an incentive to improve efficiency in the costs of capital. A disadvantage, however, is that rate-of-return regulation may give the network firm an incentive to overinvest in case the allowed rate-of-return exceeds the actual opportunity costs of capital (see further Sect. 6.7.1).

Each of the above types of tariff regulation has its own pros and cons (Table 6.2). The strength of the incentives for cost reduction is the highest in price-cap regulation, while cost-plus regulation does not give any incentive to the regulated firm to becomes more efficient. A disadvantage of price-cap regulation is the risk that the regulator sets the cap too high or too low, which harms consumers or shareholders, respectively. In addition, price-cap regulation imposes more risks on regulated firms: profits will go down when costs go up since the tariff cannot be changed, at least not

---

[6]Hence, the incentive power of this type of tariff regulation is between 0 and 1, as all costs except costs of capital are fully compensated while the compensation for the costs of capital is exogenously set.

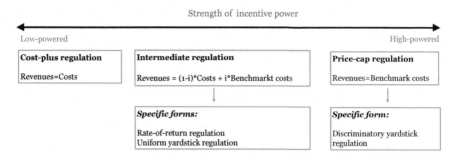

Fig. 6.5 Incentive power of various types of tariff regulation

Table 6.2 Pros and cons of main types of tariff regulation

Type of price regulation	Strength of incentives for efficiency (i.e. degree by which moral hazard is solved)	Risk of inadequate level of profits (too low or too high)? (i.e. remaining adverse selection?)	Do consumers fully benefit from realized efficiency improvements?
Cost-plus	No	No	Yes
Rate-of-return	Low	Low	To a large extent
Price cap	High	Yes	No

until the next price review (i.e. the next regulatory period). Under rate-of-return regulation, consumers would bear some of that risk, because firms would be able to increase the tariffs to protect their profits. Shareholders thus face more risk under price-cap regulation, which could make them to be less inclined to invest in the network. Regulated firms subject to price-cap regulation will only invest in its infrastructure when they expect that this will result in lower costs in the future. The other two types of tariff regulation give more certainty to investors about the return on investment, but this may result in overinvesting, gold-plating and too high tariffs for consumers. An advantage, however, of cost-plus regulation is that consumers will never pay more than the actual costs of the regulated firm, while under price-cap regulation it is possible that the regulated firm makes (high) profits during the regulatory period and before the reassessment of its costs at the end of that period. In Sect. 6.8, we will discuss how the regulator can monitor and address this risk.

6.3.3   Incentive Power and Tightness

From the above discussion of the main types of tariff regulation, it appears that incentive power and tightness are essential dimensions. The incentive power refers to the impact of realized cost reductions on the profit of a regulated firm. The other dimension is the tightness of tariff regulation, which refers to the profit level a firm is able to make in case of normal circumstances. By normal circumstances are meant, the average of the circumstances in the past, as this is generally seen as the best prediction for the future. Tariff regulation is called tight when a regulated firm makes only a small profit in these circumstances, while the tariff is called loose when the firm can make a high profit.[7]

Using these two dimensions, the regulator can design the main characteristics of tariff regulation (see Fig. 6.6). When the tariffs are based on external information (e.g. the costs of other firms operating in similar business, the so-called benchmark firms), then the incentive power is high. If the other firms which are used as a benchmark are (very) efficient, then the resulting cap may be (very) low, resulting in a (very) low profit or even a loss for the regulated firm. In this case, the tariff regulation is characterized by a high-incentive power and tight revenue level. If the firms in the benchmark are much less efficient compared to the regulated firm, then this may result in high profits under normal circumstances. Although the incentive power remains the same, the revenue level will be loose. Therefore, it is important to realize that the level of the cap does not have any effect on the incentive power.[8]

In a similar way, the tightness may vary from loose to tight while the incentive power is low. An example of a loose kind of tariff regulation with a low incentive power is where the regulator tends to overestimate the actual costs of the regulated firm in order to prevent the risk that the allowed revenues are not sufficient. When the regulator uses more conservative estimates of the realized costs, the cost-plus regulation becomes tighter.

These two different dimensions of tariff regulation become also clear when looking at the regulatory process of tariff setting (Fig. 6.7). An essential element of the process of tariff regulation is that the regulated firm decides upon its costs, while the regulator decides upon the maximum level of (i.e. the allowed) revenues. The incentive power refers to the extent the realized costs of the regulated firm play a role in the determination of the allowed revenues by the regulator. When the total allowed revenues for the firm have been set, the regulator sets the individual tariffs for the various types of products (see Sect. 6.6). These tariffs affect the realized revenues which also depend on the actual number of products which have been sold by the regulated firm. The difference between the realized revenues and the realized costs determines the realized profit of the regulated firm. The level of this profit in normal circumstances determines the tightness of tariff regulation.

As the regulator can choose different values for the incentive power and the tightness, it has to determine that set of choices it prefers the most. The regulator may have a preference for specific values of both dimensions, while the number of options may technically be restricted. Suppose, the regulator has two objectives regarding tariff regulation: (1) to enable the regulated firm to finance all investments needed for its activities (e.g. management of electricity grid) and (2) to prevent that consumers pay more than needed. Hence, this regulator wants to define the tariffs such that the risks of both financeability problems and supra-normal profits are minimized.

---

[7]Note that when tariff regulation is very loose, the cap may become ineffective. This is the case when the allowed tariffs would become higher than the profit maximizing tariffs, i.e. higher than the level of monopoly tariffs.

[8]After all, the incentive power refers to the marginal effect of changes in own costs on changes in own profits.

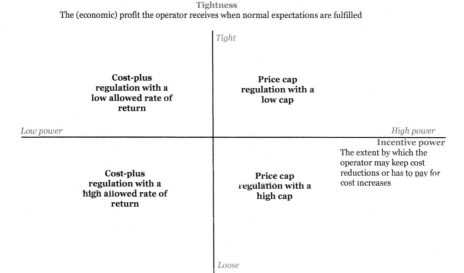

Fig. 6.6 Incentive power and tightness of various types of tariff regulation

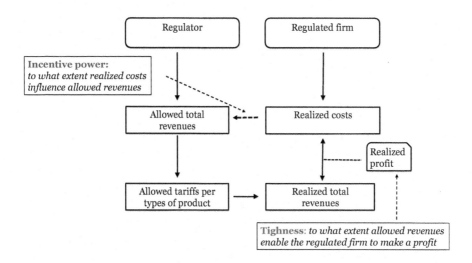

Fig. 6.7 Treatment of incentive power and tightness during the process of regulation

The optimal choice depends on the preference of the regulator regarding these two risks. These preferences can be depicted through a regulatory indifference curve (see Fig. 6.8). This curve is concave to the origin assuming decreasing marginal utility. This means that when the risk of financeability problems is high and the risk of supra-normal profits is low, the regulator is prepared to accept a

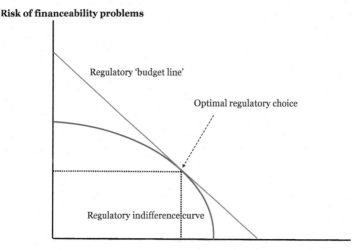

Fig. 6.8 Optimal regulatory choice regarding risks of tariff regulation

relatively strong increase of the latter risk in order to have a relatively small reduction of the former, and vice versa. As the regulator wants to minimize both risks, it prefers to be on a difference curve which is the closest to the origin. It has, however, to deal with an economic constraint on reducing both risks. Reducing the risk of financeable problems requires that the firm is allowed to charge higher tariffs, but higher tariffs increase the risk of supra-normal profits. Hence, there is a direct trade-off between both risks. This trade can be seen as the regulatory 'budget line'. The optimal regulatory choice is now determined at that point where this line is tangent with the regulatory indifference curve.

6.3.4 Definition of Costs

No matter the type of tariff regulation, regulated revenues are always related to costs, be it the costs of the regulated firm or the costs of a benchmark firm. When these costs refer to all costs, the regulation is called TOTEX regulation. This TOTEX stands for total expenditures and consists of two types of costs: OPEX and CAPEX[9]:

$$TOTEX = OPEX + CAPEX \tag{6.10}$$

---

[9]Although TOTEX means 'total expenditures', this variable refers to 'total costs', as also costs which are not expenditures like depreciation are included, while expenditures that are not costs (e.g. investment expenditures) are not included. In the remaining of this book, we will not talk about TOTEX, but use the term 'total costs' indicated by C.

OPEX stands for operational expenditures and refers to the costs of labour, energy, etc. CAPEX stands for capital expenditures and refers to costs of investments in fixed assets. CAPEX consists of two components: costs of capital (CoC) and depreciation (D):

$$CAPEX = CoC + D \tag{6.11}$$

The costs of capital are determined by the costs per unit of capital invested and the total value of the assets. The former can be determined by the weighted costs of capital (WACC) and the latter by the Regulated Asset Base (RAB).

$$CoC = WACC * RAB \tag{6.12}$$

The RAB is the valuation of the assets of the regulated firm based on Regulatory Accounting Rules (RAR). The assets included are typically the fixed assets, but sometimes also working capital and ongoing investments which have not been finished yet. The value of the RAB is commonly based on the realized, historical costs (CEER 2019). Generally, this value is indexed to compensate for inflation. If this is done, the allowed return on the invested capital (i.e. the WACC) should be deflated, which means that the real WACC should be used (see Sect. 6.5).

The RAB is a dynamic stock variable as the value at a particular moment changes as a result of investments and depreciation over time. The RAB at the beginning of year t is a function of the book value of the assets at the beginning of the previous year, the depreciation of the existing assets and the investments (I) in new assets during that year:

$$RAB_t = RAB_{t-1} + I_{t-1} - D_{t-1} \tag{6.13}$$

The treatment of investments will be further discussed in Sect. 6.7.1. The depreciation is meant to allocate the costs of using assets to specific periods. This allocation can be done in different ways. One option is to assume that every year during the useful lifetime (i.e. number of years, indicated by T) of an asset is responsible for an equal share in the costs of using the asset. This is called linear or straight-line depreciation, resulting a constant annual level of depreciation:

$$D = \frac{1}{T} * RAB_{t=0} \tag{6.14}$$

An alternative method is an accelerated depreciation method in which the depreciation is relatively high in the early years of an asset. A common approach (see CEER 2019) to realize such an accelerated depreciation is the double declining balance (DDB) method in which the annual depreciation is as follows related to the useful lifetime and the book value of the assets at the end of the previous year:

$$D_t = \frac{2}{T} * RAB_{t-1} \tag{6.15}$$

The DBB method differs in two aspects from the straight-line method: the annual percentage is twice as high, while the annual depreciation is based on the book value at the beginning of each year instead of the book value at the beginning of the period. As the annual depreciation reduces the value of the asset base every year, the depreciation declines annually as well. The speed of this decline is in the DDB method accelerated by the factor two. The economic reason to use an accelerated depreciation method is that the productivity of an asset declines with its ageing, while the straight-line method can be used when the productivity remains constant over the lifetime of the asset. As the DBB method does not result in a fully depreciated asset base at the end of the lifetime, the depreciation in the last years can be based on a straight-line method for the remaining period.

A key variable in the determination of the costs of capital is the WACC, which is the rate that the investors in the firm want to have in return for their investment. It is important to realize that the required return by the investors is determined by the market and not by the firm's management or by the regulator. The WACC represents the minimum return that a firm must earn on its existing asset base to satisfy the needs of its investors. The determination of the WACC is discussed in Sect. 6.5.

6.4   Benchmarking

6.4.1   Concepts of Productivity and Efficiency

Above, we have seen that high-powered incentive schemes need external information in order to give incentives for efficiency improvement. The next question is, therefore, which information to use. The answer to this question is determined by the objective of tariff regulation, in particular the objective of the regulator regarding the level of the tariffs. Generally, regulators want that the tariffs reflect the costs of an efficient operator. In order to determine the efficient cost level, we first need to understand the concept of efficiency.

In economics, efficiency is a key concept which is applied to several aspects of economic activity. When the concept of efficiency is used in the analysis of markets, it refers to allocative efficiency which is the efficiency of the allocation of goods (see Chap. 4). The allocation of goods is called efficient when the goods are supplied by those firms with the lowest marginal costs and supplied to those consumers with the highest willingness-to-pay (WTP). This outcome is reached when markets satisfy the conditions of well-functioning markets, and the market price is equal to the WTP of the marginal consumer. When a market is allocatively efficient, welfare is maximized, meaning that the surplus for both consumers and producers is maximized.

When the concept of efficiency is used in the analysis of the production processes of firms (or any other organization or process), it refers to productive or technical efficiency (TE). This efficiency measures the productivity of a firm in relation to the productivity of the best practice firm. Productivity (*Prodtt*) is defined by the amount of output produced per unit of input:

$$Prodtt = \frac{Output}{Input} \qquad (6.16)$$

Technical efficiency (TE) is a relative concept which is defined as the ratio of the productivity of firm *i* in relation to the productivity of the best practice *b*:

$$TE_i = \frac{Prodtt_i}{Prodtt_b} \qquad (6.17)$$

A firm that is perfectly efficient has the same productivity as the best-practice firm operating in the same business. The productive efficiency of this firm has the value of 1. This holds, by definition, for all firms that operate on the so-called technological frontier. This frontier is determined by the maximum output levels possible given a level of input and existing technologies (see Fig. 6.9). At input level I1, for instance, it is not possible to have a higher output level than O1. Lower outputs, of course, are possible. The area below the productivity frontier defines the so-called feasible production set. In this figure, both firms B and C are operating on the technological frontier, but they have a different productivity. The productivity of firm B and firm C is equal to the slopes of the ray OO1/oI1 and OO2/OI2, respectively. The lower productivity of firm C compared to firm B is due to scale

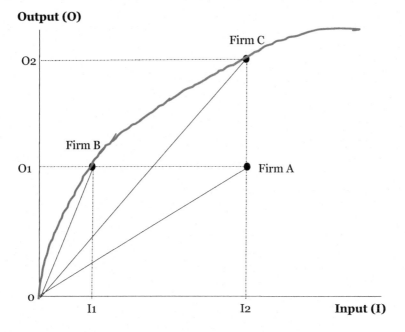

Fig. 6.9 Productivity and efficiency. *Source* Coelli et al. (2006)

inefficiencies. Although firm C is operating efficiently given its size, it could improve its productivity by operating on a smaller scale.

Firms that do not operate on the technological frontier are said to be inefficient, like firm A. The productivity of this firm is equal to the slope of the ray $OO1/oI2$. The inefficiency can be measured in two different ways since firm A can be compared to both firms B and C as both firms are operating on the technological frontier. When firm A is compared to firm C, its efficiency is defined as the ratio between the productivity $OO1/oI2$ and the productivity $OO2/OI2$. Both firms have the same input level, but a different output level. Hence, when firm C is used as a benchmark, it is implicitly assumed that firm A could improve its efficiency by realizing a higher volume given its level of inputs. This method of benchmarking is called the output-maximization approach. This method should be used when the regulator believes that a firm could reach a higher output. This movement in productivity towards the technological frontier is called catch-up.

In many circumstances, however, the output level is exogenous to a firm. In the case of energy networks, for instance, operators can hardly influence the magnitude of the utilization (i.e. the level of energy use). The only option these operators have to raise their efficiency is to increase the productivity by lowering the level of inputs per unit of output. In these cases, the input-minimization approach should be used. In this example, the efficiency of firm A should be calculated as the ratio between the productivity $OO1/OI2$ (firm A) and the productivity $OO1/OI1$ (firm B).

The two different perspectives on technical efficiency can both be related to the optimizing behaviour of firms. When the objective is to minimize the costs given a level of output, the technical efficiency can be measured as the distance of the inputs to the optimum input level depicted through an isoquant (see Fig. 6.10). When the objective is maximizing the outputs given a level of input, the technical efficiency can be measured as the distance of the outputs to the maximum possible level of outputs, depicted through the production possibilities curve.

In the above graphs, it is assumed that the technological frontier is constant, but this frontier can also move due to technological innovations. These innovations are called dynamic efficiency, while the movement of the frontier is called frontier shift (see Fig. 6.11). The change in overall efficiency of a firm is, hence, due to two factors: catching up to the frontier (i.e. moving to the frontier) and the frontier shift (i.e. the movement of the frontier).

When assessing the change in technical efficiency of a firm, the new position of the frontier has to be taken into account as well. This change in technical efficiency can be determined as follows:

$$\Delta TE = \frac{TE_{t=1}}{TE_{t=0}} \tag{6.18}$$

Hence, a firm can realize a higher productivity, as firm A has done from $t = 0$ to $t = 1$, but its technical efficiency may remain the same because the frontier has also moved. From this follows that the change in the productivity of a firm can be expressed as a function of the change in its own technical efficiency and the shift in the frontier (FS):

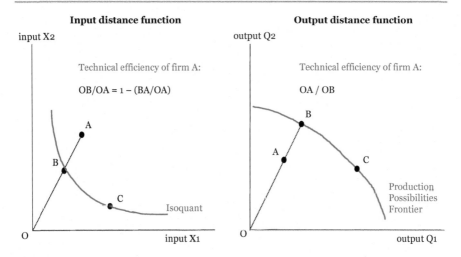

Fig. 6.10 Technical efficiency measured in two different ways. *Source* Coelli et al. (2006)

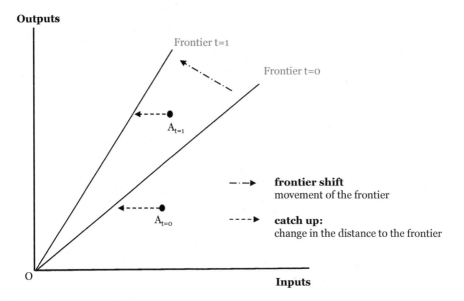

Fig. 6.11 Productivity change as function of catch-up and frontier shift. *Source* Coelli et al. (2006)

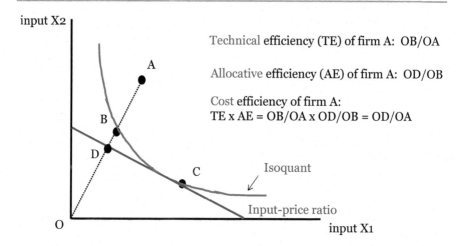

Fig. 6.12 Costs efficiency as a function of technical and allocative efficiency. *Source* Coelli et al. (2006)

$$\Delta Prodtt = \Delta TE * FS \qquad (6.19)$$

The frontier shift, i.e. the dynamic efficiency or technological innovation, can be estimated through the geometric mean of the technical efficiency of A in both t = 0 and t = 1 in comparison with the old and the new frontier:

$$fs = \sqrt{\frac{TE^{FS0}_{t=1} * TE^{FS0}_{t=0}}{TE^{FS1}_{t=1} * TE^{FS1}_{t=0}}} \qquad (6.20)$$

Operating on the frontier does, however, not imply that a firm operates cost efficiently. As the inputs may have different prices, not every point on the frontier results in the same level of costs. The relationship between the input prices can be depicted through the input–price ratio curve (see Fig. 6.12). Each point on this curve reflects the same amount of total costs used for the inputs, but not with every combination of inputs the same level of output can be realized. In this figure, point B reflects a more efficient combination of inputs than point A, but this combination is not efficient when the input prices are taken into account. The latter inefficiency is called allocative inefficiency.

Box 6.2 Decomposition of productivity change into efficiency change and frontier shift

When the development in inputs and outputs of a firm and its benchmark firm are known, then the productivity change can be determined as a function of the change in technical efficiency and the movement in the technological

Table 6.3 Example of decomposing productivity change

	Firm A		Frontier	
	t = 0	t = 1	t = 0	t = 1
Inputs	60	40	40	30
Outputs	70	70	80	80
Productivity [see Eq. (6.16)]	1.17	1.75	2.00	2.67

Technical efficiency of firm A [see Eq. (6.17)]			
	Frontier t = 0	Frontier t = 1	
t = 0	0.58	0.44	
t = 1	0.88	0.66	
Efficiency change [see Eq. (6.18)]		1.13	
Frontier shift [see Eq. (6.20)]		1.33	(check: this is equal to 2.67/2.00)
Productivity change [see Eq. (6.19)]		1.5	(check: this is equal to 1.75/1.17)

frontier. In Table 6.3, the productivity of both firm A and the frontier (based on a benchmark firm) increases from t = 0 to t = 1. Using the formulas of this section, we are able to calculate the catch-up and the frontier shift. It appears that the increase in the productivity of firm A is to a large extent related to the general improvement in technologies (i.e. frontier shift is 33%) and to a lesser extent to the catching up of the firm (i.e. efficiency change is 13%).

6.4.2 Empirical Methods

A number of empirical methods exist to estimate the productivity efficiency (Table 6.4). A generally used method by regulators is data envelopment analysis (DEA). In this method a technological frontier is estimated on the basis of data on inputs and outputs of groups of firms. The estimation technique is based on linear programming. In this method, no explicit assumption has to be made on the production processes, in contrast to the method of stochastic frontier analysis (SFA). In both methods, it is possible to determine the so-called catch-up, which is the difference between the actual productivity of a firm and the productivity of its peers, which are the firms operating on the technological frontier.

Another group of methods to analyse the productivity of a firm is ordinary least squares regression (OLS/COLS). This method does, however, not give insight into the distance of a firm to the frontier (i.e. the inefficiency). The same holds for the use of productivity indicators, in which just the productivities are compared.

Table 6.4 Pros and cons of various empirical methods to estimate productive efficiency

Method	Pros	Cons
Date envelopment analysis	• Able to distinguish catch-up and frontier shift • Not needed to assume a specific production function	Not able to correct for noise in the technical efficiency
Stochastic frontier analysis	• Able to distinguish catch-up and frontier shift • Able to correct technical efficiency for stochastic noise	A production function has to be estimated
Ordinary least squares/COLS	Gives insight into factors behind productivity	Gives no insight into the distance to the frontier
Productivity indicators	Relatively easy to apply	Not able to distinguish catch-up and frontier shift or to correct for other factors
Reference network analysis	Potentially able to fully remove information asymmetry about frontier	• Complex, time consuming • Assumptions about frontier shift have to be made
Process benchmarking	Gives insight into the costs of specific activities	Cannot be used to assess the efficiency of existing costs or to make an integrated assessment of all costs (capex and opex)

A completely different method is reference network analysis, in which the infrastructure of the network operator is simulated by constructing a small-scale version in a greenfield situation and using the latest technologies. This approach is, however, very time consuming with a serious risk that not sufficient attention is paid to inefficiencies in actual networks which cannot be removed by the operator.

The last group of methods is called process benchmarking. In this method, the benchmarking is based on assessing the processes of managing the infrastructure instead of the actual outputs and inputs.

A challenge in any benchmarking method is to find comparable firms which are acceptable for both regulator and the regulated firms. Regulators want to look for firms that are sufficiently comparable in terms of type of activities, type of clients, size and natural circumstances (e.g. mountain area, presence of rivers), while regulated firms have an incentive to emphasize their differences, either before the regulator or the court of appeal (see Sect. 3.4.3), and to lobby for a comparison that involves other firms with higher costs. If regulated firms are successful in having less efficient firms as benchmark firms, the efficient costs will be defined at a higher costs level, leading to higher regulated tariffs.

In order to address this risk that the tariffs are incorrectly increased due to successful lobbying by the regulated industry, regulators can formulate as default view point that regulated firms have to proof that benchmark firms are not similar to them. In case the regulator agrees that a firm is to some extent structurally different from the others, for instance in terms of presence of waterways which affect the costs of operating an energy network, the regulator may correct the benchmark for this firm in order to compensate for the extra costs resulting from that difference.

6.4.3 Yardstick Regulation

In the above benchmarking methods, the efficiency of a regulated firm is assessed by using external information from other firms which operate in similar circumstances. These firms are not necessarily subject to the same regulatory supervision. A method where only data is used from firms which are subject to the same regulation is called yardstick regulation (Shleifer 1985). In this type of regulation, the maximum tariffs regulated firms are allowed to charge are determined on the basis of information on the productivity of a group of comparable firms.

Three types of yardstick regulation can be distinguished: a uniform, a discriminatory and a best-practice yardstick (Table 6.5). In a uniform yardstick scheme, the tariffs of each regulated firm are based on the average costs per unit of output of all firms in the group. In a discriminatory yardstick scheme, however, the tariffs of each regulated firm are based on the average costs of all the other firms in the group. In this scheme, information on the productivity of the regulated firm itself is not used in the benchmark, while in the uniform yardstick scheme, information of all firms is used to determine the average costs. In a best-practice yardstick scheme, the tariffs for all regulated firms in the group are based on the productivity of the most efficient firm in that group.

These three types of yardstick regulation differ in the strength of the incentives for cost reductions. With a discriminatory yardstick, the tariffs are fully based on the costs per unit of output of the other firms which implies that each regulated firm is a price taker. In that sense, this scheme is similar to price-cap regulation where the tariffs are fully based on external information. Hence, the incentive power is 1 for all firms. The yardstick can be calculated in both a weighted (i.e. where each firm has the same weight, independent of its size) and an unweighted way (for instance, based on firms' output levels). With a uniform yardstick, the incentive power (i.e. the strength of the incentive to improve efficiency) depends on the market share of each firm (see also Box 6.3). This market share can also be calculated in an unweighted) or weighted manner. The higher the share of a regulated firm in the group, the lower the incentive power is. In the extreme case where one firm has a market share of almost 100%, this type of yardstick regulation comes very close to cost-plus regulation. For all other firms in the group, their market share is almost zero, which implies that for them the yardstick regulation is close to price-cap regulation.

Table 6.5 Definition and incentive power of types of yardsticks

Type of yardstick	The average revenue per unit (tariff) is based on:	Incentive power
Uniform	(Un)weighted average costs per unit of all firms	1 - (un)weighted share of a firm in the yardstick
Discriminatory	(Un)weighted average weighted costs per unit of all *other* firms	1 for all firms
Best practice	Average costs per unit of most efficient firm	0 for best practice firm; 1 for all other firms

Box 6.3 Calculating the maximum tariffs for a network operator using different types of yardsticks

Suppose a regulator wants to regulate the tariffs of a group of five network operators through yardstick regulation. The first thing to do is to collect data on costs and outputs per operator (see Table 6.6). As the costs are generally expressed in financial terms, they can be easily aggregated to one number per operator. The outputs, however, do not have a market price. Therefore, the outputs are aggregated using regulatory weights for each type of output. This results in the so-called standardized output.

In case of a weighted uniform yardstick, the yardstick (i.e. the costs per unit of output) can be calculated by summing up all costs and outputs and determine the ratio (=105/58 = 1.81). This yardstick is the tariff which every operator may charge per unit of (standardized) output. In case of a weighted discriminatory yardstick, the yardstick can be calculated per operator by summing up all costs and output of all other operators. For instance, for operator A, the yardstick = 93/52 = 1.79. In this type of yardstick, every operator may have a different maximum level of revenues.

The incentive power is the same (i.e. 1) for all operators in a discriminatory scheme, as the revenues are only related to external information. In a uniform yardstick, the incentive power is related to the relative size of operators. Operator E has a relatively high share in the total market (24/58 = 0.41), and therefore its incentive power is relatively low (1–0.41 = 0.59). This result means, for instance, that for every euro increase in costs, operator E sees a reduction in its profits of 0.59 euro, while operator A sees a reduction in profits of 0.90 euro. Hence, because of its small size, it faces the strongest incentive to improve efficiency.

Table 6.6 Example of calculating two types of yardsticks and the resulting incentive powers

Operator	Data		Weighted uniform yardstick		Weighted discriminatory yardstick	
	Total costs (x million euro)	Total standardized output (x million euro)	Yardstick	Incentive power	Yardstick	Incentive power
A	12	6	1.81	0.90	1.79	1
B	14	10	1.81	0.83	1.90	1
C	24	10	1.81	0.83	1.69	1
D	20	8	1.81	0.86	1.70	1
E	35	24	1.81	0.59	2.06	1
Total	105	58				

A key condition for the effective application of yardstick regulation is that the group of firms is sufficiently large. If the number of comparable firms is too low, yardstick regulation is likely not effective. One specific risk of having only a few firms in the benchmark is that these firms try to collude against the regulator by deciding together that no one will make efforts to improve efficiency. If that would happen, the average costs are not reduced, and, hence, tariffs remain the same (Dijkstra et al. 2017).

6.4.4 Profit Sharing

If external information cannot be used to find an appropriate benchmark for the tariffs of a regulated firm, an alternative method can be found in profit sharing. In this form of tariff regulation, the external information is derived from the past while the benefits of efficiency improvement are shared between the regulated firm and its customers. The allowed revenues of a regulated firm (R_a) in year t can be set equal to the realized revenues (R_r) in year $t - 1$ minus a fraction of the profits (i.e. realized revenues minus realized costs (C_r)) which have been realized in that year (see Fig. 6.13). This results in the following formula for the allowed revenues:

$$R_{a,t} = R_{r,t-1} - a(R_{r,t-1} - C_{r,t-1}), with\ 0 \leq a \leq 1 \tag{6.21}$$

In this formula, a is the profit-sharing parameter which determines how the benefits of efficiency improvements as well as the costs of inefficiencies are shared between the firm and its customers. If $a = 1$, the allowed revenues in t are equal to the realized costs in $t - 1$, which means that this scheme is equal to cost-plus regulation. If $a = 0$, the allowed revenues in t are equal to the realized revenues in $t - 1$, which means that this scheme is equal to price-cap regulation, provided that the revenues depend fully on exogenous factors. This implies that in that situation the incentive power is equal to $1 - a$.

Hence, incentives for costs reductions can be given without the use of information from external firms. The key condition for giving such incentives is that the revenues (and, indirectly, the profits) are not fully adapted to changes in the costs. If a regulator wants to give high-powered incentives to a regulated firm, it could simply impose the rule that the revenues remain the same (i.e. setting a equal to 0), no matter what happens with the actual costs of the firm. Such a rule creates, of course, the risk that the revenues strongly deviate from the costs, resulting in windfall profits or financeability problems. In order to reduce this risk, the regulator may set a equal to, for instance, 0.5, which implies that the firm will be compensated for 50% for any change in its costs.

Costs/revenues

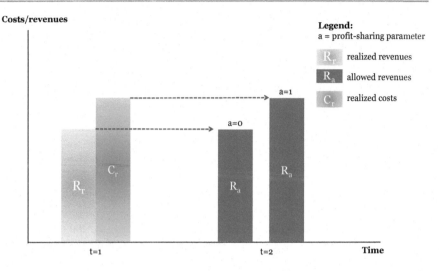

Fig. 6.13 Allowed revenues determined in a profit-sharing scheme

---

6.5   Cost of Capital

6.5.1   WACC

As regulated companies are generally capital-intensive, costs of capital form a major component of total costs. The cost of capital is also the most complicated variable in tariff regulation as it consists of a number of variables referring to the future and which, hence, have to be estimated. The cost of capital is expressed through the WACC which stands for the weighted average costs of capital. The weights refer to the shares of the two types of capital: equity (E) and debt (D). Equity is capital raised by issuing shares that entitle their owners to a share in the firm's profits, for instance by regular payments of dividend. Debt is capital borrowed from banks or other lenders with the obligation to pay them interest and to repay the loan back at some time. The key distinction between equity and debt is that equity capital faces the residual risk related to the profitability of the firm, while debt capital is (to a large extent) protected from that risk. The latter is organized by providing the lender of debt capital more or less guarantees that the borrower will pay a fixed compensation (i.e. the interest) and repay the loan after some period. In order to give the lender of debt capital these guarantees, more effort has to be done in, for instance, screening the borrower and the project, which results in transaction costs. The provider of equity capital doesn't have this kind of securities, and, hence, it faces the risk that the return on the investment is less than required, while it can also be (much) higher.

As said, the WACC is calculated by taking into account the relative weights of each source of finance. The share of debt in total capital is called the *gearing* (g):

$$g = \frac{D}{D+E} \qquad (6.22)$$

The basic formula for the WACC is the weighted average of the costs of debt (c_d) and the cost of equity (c_e):

$$WACC = c_d * g + c_e * (1 - g) \qquad (6.23)$$

In this formula, no attention is given to the role of taxes. From the perspective of investors, the taxes they have to pay on the dividends are also costs, while the firm generally can deduct interest expenses from the tax allowances. In order to make the role of tax explicit, the WACC is calculated both post and pre-tax. The formula for the post-tax WACC is as follows, with tx being the (marginal) tax rate:

$$WACC_{post-tax} = (1 - tx) * c_d * g + c_e * (1 - g) \qquad (6.24)$$

The post-tax WACC measures the cost of capital when the costs of debt are reduced with the tax deductions while the costs of equity do not include the compensation for the costs of tax payments on dividend. In the pre-tax WACC, both effects of tax are mitigated:

$$WACC_{pre-tax} = \frac{WACC_{post-tax}}{1 - tx} \qquad (6.25)$$

Regulators will use the pre-tax WACC when they want to separately deal with the impact of taxes in tariff regulation.

The essential components in the WACC formula are the estimates of the costs of equity and the costs of debt. The costs of equity refer to the compensation which is required by investors for participating in the firm and, hence, running the residual risk, while the costs of debt refer to the compensation required by lenders of debt capital. Both required compensations are partly related to the compensation needed for the inflation and the so-called time value of money. The latter basically means that current consumption is preferred above consumption in the future, which implies that everyone who postpones consumption to some moment in the future wants a compensation in return.

Besides these two common elements, both have also specific elements. For the costs of equity, the specific element is the compensation required for the financial risks related to the activities of the firm. This is called the required return on equity. For the costs of debt, the specific element is the credit risk, which is the risk that the borrower will not pay back the loan with the related interests. In addition, lenders of debt capital make transaction costs, which were mentioned above, for which they also want a compensation. Below, we discuss how the costs of debt and equity can be determined. In Box 6.4, we give an example of calculating the WACC.

6.5.2 Cost of Equity

The cost of equity (c_e) is the sum of the risk-free interest rate (r_f) and the equity-risk premium (p_e)[10]:

$$c_e = r_f + p_e \tag{6.26}$$

The risk-free interest rate is the return on an investment without any risk. In theory, this rate depends only on the time preference of investors (i.e. the above-mentioned time value of money) and the expectations regarding future income.

The equity-risk premium depends on the market-risk premium (p_m) and the sensitivity of a firm to the market risk, which is measured by β (beta):

$$p_e = \beta * p_m \tag{6.27}$$

The underlying theory behind the costs of equity is that in well-functioning (efficient) capital markets, investors only receive compensation for risks which cannot be diversified away. Non-systematic risks are project-specific risks, which can be diversified away by combining different projects with different specific risks in one investment portfolio. The so-called systematic, non-diversifiable risks are the risks related to macroeconomic developments and which cannot be mitigated through diversification.

The premium investors required for this systematic risk is called the market-risk premium. The market-risk premium (p_m) is the extra return an investor requires on top of the risk-free interest rate, i.e. it is the difference between the return on the market portfolio (r_m) and the risk-free interest rate (r_f).

$$p_m = r_m - r_f \tag{6.28}$$

An investor that invests in the market portfolio, which consists of shares of all firms active in the market, receives the market-risk premium if the capital market is efficient. Hence, the required return on equity of such an investment is equal to the sum of the risk-free interest rate and the market-risk premium. If investors invest in a different portfolio with a selection of firms, the sensitivity of this investment portfolio to the systematic risk may be more or less than the average. This sensitivity is measured through the beta. If the beta is less (more) than 1, it means that a particular investment is less (more) sensitive to the macroeconomic risk than the groups of all firms in the market. In case an investment does not face any relationship to the macroeconomic risks, the beta is zero, which implies that the required return on equity is equal to the risk-free interest rate. The relationship between the beta of an investment (i.e. the sensitivity to the market risk) and the required return on equity is expressed through the so-called security market line (see Fig. 6.14). This line shows

---

[10]The common formulation of this relationship is $c_e = r_f + \beta(r_m - r_f)$.

that the required return on equity (r_e; which is equal to the costs of equity, c_e) depends on the risk-free interest rate, the market-risk premium and the sensitivity of an investment (e.g. a regulated firm) to the market risk.

6.5.3   Market-Risk Premium

The market-risk premium can be estimated by three different types of methods: the historical method, forward-looking method and surveys.

The *historical* method is based on the idea that the expected value of the market-risk premium is equal to the average value over a number of years n in the past. Hence, the market-risk premium for the future is calculated as follows:

$$p_{m,t} = (p_{m,t-n} + \cdots + p_{m,t-1})/n \qquad (6.29)$$

Although this method seems to be straightforward, the results are sensitive to how the method is applied. If the number of years looking backwards is relatively short, the standard error is relatively large, which can be even higher than the actual risk premium, while longer periods may imply that the data series is less representative for the future. Capital markets have changed strongly over the past 100 years due to the increased use of information technologies and the resulting globalization, which have enabled investors to improve the diversification of their investments. As a result, the systematic risk in the more recent years is lower than in

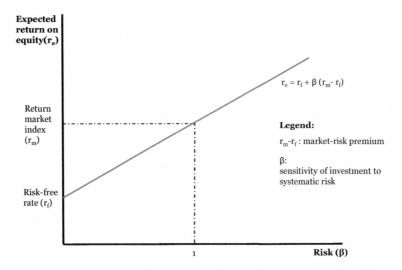

Fig. 6.14 Security market line: linear relationship between expected return on equity and systematic risk of portfolio

decades ago. Hence, when using historical data on realized market returns, one could give a higher weight to more recent observations.

The market-risk premium is also sensitive to the choice of the type of the risk-free investment. The risk-free interest rate is generally measured through the interest rate on government loans, but these loans exist in different types. When short-term government securities (so-called treasury bills) are used, different results for the risk-free interest rate are obtained than when long-term government securities (so-called treasury bonds) are used. The relationship between the interest rate and the duration of loans can be expressed through a yield curve. If this yield curve is upward sloping (i.e. higher rates for higher duration of securities), then a lower market-risk premium is found when (long-term) bonds (with relatively high interest rates) are used. In general, treasure bills can be viewed as the best proxy for risk-free investments, since no long-term risks are included.

What also affects the outcome of the calculation of the market-risk premium using historical data is how the average is calculated. This can be done in an arithmetic way (like in the above formula), but it could also be done in geometric way (taking the square of the product). In the former method, it is assumed that the annual returns are independent from each other, while in the latter method, it is assumed that the annual returns depend on each other through the interest-on-interest effect.

Another method to estimate the market-risk premium is the *forward-looking* method. This method is based on the theoretical insight that the actual value of stocks is related to the expected future revenues from holding the stocks, i.e. that the current value (V) of stocks is equal to the present value of the future dividends (Div):

$$V = \sum_{t=1}^{t=\infty} \frac{Div_t}{(1+r)^t} \tag{6.30}$$

The expected future dividends on equity can be based on expectations of analysts or forecasts of macroeconomic growth (e.g. current dividends times annual growth rate). The required return on equity investments is equal to the discount factor r which is needed for making the present value of the expected dividends equal to current value of stocks. From this, the implied market-risk premium (p_m) can be calculated as the difference between this discount factor and the risk-free rate (r_f):

$$p_m = r - r_f \tag{6.31}$$

The third method to estimate the market-risk premium is to conduct *surveys* among investors and ask them about their assessment of risks related to the market portfolio and the required return. This method appears, however, to be sensitive to a number of factors, in particular to the current stock value and to whom is included in the survey. It appears that the surveys result in higher estimates for the market-risk premium, when stock prices have increased strongly and v.v. In addition, it appears that individual investors seem to have higher expected returns than institutional investors, while academics appear to use higher market-risk premium estimates than analysts.

6.5.4 Asset and Equity Beta

When the market-risk premium has been determined, the next step is analysing to what extent a particular firm is sensitive to the systematic risk and to what extent the shareholders of that firm are confronted with it. The former aspect is measured through the so-called asset beta (β_a), while the equity beta (β_e) measures the exposure of shareholders to the systematic risk.

The asset beta of a particular firm can be estimated by analysing the relationship between the returns on the market portfolio and the returns of that firm. The asset beta of energy network firms is generally around $0.3 - 0.4$, which means that these firms are not very sensitive to macroeconomic changes (see Fig. 6.15). This limited sensitivity is related to the fact that the revenues are regulated, while the usage of electricity and gas is not strongly related to the economic business cycle.

As discussed above, the systematic risk of a company is allocated to the shareholders. Their exposure to the systematic risk depends on three factors: the asset beta, the gearing and the tax rate. The higher the gearing (i.e. the higher the share of debt in total financing), the lower the share of equity and the higher the sensitivity of each unit of equity to the systematic risk. The equity beta is lowered if the tax rate is higher, because taxation on dividends implies that the risks are shifted away from the shareholders to the government (as collector of tax revenues). The influence of tax on the equity beta can be determined in different ways.

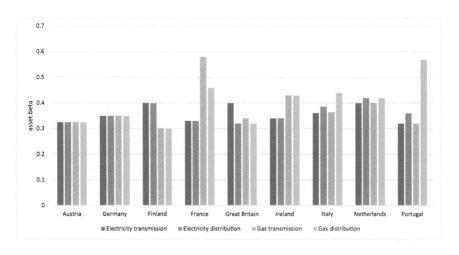

Fig. 6.15 Asset betas recently set by regulators for electricity and gas transmission and distribution networks in a number of European countries. *Source* NERA (2016)

One approach is Modigliani–Miller approach. Using this approach, the equity beta can be calculated as follows:

$$\beta_e = \frac{(1-g) + g(1-tx)}{1-g}\beta_a \tag{6.32}$$

6.5.5 Cost of Debt

The costs of debt depend on the risk-free interest rate, a debt premium (p_d) and non-interest fees (t_d):

$$c_d = r_f + p_d + t_d \tag{6.33}$$

The debt premium is the required return on a loan above the risk-free rate as compensation for credit risks. This premium depends on risks related to lending capital to a firm such as the risk of default and the risk of prepayment. The debt premium can be determined by comparing actual interest rates of bonds issued by comparable firms with the interest rates on government bonds (with the same duration). This difference is called 'the spread'. The non-interest fees are related to the transaction costs of lending capital to a firm.

Box 6.4 Calculating the WACC

When a regulator wants to determine the WACC, it has to determine the value of a number of variables. When it has the following information:

Parameter	Value
Risk-free rate (r_f)	0.5%
Asset beta (β_a)	0.4
Gearing (debt/assets) (g)	0.4
Tax rate (tx)	20%
Market-risk premium (p_m)	5%
Debt premium (p_d)	0.7%
Non-interest fees (t_d)	0.1%
Inflation (Infl)	2.0%

The WACC can be determined as follows:

[1] Equity beta $= \beta_e = \frac{(1-g) + g(1-tx)}{1-g}\beta_a = 0.61$

[2] After-tax cost of equity $= r_f + \beta_e * p_m = 3.57$

[3] Pre-tax cost of debt $= r_f + p_d + t_d = 1.3$

[4] Nominal after-tax WACC = $(1 - g * [2]) + (1 - tx) * g * [3])$
= 2.56
[5] Nominal pre-tax WACC = $[4]/(1 - tx)$ = 3.20
[6] Real pre-tax WACC = $100 * ((1 + [5]/100)/(1 + Infl/100) - 1)$
= 1.17
When the latter variable is used to determine the compensation for the costs of capital, separate arrangements are needed to deal with inflation and the effects of taxation.

6.6 Tariffs and Revenues

6.6.1 Tariff Structure

While incentive power and tightness refer to the total revenues of a regulated company, another fundamental aspect of tariff regulation is the design of the tariff structure. The regulatory challenge here is to find a tariff structure which is allocatively efficient, which means that the tariffs are related to the marginal costs of network usage, under the condition that the average overall revenues should be at least equal to the average costs of the network operator. The latter condition is called the financeability constraint.

Theoretically, the first best method to deal with this challenge is to set the individual tariffs equal to the respective marginal costs and to use the consumer surplus to give the firm compensation for the fixed costs. A regulatory solution which follows this approach is the so-called Loeb–Magat proposal. In this approach, the regulated firm is free to set the tariffs, but it gets a subsidy equal to the consumer surplus. This gives the firm the incentive to set tariffs equal to the marginal costs as that maximizes the consumer surplus, as was explained in Sect. 6.2.1. This solution requires that the regulator is able to estimate the consumer surplus, which means that he is perfectly informed about the demand curve, which is often not the case. Another caveat of this solution is that it enables the regulated company to fully capture the consumer surplus. This adverse effect can, however, be addressed by imposing the obligation on the company to pay for the licence to operate, for instance through a process of competitive tendering (see Sect. 6.2.3).

Another theoretical solution which minimizes the deadweight loss and maximizes the revenues for the firm is Ramsey pricing. In this approach, the tariffs are inversely related to the price elasticity of the various consumer groups (see Fig. 6.16). Suppose a company sells two different products in two different markets: A and B. Suppose further that the demand curve for product A is steeper than for product B, which means that the consumers of product A are less price sensitive. When the consumers would face a similar price increase (from P1 to P2),

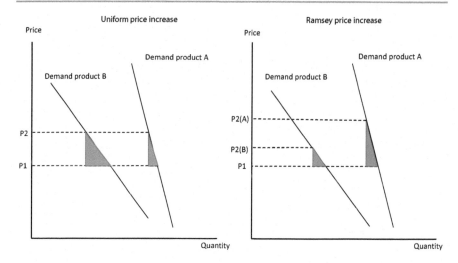

Fig. 6.16 Uniform and Ramsey prices in markets with two different demand curves

consumption of product B will be reduced more strongly than consumption of product A. If this price increase is not related to marginal costs, but meant to recoup fixed costs, the price increases result in a deadweight loss. This effect on deadweight loss can be reduced by differentiating the price. Consumers which are more sensitive should only pay a relatively small margin on top of the marginal costs, while less price-sensitive consumers should pay a higher margin on top of the marginal costs. As this latter group is less price sensitive, there is less reduction in consumption and, hence, less deadweight loss. Consequently, differentiating tariffs by making them inversely related to the price elasticities is an effective tool to reduce the deadweight loss. By charging higher tariffs for those products where consumers are less price sensitive enables a company to recoup all its (fixed) costs with lower tariff increases for other products and consumers.[11]

Although being efficient from a theoretical–economic point of view, Ramsey prices may not be perceived as fair by consumers. Those consumers who really depend on the use of energy, which means that they have a low-price elasticity, will pay more than those who have more alternatives. From the perspective of fairness, this tariff design may therefore be less appreciated (see further on this issue, Sect. 11.4).

The application of Ramsey prices requires that the regulated company as well as the regulator are well informed about the price sensitivity of various types of users for the various types of products. Often this information is not present, while the collection of information and calculation of the tariffs would be fairly complicated. In that case, an alternative solution can be used.

---

[11]The Ramsey prices can be calculated by searching for these product prices which maximize the consumer surplus under the condition that the overall revenues are equal to the overall costs (i.e. the financeability constraint).

A less complicated option to solve the regulator challenge to reduce the dead-weight loss of tariffs while taking into account the financeability constraint is nonlinear tariffs. Examples of such a structure are two-part tariffs or multi-part tariffs. One part is a fixed fee, for instance a fixed amount per unit of capacity per year, while the other part(s) are related to the marginal costs of actual network usage (see Fig. 6.17). If the fixed fee is related to the fixed costs, and the variable parts to the marginal costs, then both objectives are realized: sufficient revenues to recoup the fixed costs, while minimizing the deadweight loss.

A bit more advanced scheme is a so-called self-selecting two-part tariff scheme. Here customers can choose from a number of alternative two-part tariff schemes. As they have more information on how intensive they are going to use the infrastructure, they may prefer to pay more or less through fixed or variable fees. In Fig. 6.17, tariff structure A has a lower fixed fee and a higher variable fee compared to tariff structure B. Consumers who expect to be heavy users (i.e. consume more than quantity Q^*) will choose tariff structure B, while consumers who expect only incidentally to use the infrastructure (i.e. with a consumption level lower than Q^*) will choose tariff structure A. As a consequence, heavy users contribute more to the recovery of the fixed costs, while their marginal decisions can be more efficient as their marginal tariff is lower and, hence, closer to the marginal costs. In addition, the two-part tariff scheme allows light users to consume the product because of the low fixed fee. In general, a two-part tariff scheme can be seen as a Pareto improvement compared to a linear scheme with a fixed fee for everyone, as both consumer groups can benefit, while also the supplier realizes a higher level of revenues (see also Decker 2015).

Another pricing solution to obtain both efficient prices and recovery of fixed costs is called peak-load pricing. This principle can be applied when the commodity is not storable while the capacity to supply to the market is fixed during a period of time, such as in the case of transportation infrastructure and electricity generation (see further Sect. 7.5). In a system of peak-load pricing, the price of the commodity is only determined by the marginal costs when demand is less than the capacity constraint. This price is called the off-peak price. When, however, the demand is higher than the capacity constraint, the price has to go up in order to clear the market. The resulting prices are called peak prices. These peak prices form the source for the remuneration of the fixed costs. Hence, users who use the infrastructure during hours of scarcity pay more than the marginal costs, while users who only consume during off-peak hours, when there is sufficient capacity, only pay the marginal costs.

As long as the price is below the long-run marginal costs of extending the capacity (i.e. the sum over operational and investment costs per unit of output), a network company will not invest and consumers pay a price which is determined by the marginal willingness-to-pay (i.e. the intersection of demand curve and vertical part of the supply curve). Only when this price exceeds the long-run marginal costs of network capacity, a network company has an incentive to extend its capacity. The break-even condition for the investment decision is that the investment costs per unit are equal to the expected peak-demand revenues minus operational costs during the lifetime of the investment. Hence, also the peak prices are related to the marginal costs, as the long-run marginal costs of extending capacity form a cap on the peak prices.

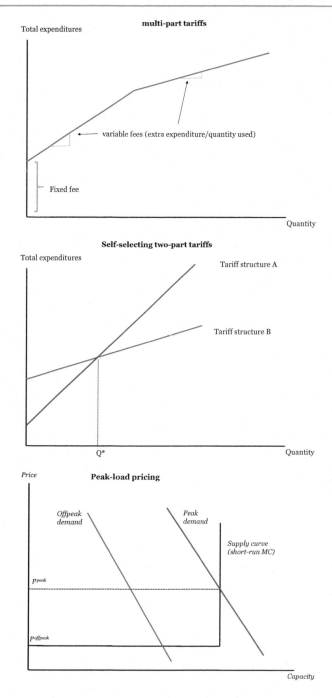

Fig. 6.17 Alternative tariff structures to minimize deadweight loss while taking into account financeability constraint

The peak-load pricing principle can be applied to make energy network tariffs depending on time and location since bottlenecks in network capacity also have these two dimensions. When a particular part of a network is not congested during a particular period of time, the tariffs for using that part of the network could be set equal to the marginal operational costs. In case of congestion in some parts of the network during some periods, the tariffs could be raised in order to give incentives to network users to reduce consumption until demand equals the available capacity. Such locational-specific and time-varying tariffs give also incentives to the network operator to invest in network extension when the tariffs are sufficiently high to recover the investment costs. See further on this issue Sect. 7.3.

The above pricing methods are based on economic principles to foster efficient usage of the infrastructure. In practice, regulators and network companies also use more administrative pricing methods. One of such methods is called fully distributed costs (FDC). This method entails that the common costs of an infrastructure are allocated over various products and users by using administrative rules of thumbs, such as the contribution of a product to the total variable costs.

6.6.2 Allowed Revenues and X-factor

When the regulatory choices regarding incentive power, tightness and tariff structure have been made, the levels of allowed revenues during the next regulatory period can be determined. The notion of allowed revenues is used, as generally regulated companies are free to have a lower level of revenues. The allowed revenues can, therefore, be seen as maximum level of revenues. Before discussing the determination of the individual tariff levels in Sect. 6.6.3, we first discuss the determination of the annual allowed revenues during each year of the next regulatory period.

Box 6.5 Tariff structure applied by an electricity distribution grid operator

Operators of electricity grids differentiate their tariffs to a number of dimensions, such as the voltage of the network part a user is connected to (e.g. high, medium or low voltage), the capacity of the connection (in ampere), the annual number of hours of usage, the timing of use and the actual level of transported energy. The Dutch distribution operator Stedin, for instance, distinguishes the users that are connected to the low-voltage part of the grid in three types: residential users with only a capacity tariff, other small-scale users and other users. For each group of users, the tariff structure is different (see Table 6.7). The residential users with a capacity tariff pay a fixed tariff per year depending on the individual capacity of their connection. The other small-scale users pay a fixed transport fee that is only differentiated between two groups with different capacities. The other users to the low-voltage grid pay three types of tariffs: a fixed fee per year, a capacity tariff that depends on the individual capacity, and a transport fee that depends on the amount of energy

Table 6.7 Tariff structure of Dutch electricity distribution grid operator

Group of users	Tariff types	Unit
Residential users with capacity tariff	Capacity tariff	Euro/capacity/year
Other small-scale users	Fixed tariff if capacity is $\leq 1 * 6A$	Euro/year
	Fixed tariff if capacity is $\leq 3 * 80A$	Euro/year
Other users	Fixed tariff transport services	Euro/year
	Capacity tariff	Euro/capacity/year
	Transport tariff off-peak hours	Euro/kWh
	Transport tariff normal hours	Euro/kWh

Source ACM, https://www.acm.nl/nl/publicaties/tarievenbesluit-stedin-elektriciteit-2019

transported and the time of the day. For network users connected to other parts of the grid (i.e. medium and high voltage), similar types of tariffs exist.

When these revenues are related to the own realized costs (i.e. pure cost-plus regulation), it is fairly straightforward to determine them. For instance, the maximum level of revenues in year t is equal to the realized costs by the regulated company in year t − 1. In most cases, however, regulators prefer to use external information on appropriate cost levels as benchmark for the allowed revenues in order to give some incentives for efficiency, as discussed in the previous sections. The issue is then how to translate this external cost information to the annual levels of allowed revenues.

Suppose that the allowed revenues of a regulated company may not be higher than the costs of an efficiently operating firm at the end of the regulatory period. These costs can be determined through various types of benchmarking methods, as we have seen in Sect. 6.4. In that section, we have seen that there are two aspects in determining the level of efficient costs of the benchmark firm: catch-up effect and frontier shift. The former effect refers to the actual difference between the costs (per unit) of the regulated company and the benchmark firm, while the latter effect refers to the future change in efficiency of the benchmark firm.

To determine the annual levels of the allowed revenues of the regulated company, the results of benchmarking regarding the catch-up effect and frontier shift can be used in several ways. One way is to first calculate an estimate for the current costs when the firm would have addressed actual inefficiencies (i.e. controlling for the catch-up effect) and next to estimate what the expected changes in technological developments during the new regulatory period will be (i.e. controlling for the frontier shift) (see Fig. 6.18).

The estimate of the costs of the regulated company after controlling for the catch-up effects (C_e) is based on its actual realized total costs (C_r), the technical efficiency score (i.e. the difference in productivity between the regulated company

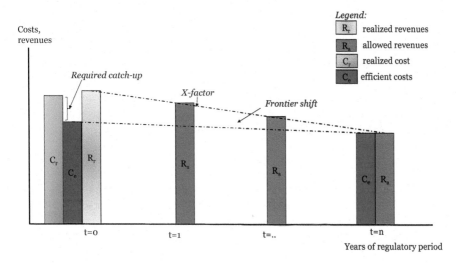

Fig. 6.18 Translating results of benchmarking to the allowed levels of annual revenues

and the benchmark firm) and the number of years the firms is given to catch up with the benchmark firm, i.e. to reach the frontier (N):

$$C_{e,t=0} = C_{t=0} * TE^{\frac{1}{N}} \tag{6.34}$$

If, for instance, a regulated firm has an efficiency of 60% and the regulator gives this firm 10 years (i.e. N = 10) to catch up to the frontier, then the level of efficient costs at the start of the regulatory period is set at $0.60^{1/10}$, which gives a level of 95% of the actual level of costs. This implies that instead of requiring that the regulated firm immediately reduces its costs to 60%, the costs only have to reduce by 5%. Hence, if a firm is highly inefficient (compared to the benchmark firm) and the regulator believes the firm needs a long period to catch up, it can simply use a high number for N. The higher this number of years, the closer the value of the efficient costs (C_e) is to the value of the realized costs C_r in the last year of the previous regulatory period. It is generally up to the discretion of the regulator how quickly the revenues of the regulated company will be set on the level of the benchmark (Baldwin et al. 2012).

Next, the regulator may decide also to take into account the expected shift in the technological frontier as a result of future technological improvements. The efficient cost level (C_e) at the end of the new regulatory period of n years can be calculated as the product of the current efficient cost level and the annual percentage frontier shift (fs):

$$C_{e,t=n} = C_{e,t=0} * (1 - fs)^n \tag{6.35}$$

If, for instance, the frontier shift is set at 2% per year and the length of the new regulatory period is 3 years (i.e. n = 3), then the level of efficient costs at the end of the regulatory period is equal to 0.98^3 (i.e. 94%) of the level of efficient costs at the start of the regulatory period.

When the level of efficient costs at the end of the regulatory period has been determined, the next step is to translate this level into levels of annual allowed revenues. The relationship between the allowed level of allowed revenues (R) and the costs is expressed in the so-called CPI-X formula:

$$R_{a,t} = (1 + CPI_t - X) * R_{a,t-1} \tag{6.36}$$

In this formula, CPI stands for the consumer price index and X is the so-called x-factor. This formula says that the annual level of allowed revenues may increase with the annual (expected) inflation and decrease with the annual efficiency factor. The x-factor depends on two factors: (a) the productivity change the regulated firm has to realize based on the above discussed regulatory decisions regarding catch-up and frontier shift and (b) the starting point for the revenues. When the starting point is determined by the total revenues in the last year of the previous regulatory period, the x-factor is the annual percentage change which takes that level of revenues (R_r) to the level of efficient costs (C_e) at the end of the new regulatory period of n years (Fig. 6.18). This x-factor is calculated as follows:

$$X = 1 - \left(\frac{C_{e,t=n}}{R_{r,t=0}}\right)^{1/n} \tag{6.37}$$

If, for instance, the level of efficient costs at the end of the new regulatory period of 3 years (i.e. $C_{e,t=3}$) is 90 and the actual level of revenues in the last year of the previous regulatory period (i.e. $C_{r,t=0}$) is 100, then the x-factor is $(1-90/100)^{1/3}$, which is equal to 0.035. Applying this number in Eq. (6.36), while ignoring the CPI, gives an allowed level of revenues of 96.5, 93.2 and 90 in the years 1, 2 and 3 of the new regulatory period.

The x-factor is often called the efficiency factor, but it is important to realize that the level of this factor only says to what extent the current level of revenues has to change from one year to the other. The incentive power to improve efficiency is not determined by the x-factor, but by the degree that the own costs of the regulated firm affect the costs of the benchmark, as discussed in Sect. 6.3.

6.6.3   Maximum Tariffs

When the total level of the annually allowed revenues during each year of the next regulatory period and the tariff structure have been determined, the next and final step in tariff regulation are to allocate these allowed revenues to maximum tariffs for various products a network operator sells. This set of tariffs is called the tariff basket. These tariffs refer to the products of the network operator which consist of the rights to use the network at the various (entry and exit) points (see Box 6.6).

Box 6.6 Number of different tariffs for using the Dutch high-pressure gas transport network

Network operators sell access to their network. Their product is, hence, capacity, which can be expressed in MWh/h, so the ability to inject or withdraw energy (MWh) during a period of time (hour). Gas network operators typically have various types of customers, such as local producers, industrial users, local distributional grids, storages and borders. For each of these users, they may define different tariff levels. The Dutch gas transmission system operator GTS, for instance, uses about 1100 different types of tariffs. Within each group, there can be a large spread in tariff levels; see for instance Fig. 6.19 with a histogram of the tariffs used for the exit to the local distribution networks.

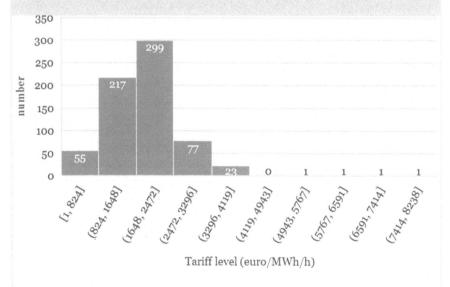

Fig. 6.19 Distribution of exit tariffs of Dutch gas transmission network for the local distribution grids. *Source* Website GTS

Energy transportation and distribution networks generally offer many different types of products: every connection to the grid is a product and there are usually many types of connections (e.g. varying in size and location). In gas markets, for instance, the gas network operator gives access to various types of domestic producers, storage operators and consumers, while it also may have international connections. The access to the network is the basis for the functioning of the gas market. In case of an entry–exit system, gas that is injected somewhere in the grid can be traded in the wholesale market, as is discussed in Sect. 3.3.3.

Generally, network operators are free to set individual tariffs for using the networks (tar$_i$), provided that the expected revenues (R$_e$) are below the allowed level of revenues (R$_a$)[12]:

$$R_e = \sum_i tar_i * V_i \leq R_a \qquad (6.38)$$

In order to check whether this condition holds, the regulator has to assess both the proposed tariffs and the underlying assumptions regarding the expected volume of the sales of each product (V$_i$). When the above condition is satisfied, the proposed tariffs of the network operator can be approved.

The expectations regarding the future volume of sales of products can be based on data on historical usage in combination with new information on changes in the network size or within the group of network users (such as a recently realized extension of the grid to a new group of households). In the above, it is assumed that the network operator faces volume risk, which is that the consequences of any deviation between the realized volumes and the assumed volumes are for the network operator. When, however, the network operator cannot influence the utilization of its network, it makes sense to remove this risk from the revenues. This may, for instance, be relevant when the actual utilization of a grid is strongly dependent on external circumstances such as outside temperature (which in particular holds for gas and heat networks) and the economic business cycle (which is in particular relevant to electricity networks).

When the volume risk is removed from tariff regulation, it is called revenue regulation. This can be implemented by determining a new level of allowed level of revenues (R_a^*) in year t on the basis of the initially determined level of allowed level of revenues and the deviation between realized (R$_r$) and allowed revenues in the previous year:

$$R_{a,t}^* = R_{a,t} - (R_{r,t-1} - R_{a,t-1}) \qquad (6.39)$$

As a consequence, the firm will be perfectly compensated for any deviation between realized and assumed utilization of its network. The only remaining risk

---

[12]Some regulators also require that each separate tariff is related to the costs of using a specific product. This is, however, a difficult objective to realize as most costs of network operation are so-called common costs, which means that they are shared by all users and cannot be attributed to specific products. In practice, several rules-of-thumb are used for this kind of cost allocation, such as the method of fully distributed costs, as discussed in Sect. 6.6.1.

for a firm subject to revenue regulation refers thus to changes in its costs. Such a reduction in risks for the regulated company may imply that the systematic risk for the investors decreases as well, which should result in a lower allowed return on capital (i.e. a lower WACC, see Sect. 6.5).

6.7 Performance of Networks

6.7.1 Tariff Regulation and Investments

Although tariff regulation is primarily meant to prevent that consumers pay more than what would be needed in competitive circumstances, this does not imply that the tariffs should be reduced to the level of the short-run marginal costs. It also holds for competitive markets that prices, at least in the long term, are related to the fixed costs. This implies that the tariff regulation also needs to control for the financeability constraint. This constraint does not only refer to the actual costs of capital, but also to the future costs as the regulated firm needs to be able to finance investments to maintain or upgrade the network. The energy transition from fossil fuels to renewables, for instance, requires network firms to adapt their grids. Increasing amounts of solar and wind energy need to be able to be fed into the electricity grid, while gas grids may need to facilitate the injection of biomethane. Moreover, offshore wind parks have to be connected to the grid, while more and more households and firms put solar panels on their roofs, which may create bottlenecks (congestion) in the networks. Hence, in order to facilitate the energy transition, energy network operators need to be able to finance the required investments. This means that the regulated revenues need to take the extra costs of investments into account.

The general method to deal with investments within tariff regulation is that the Regulated Asset Base (RAB), which is the regulatory book value of invested capital, is corrected with investments and depreciation (see Sect. 6.3.4). This means that if a regulated firm invests in extending its network, the value of this investment is added to the RAB. As the RAB increases, the CAPEX increases as well. Subsequently, the allowed revenue in the new regulatory period increases. In this manner, the reimbursement of the investment costs is delayed as the investments have to be done before the allowed revenues can increase. The regulator may, however, give a compensation for the costs of this delay in the reimbursement of costs.

It is important to understand, though, that the increase in revenues from tariffs is not meant to finance investments, but only meant to compensate for extra costs. Note that investments are not costs, but just expenditures. The costs of an investment consist of the depreciation costs and the required compensations for the providers of debt and equity. Hence, tariffs to be paid by network users are only meant to give compensation for the costs of network operators. When a firm invests, its future revenues may increase if the investment results in a higher network usage, and this will raise the willingness of banks and shareholders to provide extra capital to finance the investment. Generally, one may assume that when a regulatory framework for tariff regulation exists which gives the regulated firm certainty that it will get reimbursement of its efficient costs resulting from a new investment, it will also be able to attract sufficient capital from investors for efficient investments.

There might, however, be a few hurdles. If a regulated firm is uncertain to what extent the regulator will view the investment as necessary or efficient, it will also be uncertain about the extent the costs can be compensated through the regulated tariffs in the future. Because of that regulatory uncertainty, financing investments may be problematic. For this reason, it is crucial that the regulator provides clarity beforehand about the assessment method that will be used to assess investments. By being transparent about the future design of tariff regulation, the regulator can mitigate this risk of hold up.

In addition, a firm may face financing difficulties related to other constraints investors are imposing, such as the common requirement by lenders of debt capital that the share of equity should have a certain minimum value. If shareholders are not able to provide additional equity, an investment might not be financeable at all. Such a problem is, however, not related to tariff regulation, but to how the capital market functions.

The design of tariff regulation may also have an effect on the incentives for investments. Rate-of-return regulation, for instance, may give the regulated firm an incentive to overinvest in case the allowed rate-of-return exceeds the actual opportunity costs of capital (Viscusi et al. 2005). To illustrate this, suppose a firm uses two inputs: capital and labour (see Fig. 6.20). The isoquant describes by which combinations of inputs a specific number of outputs can be produced, while the isocost line describes by which combinations the firm has the same total costs (see Sect. 4.2.2). In order to minimize their costs, a firm wants to be on the isoquant which is closest to the origin. Any point on this curve is technically efficient, but the relative prices of the inputs determine which point results in the lowest costs. The optimal mix of inputs is where the isocost line is tangent with the isoquant. If the costs of capital reduce, then the slope of the isocost line changes (in this figure, from $I1$ to $I2$). The new optimal input mix changes from point E to point A. In the new optimum, more capital is used (C_2 instead of C_1) and less labour. Hence, reducing the costs of capital gives an incentive to firms to become more capital intensive.

In the case of rate-of-return regulation, the reduction in costs of capital occurs when the rate-of-return given to the regulated firm exceeds the actual costs of capital in the capital market. In such a case, a regulated firm will invest more which may result in so-called gold-plating within the firm as well as higher tariffs for customers. This effect is called the *Averch–Johnson effect*. Empirical evidence has indeed confirmed that tariffs under rate-of-return regulation may give firms an incentive to invest in capital-intensive technologies which results in higher costs (see, e.g. Cicala 2015).

Applied to the field of energy transition, a network operator that is incentivized to search for capital-intensive solutions may prefer to solve congestions created by increased supply of renewable energy by investing in network extension, while a network operator that is incentivized to improve overall efficiency may prefer to

Labour

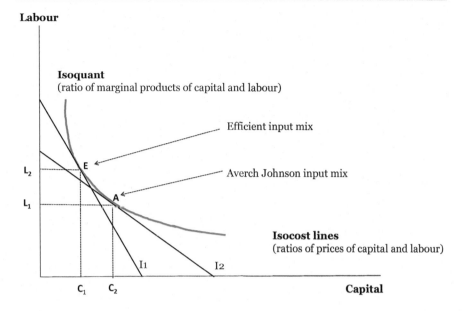

Fig. 6.20 Averch–Johnson effect of rate-of-return regulation

stimulate network users to provide flexibility through, for instance, dynamic network tariffs (see further on this issue Sect. 7.3.3).

6.7.2   Incentives for Quality

When the regulation is only directed at the tariffs and revenues of the regulated company, the risk exists that the quality of performance by this company deteriorates. This risk is in particular relevant when tariffs are based on incentive regulation. In this type of regulation, the regulated company receives an incentive to increase its productive efficiency, but this may go at the expense of quality. This adverse effect of incentive regulation may occur when the network users don't have an outside option, which is generally the case when the network operator has a natural monopoly.

The quality of performance of an energy network operator refers to the amount of energy that constantly is transported in a specific way. For gas networks, this refers to the pressure of the gas flow and the physical characteristics of the gas measured through the so-called quality spectrum (see Sect. 2.3.3). For electricity networks, quality refers to voltage levels, frequency and shapes of the electric waves (see Creti and Fontini 2019). In order to measure the quality performance of networks, various indicators can be used referring to different aspects of interruptions in supply (see Table 6.8). Using such quality indicators, regulators can give incentives to the regulated companies. This can, for instance, be done through

Table 6.8 Indicators for quality of electricity networks

Quality index	Meaning
SAIDI	System Average Interruption Duration Index
SAIFI	System Average Interruption Frequency Index
CAIDI	Customer Average Interruption Duration Index
MAIFI	Momentary Average Interruption Frequency Index

See: https://www.pge.com/en_US/residential/outages/planning-and-preparedness/safety-and-preparedness/grid-reliability/electric-reliability-reports/electric-reliability-reports.page

regulatory standards to which the regulated companies have to commit. If they fail to do so, regulators can issue fines.

Instead of standards, regulators can also choose for a financial instrument, such as a bonus-malus system. If the quality (according to one or more indicators) is above a certain threshold, then the company is entitled to a bonus (implemented as an increase of the allowed level of revenues), otherwise it has to pay a malus (implemented as a discount on the allowed revenues). This bonus-malus scheme can be implemented in the tariff regulation through a so-called q-factor. The formula for determining the level of allowed revenues (see Eq. 6.36) then changes into:

$$R_{a,t} = (1 + CPI_t - x + q) * R_{a,t-1} \qquad (6.40)$$

Another type of financial measure to incentivize network operators to maintain quality levels is to oblige them to compensate network users for any disturbance in network services.

6.8   Ex Post Evaluation

6.8.1   Monitoring Profits of Regulated Firms

As regulators act on behalf of societies, based on legal frameworks, they usually have to report to society to what extent the objectives of regulation have been realized. One of the objectives of regulating the tariffs of a company operating a natural monopoly is to prevent that consumers pay too much. When the tariffs are regulated on the basis of the cost-plus approach, then the consumers will never pay more than the actual costs. As the caveat of this type of regulation is that the regulated company does not have any incentive to reduce costs, regulators may choose a kind of price-cap regulation in which the tariffs are related to other sources of costs. A potential caveat of this type of regulation, however, is that consumers pay more for the services provided by the regulated company than what would be needed looking at the realized costs of the company (see Hauge and Sappington 2012). After all, the regulated company may have become more efficient than what was assumed initially (i.e. at the start of the regulatory period).

Therefore, when regulators report the realized profits by regulated companies, they have to be careful how to define and how to assess the profits. Paying attention to these differences is important to understand to what extent tariff regulation has been effective in preventing that the regulated company makes supra-normal profits. The realized profits can be calculated in four different ways: (a) by looking at the difference between realized revenues and realized costs, (b) by looking at the difference between realized revenues and efficient costs, (c) by looking at the difference between allowed revenues and realized costs and, finally, (d) by looking at the difference between allowed revenues and efficient costs. Each of these alternative definitions of profits can result in different outcomes (see Fig. 6.21). As a result, they also have a different meaning in the assessment of the effectiveness of tariff regulation (see Table 6.9).

1. Difference between *realized* revenues and *realized* costs

The value added of looking at the realized values of both revenues and costs is that this enables the regulator to explain the published profits by a regulated company, by pointing at the influence of the accounting rules on the profits. Regulators generally prescribe a specific accounting scheme (the so-called Regulatory Accounting Rules, RAR), while companies may use their own accounting scheme (the so-called commercial accounting rules). As a result, the realized profits can have different values depending on which accounting scheme is used. The reason for regulators to prescribe their specific accounting rules is to have a standard approach in determining costs and revenues, while they also may have a preference for specific rules, such as how to determine the financial value of an asset or regarding the depreciation method (see Sect. 6.3.4). Hence, when analysing the realized profits of a regulated company, it is important to use the relevant (i.e. regulatory) accounting rules.

The level of the realized profits itself does not say anything, however, about the effectiveness of tariff regulation as the realized revenues may deviate from the allowed revenues, while the realized costs can deviate from the efficient costs. This is due to the fact that the allowed revenues and efficient costs are elements in ex-ante regulation, which means that the values of these quantities are set before the start of a new regulatory period. The allowed revenues are based on expected volumes of utilization, while the realized revenues depend on the actual volumes of utilization. As the actual volumes of utilization may be higher than the expected volume, the realized revenues can also be higher than the allowed ones. For instance, in the case of a gas network operator, the expected utilization of the gas network can be based on normal (i.e. long-term average) outside temperatures, while the actual temperatures in a specific year can strongly deviate from the normal values. As discussed in Sect. 6.6.3, regulators may choose to control for the volume risks, by adapting the allowed level of revenues for deviations between actual and expected utilization in the previous year.

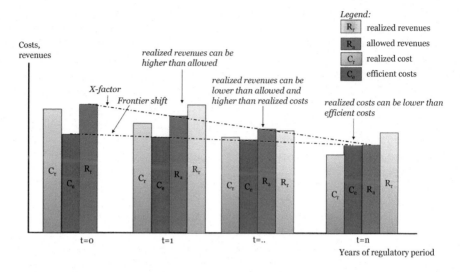

Fig. 6.21 Monitoring profits of a regulated company

Table 6.9 Assessment of profits of a regulated firm, using different profit definitions

Costs	Revenues	
	Allowed	Realized
Efficient	Profits/losses result from regulatory choices. Regulator can apply a glide path to bring profit to zero at the end of the regulatory period	Profits/losses should be possible in case of incentive regulation, as a firm may make more/less revenues than what was assumed at start of regulatory period
Realized	Profits/losses should be possible in case of incentive regulation, as firm may become more/less efficient than what was assumed at start of regulatory period	Profits/losses reflect actual financial performance; value does not give any information on effectiveness of tariff regulation, but is relevant to assess financial strength of firm

2. Difference between *realized* revenues and *efficient* costs

Comparing the realized revenues (based on Regulatory Accounting Rules) with the efficient costs gives insight into what extent the revenues are sufficient to compensate for the costs of an efficiently operating company. Users of the services of the regulated company may want that the revenues are never higher than the efficient costs and may use this definition of profit to analyse to what extent they are paying too much.

It is, however, essential to realize that price-cap regulation is a kind of ex-ante regulation, which means that the allowed level of revenues is determined at the start of a regulatory period, and that, hence, afterwards, realized revenues can be different for many reasons, as indicated above. If network operators face this volume risk, then it belongs to the rules of the game that they may make some higher profits in some years.

In addition to this, price-cap regulation has also as an objective to give incentives to the regulated company to become more efficient, which means that this company needs to have the possibility to make excess profits. In other words, price-cap regulation implicitly assumes that the possibility of making excess profits is also part of the game.

3. Difference between *allowed* revenues and *realized* costs

Similar remarks can be made for the difference between allowed revenues and realized costs. A regulated firm may be allowed to make more revenues that what is needed to compensate for its costs, just because the firm has the incentive to become more efficient, so to reduce its costs even below what is assumed to be the benchmark. Moreover, the level of efficient costs is based on calculations and assumptions made at the start of new regulatory period and during this period, the costs of an efficiently operating firm may reduce more than expected at the start of the period.

4. Difference between *allowed* revenues and *efficient* costs

The final type of comparison is between allowed revenues and efficient costs. Both are set at the start of the regulatory period when setting the annual allowed revenues (see Sect. 6.6.2). Any differences between these quantities are fully due to choices made by the regulator. The regulator may, for instance, decide that the realized revenues in the past form the starting point for the future revenues and when the realized revenues are above the efficient cost level at the start, the regulated company will make extra profits compared to the efficient cost level during the regulatory period (as indicated in Fig. 6.21). This is known as a glide path by which the excess profits (i.e. differences between allowed revenues and efficient costs) are gradually eliminated (Baldwin et al. 2012). The alternative for this approach would be to immediately take the allowed revenue to the level of efficient costs in year $t = 1$, which is called the P0-adjustment.

6.8.2 Monitoring Financial Strength of Regulated Firms

While in normal markets, the risk of adverse financial situations and bankruptcy belongs to the key mechanisms behind productive efficiency, in regulated markets it is often not acceptable to let the regulated companies go bankrupt because these companies fulfil key activities in an economy, while alternative suppliers are not present. Because of these factors, regulators want to prevent that a regulated company is not able to conduct its (legally embedded) tasks, such as the transport of gas or electricity. This implies that the regulatory objective to prevent that the regulated companies make supra-normal profits (i.e. realizes monopoly profits) is often subject to the condition that the company should be financially stable. This means that the regulated companies should have a sufficiently strong financial position to continue their activities and to make the necessary investments.

The financial position of a regulated company is affected by tariff regulation. If the regulator has chosen for a tight regulatory scheme, for instance, the allowed revenues give less room for financial buffers. Since price-cap regulation is a kind of ex-ante regulation, the regulator has to make assumptions on future conditions, which implies that these assumptions may appear to be too optimistic afterwards. When the regulated company experiences, for instance, a lower utilization of its infrastructure and higher costs than expected, this may result in negative cash flows. When such negative cash flows occur for a number of years, the regulated company may face problems in financing new investments or even enter into situations of default on existing financial obligations. If a regulator wants to minimize these risks, it can make the regulation looser.

The financial position of a company can be monitored through a number of financial indicators, such as earnings before interest and taxes (EBIT)/interest, net operational cash flow/interest, net operation cash flow/debt and debt/total assets. These financial ratios measure to what extent the net revenues from operations enable the company to meet its financial obligations, in particular those regarding the providers of debt. Another method to assess the financial position of a company is to use financial rations by agencies like Moody's, insofar these ratings are present.

Exercises

6.1 Why is subadditivity of costs a more general characteristic of a natural monopoly than the presence of economies of scale?

6.2 In which circumstances can antitrust be an effective policy measure to address market power?

6.3 A regulator does not know precisely what the cost structure of a TSO is. Does this information problem result in moral hazard or adverse selection?

6.4 What is the definition of incentive power in tariff regulation?

6.5 What is difference between tightness and incentive power?

6.6 What is meant by TOTEX regulation?

6.7 What is difference between productivity and efficiency?

6.8 Define the relationship between productivity change (p), catch-up (c) and frontier shift (fs).

6.9 What is the incentive power of the most efficient firm in case of best-practice regulation?

References

Baldwin, R., Cave, M., & Lodge, M. (2012). *Understanding regulation; Theory, strategy and practice* (2nd ed.). Oxford University Press.

Cicala, S. (2015). When does regulation distort costs? Lessons from Fuel Procurement in US Electricity Generation. *American Economic Review, 105*(1), 411–444.

Council of European Energy Regulators (CEER). (2019). Incentive regulation and benchmarking. CEER Report C18-IRB-38-03.

Coelli, T. J., Rao, D. S. P., O'Donnel, C. J., & Battese, G. E. (2006). *An introduction to efficiency and productivity analysis*. Berlin: Springer.

Creti, A., & Fontini, F. (2019). *Economics of electricity: Markets, competition and rules*. Cambridge University Press.

Decker, C. (2015). *Modern economic regulation; An introduction to theory and practice*. Cambridge University Press.

Dijkstra, P. T., Haan, M. A., & Mulder, M. (2017). Design of Yardstick competition and consumer prices: Experimental evidence. *Energy Economics, 66*, 261–271.

Hauge, J., & Sappington, D. (2012). Pricing in network industries. In R. Baldwin, M. Cave, & M. Lodge (Eds.), *Understanding regulation; Theory, strategy and practice* (2nd ed.). Oxford University Press.

Mulder, M., Perey, P., & Moraga, J. L. (2019). Outlook for a Dutch hydrogen market; Economic conditions and scenarios. CEER Policy Papers 5, March.

NERA. (2016). The beta differential between gas and electricity networks—A review of the international regulatory precedent, London, 22 March.

Shleifer, A. (1985). A theory of yardstick competition. *RAND journal of Economics, 16*(3), 319–327.

Varian, H. R. (2003). *Intermediate microeconomics; A modern approach* (6th ed.). New York/London: W.W. Norton & Company.

Viscusi, W. K., Harrington, J. E., & Vernon, J. M. (2005). *Economics of regulation and antitrust*. The MIT Press.

Reliability of Energy Supply as Semi-public Good

7

7.1 Introduction

Gas, electricity and heat markets depend on the reliability of the physical networks for transport. This reliability can be seen as a semi-public good and, as a consequence, its provision is not fully secured by the market. This chapter first explains how the functioning of energy markets and networks are interrelated before discussing regulatory measures which can be taken to ensure reliability of energy supply. Section 7.3 goes into the organization of financial incentives to network users related to network capacity; Sect. 7.4 discusses how financial incentives can be given to market participants to contribute to system balancing, while Sect. 7.5 discusses how to assure that there is always sufficient production capacity and that all production is priced even if the market is not able to clear.

7.2 Energy Networks and Markets

Just as in any other market, producers and consumers enter into commercial transactions in energy markets. These transactions can be settled bilaterally, via brokers or on exchanges as we have seen in Sect. 3.2. A key difference between transactions in electricity, gas and heat markets and those in other markets is that the former are based on the presence of a physical infrastructure for transport and distribution. Suppliers can only sell and deliver their commodities to consumers when both are connected to the same infrastructure for transport. Hence, these markets are built on the presence of a well-functioning physical infrastructure (see Fig. 7.1).

© Springer Nature Switzerland AG 2021
M. Mulder, *Regulation of Energy Markets*, Lecture Notes in Energy 80,
https://doi.org/10.1007/978-3-030-58319-4_7

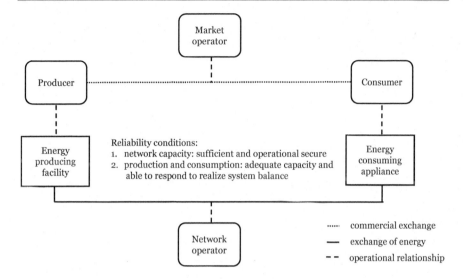

Fig. 7.1 Relationship between energy markets and energy networks

The functioning of these networks depends on a few conditions.[1] These conditions refer to the reliability of both the grid and the assets of producers and consumers. Regarding the grid, its capacity should be adequate in order to facilitate all commercial transactions. This means that the capacity of the infrastructure should be sufficient to transport the amounts of energy resulting from the commercial transactions. Note, that the network operator does not transport all the flows resulting from the individual production and load decisions, but it only needs to transport the netted value of all these decisions regarding a particular line (see Sect. 2.4.4). In addition, the operation of the grid should also be secure, which basically means that a malfunctioning of a particular grid component does negatively affect the functioning of the grid. Regarding the role of producers and consumers, they should be able to respond to changes in system circumstances in order to realize permanently system balance. This balance basically means that the injected power should be permanently equal to the power withdrawn from the grid. This condition also requires that there is adequacy of production capacity in order to be able to permanently provide the demanded energy. It belongs to the key tasks of network operators to contribute to the realization of these conditions (see Box 7.1).

A fundamental problem in markets based on a physical grid is how to coordinate the usage of the grid. This issue is in particular relevant in electricity systems because of the physical peculiarities of electricity, which is that this commodity itself can hardly be stored but only through conversion to other types of energy, while the transportation infrastructure cannot provide any flexibility itself to

---

[1]These conditions are generally summarized as security and adequacy. Security refers to the ability of a system to counteract sudden disturbances, while adequacy refers to the ability of a system to supply the needed energy for the demand at all times (Stoft 2002).

balance production and consumption. In gas and heat markets, this is less of an issue, because these commodities can be stored, while the infrastructure is also able to handle temporary deviations between injected and withdrawn energy (in gas markets, this is called line pack).

The problem how to coordinate the usage of an electricity grid is part of the so-called unit commitment problem. This problem entails the question how to utilize the various generation units in order to realize an objective (such as minimizing overall generation costs) while meeting the network constraints. This is a complex problem as there are generally many generation units, which vary in characteristics (such as regarding start-up and shut-down times, ramping rates and minimum down times) and dependence on external (weather) circumstances), while the load generally is highly volatile. In addition, the generators and the electricity consuming appliances are generally located at various locations within the grid, while the flows depend on the capacity of network parts.

A particular characteristic of transport of electricity is that the energy is not transported according the shortest distances between production and consumption, but the flows of energy are related to the resistance within the grid. This follows from the physical (Kirchhoff's) law that electric energy will choose the path of the lowest resistance (see Sect. 2.3.4). When one part of a network is constrained, the power flows are constrained in all network parts (see Sect. 2.4.4). Hence, bottlenecks in one region of a network have effects on the available capacity in other regions.[2]

Box 7.1 Tasks of network operators

The core tasks of network operation can be distinguished in five groups: (1) giving access of producers, consumes and traders to the network, (2) maintenance of the physical quality of networks (which is called asset management), (3) management of the physical properties of the networks (which refers to operational security and voltage control), (4) management of the power balance in the system (frequency control) and (5) facilitation of the wholesale market, in particular the settlement of imbalances. The last two tasks only hold for transmission system operators, not with distribution system operators.

1. Operators facilitate the energy market by giving access to produces and consumers to the grid. In order to prevent that a network operator restricts the available capacity for other producers, it is crucial to remove any incentives for the operator to do so. This can be done by unbundling the network-operation task from other, commercial tasks in production and supply. A fully unbundled network operator does not have any interest anymore in restricting the availability of network to other producers.

---

[2]Although it is to some extent possible to use equipment, like phase shifting transformers, to change the flow pattern in the meshed grid with the aim to achieve a better overall utilization.

Instead, when such an operator wants to maximize its own profits, it will pursue maximum utilization of the infrastructure. Because of the importance of unbundling for the creation of competitive wholesale and retail markets, this measure is widely adopted as we have seen in Sect. 3.4.2.

2. Maintaining the physical quality of the grids is important for the quality of the services provided by the network operator. Without any specific regulation of the quality, network operators may choose for suboptimal quality levels. This risk exists in particular when the operators receive high-powered incentives to improve efficiency. In order to address this risk, regulators implement quality regulation, as was have discussed in Sect. 5.7.2. In addition to this quality regulation, network operators may also be subject to specific regulation directed at investments in grid capacity.

3. Besides maintaining the physical quality of the assets, also the physical properties of the system has to be managed. Voltage control is a fundamental condition for the functioning of in particular electricity markets. This is relevant in all parts of the electricity grid, including the high-, medium- and low-voltage network parts.

4. Another aspect of the reliability of the network is the power balance of the system, which is managed through frequency control. This activity is conducted by the system operator. This is discussed in Sect. 7.4.

5. In order to facilitate the wholesale market, balancing responsibility has to be organized, which is further discussed in Sect. 7.4. In addition, network operators contribute to the functioning of the wholesale markets by creating and maintaining cross-border connections, providing access to these connections and using them as efficiently as possible. The latter activity is done through the allocation of cross-border capacity through measures like auctioning of capacity, including integration of balancing markets and integration of markets on other time frames (forward, day-ahead and intraday). This topic is further discussed in Chap. 10, where we pay attention to the international dimension of energy markets.

These effects can be seen as externalities between network parts which may result in suboptimal investments when these network parts (i.e. different regional networks) are operated by different network operators and when there is insufficient coordination between these operators. A network operator in a specific region will invest less in network capacity than what would be optimal from a social perspective when other network operators in other regions benefit from such investments without the need to compensate the first mentioned operator for the investment costs. So, investments to prevent congestion in one part of the network are semi-public goods[3] as users of other network parts benefit while the investing network operator may not be able to force them to pay.

7.3 Pricing of Network Capacity

7.3.1 Organization of Market and Network Usage

Because of the interrelatedness of all elements in an electricity grid, it is evident that real-time operation of an electricity grid needs to be centrally coordinated. A fundamental organizational issue, therefore, is how to combine the required central coordination of real-time network usage with decentralized commercial decision-making by producers, consumers and traders. It is a generally accepted view that these market parties can enter into commercial contracts a long time before real time, but a more debated issue is to what extent the decisions a day before real time should be centrally coordinated. Electricity markets around the globe vary in how they have organized the short-term (i.e. the spot) market. In some systems, such as in several States in the USA, this is centrally coordinated by the system operator, while in others, such as in most European countries, these spot markets are based on decentralized decision-making where the transactions are coordinated by commercial entities.[4] These two different ways of organizing the spot market have consequences for how network capacity is priced. Table 7.2 summarizes the main differences between these two alternative ways of organizing electricity markets.

7.3.2 Pricing of Network in Nodal Pricing System

In centrally coordinated real-time markets, network characteristics play a role in commercial transactions. These markets are called locational systems as the spot market is cleared for every node within the network, based on the bids of suppliers and buyers in each node (see Fig. 7.2). This type of system results in equilibrium prices for every node. These prices do not only reflect the marginal costs of the suppliers as well as the marginal willingness to pay of buyers on a node, but also the presence of grid scarcity. In case the network capacity would be abundant, the prices on all nodes are the same. When, however, capacity of the network is limited in some regions, prices will differ between various nodes.

---

[3]These types of goods have some characteristics of public goods, which are goods that cannot be exclusively made available to a limited group of users (i.e. the goods are non-exclusive) and that can be consumed without affecting the availability to others (i.e. the goods are non-rivalrous). In case of semi-public goods, one of these characteristics hold.

[4]In relation to this, in markets with centrally coordinated dispatch, the operation of the network and the operation of the spot market are conducted by one agency, the so-called independent system operator (ISO), while in countries with decentralized dispatch, the operation of the network is not (necessarily) integrated with the management of the network. In the latter markets, the transport and network management is done by the so-called transmission system operators (TSO).

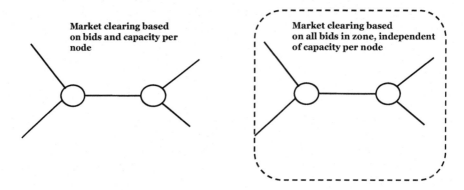

Fig. 7.2 Market clearing in a nodal (left) and in a zonal (right) system

Another consequence of a nodal pricing system is that market participants (i.e. producers and consumers) receive locational price signals because of the regional price differences depending on network availability (see Box 7.2). In addition, in these centrally coordinated spot markets, prices for users are increased by the coordinating party with surcharges which are used to give the producers a capacity payment (to ensure availability of supply) and to compensate for the costs of the network operator (Frontier, 2009). This is related to the fact that the real-time prices per node result in less inframarginal profits for producers and, as a result, the real-time prices are less equipped to incentivize producers to invest in new generation capacity. Hence, in this system, the electricity prices do not only reflect the costs and (locational) scarcity of producing electricity, but also the costs and (locational) scarcity of network capacity.

Box 7.2 Nodal prices and transmission rights

In a nodal pricing system, the electricity market is cleared for every node in the system. The electricity prices depend on the supply and demand conditions on the various nodes and the availability of transport capacity between nodes. Based on Stoft (2002), we can make the following stylized example of an electricity system consisting of two nodes (see Fig. 7.3). The nodes differ in the size of the demand, which is assumed to be perfectly price inelastic and the marginal costs of production. In Node 1 demand is 100 MWh and in Node 2 it is 1000 MWh. In Node 1, the marginal costs start at 40 euro/MWh and in Node 2 they start at 60 euro/MWh, and in both nodes the marginal costs increase linearly with production volume. Between these two nodes, there is power line with a maximum capacity of 500 MW. Because the marginal generation costs are higher in Node 2, consumers would like to buy electricity from producers in Node 1. Hence, the total demand in Node 1 is equal to the local demand (of 100 MWh) and the maximum demand from Node 2, which

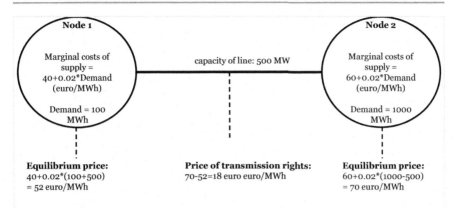

Fig. 7.3 Locational power prices in simplified case of two nodes. *Source* Stoft (2002)

is equal to the constraint of line (500 MWh). This results in an equilibrium price of 52 euro/MWh in Node 1 (assuming competitive market circumstances). In Node 2, the residual demand is (1000–500=) 500, resulting in competitive market price of 70 euro/MWh.

The price difference between the two nodes can be seen as the value of the transmission capacity. When for this capacity, transmission rights have been issued and allocated to someone, the holder of this right receives this value. If this initial holder is the network operator, it can, for instance, sell the rights to the producers in Node 1. The maximum price these producers are prepared to pay for these rights is precisely the extra profits per unit they can earn by selling electricity to consumers in Node 2 (in this example 18 euro/MWh). If the network operator sells the transmission rights to the consumers in Node 2, the maximum price they are prepared to pay is also this same price because by importing electricity from Node 1, they can have access to cheaper electricity.

7.3.3 Pricing of Network Capacity in Zonal System

In electricity systems with decentralized decision-making on spot markets, which is common in Europe, a zonal pricing system is applied. In Europe, many zonal markets are determined by the borders of countries, but in a number of countries, several zones exist. Denmark, for instance, is split into Eastern and Western Denmark, while Sweden is split into four regions.

In zonal systems, the market is cleared for a region (i.e. a market zone), not taking into account the physical characteristics of the network within a zone. Hence, a zonal market means that in a particular region, the market is cleared only on the basis of the bids of suppliers and buyers, independently of the characteristics of the network within that zone. This implies that the market parties may assume that there is sufficient capacity within the zone to transport the outcome of all transactions within the zone. This is the so-called copper-plate assumption.

As the wholesale market is cleared based on the bids of marginal costs of the suppliers as well as the willingness to pay of buyers within the zone without taking into account whether sufficient transport capacity exists within the zone, all market participants face the same electricity price. When after concluding the commercial transactions in, for instance, the day-ahead market, the network operator expects that not all suppliers can actually produce in real time due to the presence of physical bottlenecks within the grid in the zone, he can prevent the congestion by applying redispatch as a kind of congestion management. Regulators may force network operators to do this is as efficiently as possible (see Box 7.3).

As electricity prices in a zonal system do not give any locational signals to network users, network operators may want to take other measures to influence the locational decisions. After all, when the network operator knows that at some locations in the network the capacity is limited while on other locations more capacity is available, he can mitigate the costs of network management to stimulate network users to instal their production facility or consumption appliance at the latter locations. To what extent this is allowed is determined by the regulatory framework. Generally, network operators have to honour all the requests made by network users and, in addition, they are not allowed to influence them to take other decisions. This regulatory rule follows from the importance of having fully independent network operators for the functioning of markets.

An option that network operators may explore is to design the structure of network tariffs in such a way that they take the marginal costs of network extension into account. In Chap. 6, we saw that network tariffs are meant to let network users pay for the (efficient) costs of the infrastructure. Within these tariffs, network operators may introduce dynamic or locational variations to give network users incentives to take the network scarcity at particular moments of time or locations into account. These locational and dynamic variations in network tariffs can be seen as a kind of scarcity prices. Such dynamic network tariffs may induce network users to, for instance, postpone the consumption of energy (which is called load shifting) or to reduce their consumption of energy (which is called load shedding). Economically, such dynamic and locational prices are efficient as they reflect the scarcity at particular times and places, but network users, in particular residential users, may view them as unfair (see e.g. Neuteleers et al. (2017)). In Sect. 11.4, we further discuss the fairness of such regulatory measures from a behavioural-economics perspective.

Adding locational and dynamic variation to grid network tariffs should, of course, not result in supra-normal profits for the network operator, as the other objectives of tariff regulation remain valid. Hence, when a network operator receives higher revenues because of the presence of scarcity prices, the spending of these revenues should be regulated such that the network users benefit, either by obtaining lower tariffs at not-congested places or by having less congestion in the future (see Jafarian et al. 2020). A regulatory measure to assure this is to require that the network operator uses the scarcity revenues to pay the costs of investments which are directed at solving congestion and, otherwise, that these revenues are used to lower network tariffs in the next regulatory period.

Box 7.3 Redispatch in a zonal electricity market

Suppose that producer A has sold 100 MWh electricity in the day-ahead market for hour h, the next day at a market price of 50 euro/MWh, which exceeds its marginal costs, which means that it is a inframarginal producer (see Table 7.1). If this producer is located in a part of the grid that does not have sufficient capacity that particular hour, the network operator may ask producer A not to produce. Instead, producer B located in another part of the grid without a congestion can be asked to produce 100 MWh more that particular hour the next day, so that on a system basis the same amount of electricity is produced.

Table 7.1 Example of redispatch in an electricity grid

Variable	Producer A (congested region)	Producer B (non-congested region)	Network operator
Marginal production costs (euro/MWh	40	60	–
Sold at day-ahead market for next day, hour h (MWh) at market price (euro/MWh)	100 MWh at 50 euro/MWh	0	–
Production because of congestion management (MWh)	0	100	–
Compensation required (euro/MWh)	−40 (i.e. savings on production costs)	60 (i.e. extra production costs)	20 (difference between price received from A and price paid to B, which is equal to difference in marginal generation costs)

Producer B, that produces more, will need compensation which is, in theory, equal to the marginal costs of its production (say 60 euro/MWh). This compensation is paid by the grid operator. Producer A, in the congested region, will be required to pay the network operator as this producer saves on its generation costs, while it already has sold electricity in the day-ahead market (against 50 euro/MWh). The network operator wants to receive a compensation equal to the saved marginal costs of that producer (which is 40 euro/MWh), and as a result, producer A is still making the profit of 10 euro/MWh.

As the marginal costs of producer B are higher than those of producer A (after all, otherwise producer B would have sold the electricity in the day-ahead market), the network operator has to pay more to producer B than it receives from producer A. This difference in production costs is part of the system costs of the network operator. These system costs are generally socialized among all network users.

This example shows that a lack of available transport capacity within a bidding zone can be solved while honouring the commitments in the wholesale market and maintaining system balance as well. In practice, it may be complicated to determine the marginal costs of the producers that have to produce more because of the information asymmetry between network operator and producers. A market-based method to determine the price for these producers is to organize a market in which producers can submit bids for providing flexibility to the network operator.

An alternative pricing method that can be used to reflect network scarcity is the use of interruptible and firm contracts. In the former case, network users get conditional access to the grid, depending on available capacity, while in the latter case the network users receive the certainty that a specific amount of network capacity is always available to them. Because interruptible contracts offer less certainty to network users, the price (i.e. tariff) for using the grid is lower than when the capacity is firm. When network users can choose from these two types of contracts, only those users who highly value the permanent availability strongly will buy firm contracts, which makes this pricing structure efficient. In addition, when a number of network users have bought interruptible access contracts, the network operator has obtained the flexibility to reduce the network use in case of congestion (Table 7.2).

Domestic network congestion may spillover to neighbouring countries because of the physical laws determining how energy flows within electricity grids (see Sect. 2.4.4). In order to prevent that network operators do not invest enough to solve bottlenecks in their network, regulators of related market zones may force them to cooperate and make joint investment plans. This is the reason why in the EU, the electricity grid operators are forced to make joint long-term investment plans in which they coordinate the investments in the high-voltage grid.[5]

7.4 Pricing of System Balance

7.4.1 Public-Good Character of System Balance

Network balance is relevant in all energy markets where the delivery is based on physical transportation networks. This holds for gas, electricity and heat networks. These networks are in balance when the injected energy is equal to the amount of energy that is withdrawn or that is lost through the transmission (see Fig. 7.4).

In gas networks, the commodity can only be transported when the gas is put under pressure.[6] This pressure can be controlled by keeping the amount of injection

---

[5]See https://tyndp.entsoe.eu/.
[6]The transport networks of gas are, therefore, also called high-pressure networks, while the distribution networks are called low-pressure networks.

Table 7.2 Main differences between nodal and zonal pricing systems

Nodal pricing	Zonal pricing
Electricity price is determined per node of the grid	Electricity price is determined per bidding zone, in which market participants may assume the presence of a 'copper plate'
Spot (real-time) market is cleared by central operator (independent system operator, ISO); centralized dispatch of power plants	Day-ahead market is operated by various parties (exchanges, OTC); market participants have programme responsibility regarding real-time balance; separate balancing market to give incentives to market parties and to financially settle imbalances
Products in forward markets are based on system-wide reference price representing all nodal prices, in order to obtain a hedge against differences between nodal and reference prices, financial transmission rights can be used	Products in forward markets are related to spot markets products and, hence, offer a hedge against price risks in these markets
Congestion is solved through nodal prices	Intrazonal congestions are solved through redispatch or interruptible/firm contracts; interzonal congestion is solved through cross-border transmission capacity allocation mechanisms
Electricity price per node reflects: marginal generation costs and WTP, costs of infrastructure (transportation costs), network scarcity, and margin for investments in generation capacity	• Electricity price is only related to demand and supply (i.e. WTP and marginal generation costs) • Transportation costs are reimbursed through regulated network tariffs
Incentives for investments in generation capacity are sometimes given through capacity mechanisms as the real-time pricing per node results in less inframarginal profits	Incentives for investments in generation capacity are given through scarcity prices on market area level
Use of market power is monitored through the bids submitted to the ISO	Use of market power is monitored by the competition authority based on ex post antitrust (e.g. analysis of physical withholding) and specific regulation, such as REMIT (see Sect. 3.4.2)
No need to calculate available cross-border capacity as the physical capacity is already allocated to the market	Cross-zonal capacity is allocated to market parties

of gas into the network equal to the amount of gas that is subtracted from the network (van Dinther and Mulder 2013). Hence, in gas networks an imbalance occurs when more gas is extracted than is injected into it, or vice versa. In the first case, the gas pressure drops, in the latter it rises. When the pressure drops (too much), it will not flow sufficiently, which means that consumers will not get the amount of gas per time period which they require. When the pressure increases (too) much, this may result in physical incidents. Gas networks, however, have some flexibility to handle with minor deviations between the amount of injected and withdrawn gas. This flexibility is called line pack.

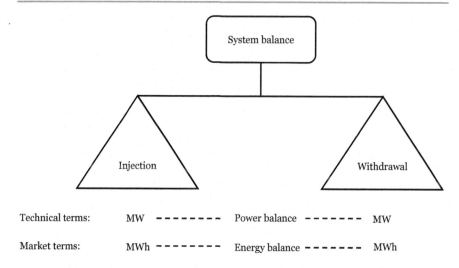

Technical terms: MW - - - - - - - -   Power balance - - - - - - -   MW

Market terms: MWh - - - - - - - -   Energy balance - - - - - -   MWh

Fig. 7.4 Definitions of system balance in technical and market terms

In heat networks, the water that is used to transport the heat also needs to be put under a constant pressure in order to let the water flow. As the water itself is not injected or withdrawn by the users of heat, this pressure is only affected by the evaporation of water. More important in heat networks is to manage the temperature of the heat at the different places within the network. By adding energy to the water, the temperature increases. When users use the heat from the network, the temperature of the water in the network drops as the energy from this water is transported to the water in the infrastructure of the users. By transporting the water through the pipeline infrastructure, the temperature of the water drops further because of the loss of energy, which is the reason that heat networks can only be profitably exploited at a limited spatial scale, as discussed in Sect. 3.3.5. Maintaining the balance in heat networks, therefore, refers to the amount of energy added to the water in the system versus the amount of energy withdrawn from the system. As heat networks store energy, the consumption of heat does not need to be equal, though, to the production of heat at any moment of time.

The necessity to keep the system in balance is particularly relevant for electricity markets. What is special about this type of market and what makes it different from gas and heat markets is that the consumption of electric power occurs in the same second as the production. Only when consumption equals production, the voltage remains constant. In an AC system, this results in a constant frequency. In such systems, the frequency may only slight vary within a bandwidth of about 200 mHz. (i.e. 0.2 Hz), around a frequency of 50 or 60 Hz. When the frequency moves outside this range, turbines will turn off, which may result in a black out. Hence, it is important to maintain the frequency as all generation units and load devices are designed at a specific frequency. Balancing the electricity grid is, therefore, equal to frequency control. A grid operator responsible for frequency control has to ensure that the domestic production always equals domestic demand. Because of this task, these network operators are also called system operators. When a network operator is responsible for both the transport of energy and the management of system balance, it is called a transmission system operator (TSO).

The condition for the constant frequency is that the total generation is constantly equal to the total load. This is called the power balance, which is that the flow of energy per second (=Joule/second = Watt) of generation and load is equal. When this condition is not met, the frequency in an AC network increases, and the other way around, damaging the turbines and appliances may be hurt. This power balance is the technical requirement for the functioning of the power system.

When this technical condition is translated to market participants, the balance can be formulated in energy terms, which results in the energy balance. This energy balance says that the energy produced and consumed over a period of time (i.e. Watt hours) needs to be equal to each other. This period of time may be, for instance, 15 min.

Maintaining the balance is important in any of these markets. In gas markets, controlling the appropriate gas pressure in the network is important to enable the gas market to function; in heat markets, the temperature of the water needs to be constant at the delivery points where users extract heat from the network, while in electricity systems the electric voltage and frequency need to be kept constant within margins. Individual market players will, however, not take care of these system requirements, because of the semi-public-good character of this activity. Other players will benefit from this activity without the incentive to pay for it, which results in free-riding behaviour. This is a clear market failure in these markets which asks for a kind of central organization, such as the creation of a balancing market.

Until the liberalization of markets, the frequency control was completely done by one actor, the network operator, being fully responsible for all actions needed to maintain system balance. These central operators not only managed the grid and were responsible for the frequency control but also managed the supply of energy to the network, for instance, from gas storages, or through the dispatch (i.e. utilization) of the generation plants (i.e. this was a centralized dispatch).

In particular in electricity markets, these centrally organized systems have been replaced by decentral systems in which many different economic agents (i.e. market participants) make their own decisions regarding the production and sales of electricity. In addition, because of the unbundling of the network operators, these operators are not allowed any more to produce or supply themselves, which means that for the frequency control the operator depends on the actions taken by other agents in the market. The decentral decisions taken by these agents need to be coordinated in one way or the other, to ensure that the system constantly meets the physical requirements to keep running. This coordination is done through so-called balancing mechanisms. Because these market-based balancing regimes are most strongly developed in electricity markets, we focus on this market and how it is designed in Europe, in the remaining of this section.

7.4.2   Definition of Balance on System Level

The imbalance of an electricity system is determined by the extent the total injection of power in the grid deviates from the total withdrawal of power. This refers to the so-called active power balance, which means that the input to the network must be permanently equal to the output (Nobel 2016). For a completely independent grid, the power balance at any moment of time is defined by the equalization of domestic power (G) and domestic load (L):

$$Power\,balance\,of\,isolated\,grid: \\ G = L \tag{7.1}$$

In practice, most grids are connected to each other. In Europe, for instance, the AC networks constitute a meshed network in which electric energy flows within and between grids driven by physical laws. Consequently, the power balance can only be maintained by the joint effort of all European system operators. These operators together have the legal task to maintain the frequency of the European electricity grid.[7] In order to fulfill this task, they have delegated this joint responsibility in terms of responsibilities for each individual system operator operating a particular load–frequency control (LFC) area. Each single operator of a high-voltage network (i.e. a control area) is responsible for securing its power balance. This entails that, at any point in time, the total domestic power plus import (I) must be equal to the total domestic load plus export (E) (plus energy losses during transmission):

$$Power\,balance\,of\,grid\,(market\,zone)\,in\,international\,network: \\ G + I = L + E \tag{7.2}$$

This power balance in a control area can be monitored by measuring the flows on the border with the neighbouring areas.

The management of the power balance can be completely done by the system operator, as was the case until the liberalization of electricity markets. In that case, this operator takes all the necessary measures itself to maintain the power balance by using so-called operating reserves which have been contracted (see Box 7.4). As the utilization of the operating reserves is managed by the system operator, the costs for this centralized activity can be covered through public funding or levies imposed on all network users (which is called socialization of costs). The method to determine these levies can be based on a kind of cost-plus regulation, which means that all costs are just passed on to the network users, without giving any incentive to the system operator or network users to behave more efficiently.

---

[7]In Europe, the AC network frequency is 50 Hz, while in the USA, for instance, it is 60 Hz.

7.4.3 Delegation of Balancing Responsibility to Market Participants

Because of the inefficiencies of such centrally organized balancing schemes, regulators have looked for other options to reorganize this. A more efficient option to organize the management of grid balancing is that the system operator makes use of market mechanisms to give incentives to network users to stay in balance themselves as well as to provide support to restore system balance in case of imbalances. The latter means that network users may deviate from their own programme in order to help the system to restore frequency. The way market parties experience financial consequences of contributing to system imbalance affects the efforts these parties want to make to prevent this.

Box 7.4 Operating reserves for upward and downward regulation

Operating reserves are the assets which the system operator has contracted in order to restore system balance within a short interval of time. These reserves can be supplied as upward regulation or downward regulation.

These reserves can be distinguished in spinning and non-spinning reserve (Stoft 2002).

- Spinning reserves consist of capacity that is already synchronized with the grid (i.e. spinning) and, therefore, it can respond very quickly. The change in output can be provided by both generation plants and load devices. Steam power plants, for instance, can generally ramp up (i.e. increase their output) at a rate of 1% per minute. Some load devices, such as those use to produce heat, can supply spin by reducing the load in short period of time (e.g. 10 min).
- Non-spinning reserves consist of capacity which is not jet connected to the system, but which can start up quickly and increase production or reduce load. These reserves need a bit more time than the spinning reserves, 30–60 min to supply spin. This type of reserves can be supplied by, for instance, gas turbines.

The spinning reserves are used to function as Frequency Containment Reserves (FCR) and Frequency Restoration Reserves (FRR).

- In case of a change in the frequency, for instance due to a sudden technical problem in a power plant, the FCR is used to prevent any further changes in the frequency. These reserves are used to stop any further change in frequency within 30 s, and they act for the total synchronous systems (i.e. of several interconnected market with the same frequency). Until recently, the FCR was called primary reserves.

- When the change in the frequency has stopped, the FRR is used to bring the frequency back to the standard level. This FRR is used in that zone (control area) where the change in the frequency was initiated. The FRR consists of operating reserves with an activation time typically between 30 s and 15 min. Until recently, FRR was called secondary reserves. FRR consists of two types: automatic FRR (aFRR) and manual FRR (mFRR).

Non-spinning reserves can act as replacement reserves (RR). These consist of reserves to be activated in a period longer than 15 min and were called tertiary reserves in the recent past. These reserves are meant as back-up for the FRR. For instance, in case all FRR is in use to restore the frequency and another incident happens, then the RR can be used. In some systems, the system operator does not contract RR, but market participants are incentivized to provide these reserves.

In this respect, it is important to distinguish the (technical) system (im)balance, which is defined in terms of power (MW), and the (market) imbalance on the level of individual market participants, which is defined in terms of energy (MWh). On the level of market participants, imbalance is defined by the extent a market participant acts in real time according to its commitments made earlier. As we have seen in Sect. 3.3.4, both producers and consumers in electricity markets make use of forward markets in order to reduce their risks. By concluding long-term contracts for delivery of electricity in the future (such as yearly or quarterly ahead), they obtain certainty about the prices for their future supply or consumption. This is in particular important in electricity markets as trading electricity in or close to real time is riskier as the market prices then are fully determined by the real-time market circumstances for production and load because of the impossibility to store electricity and the relatively high inflexibility of demand and supply. As a consequence, electricity markets mainly comprise forward markets, in which trading takes place before the electricity is physically delivered and consumed (see Box 7.5).

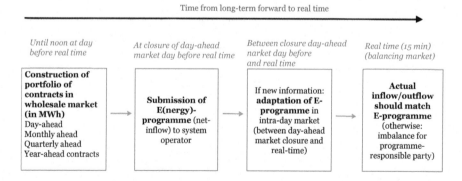

Fig. 7.5 Relationship between wholesale market and imbalance market: activities and responsibilities of Programme Responsible Parties (BRP)

The result of all transactions in the various forward markets is that for each moment of time, market participants have commitments to supply or consume based on a number of contracts concluded in the past (see Fig. 7.5). The commitments are always between two market participants (seller and buyer) with opposite positions in each contract, which means that the sum of all contracts in an electricity system for a specific moment of time should be equal to zero. This means that all transactions concluded in the electricity market should be in balance.

Box 7.5 Constructing portfolios of various types of forward contracts in electricity markets

In electricity markets, market participants typically conclude various types of forward contracts for future delivery. Some contracts are traded a long time (a few years) in advance; other type of long-term contracts are on a quarterly or monthly basis. Retailers, for instance, can use long-term contracts to supply the (expected) baseload of their customers, which is the minimum level of load over a longer period of time (see Fig. 7.6). On top of these contracts, retailers buy other types of forward contracts to adapt their portfolio of contracts to the expected load profile of their customers. For expected daily variations within the total load, the traders can go to the day-ahead market where electricity is traded on hourly basis for the next day, while the intraday market can be used to adapt portfolios to changes in expected load within the current or next day. By using these different types of products, market participants can adjust their positions up to real time. To what extent market participants have an incentive to do so depends on the design of the balancing market, i.e. the financial consequences of being in imbalance in real time.

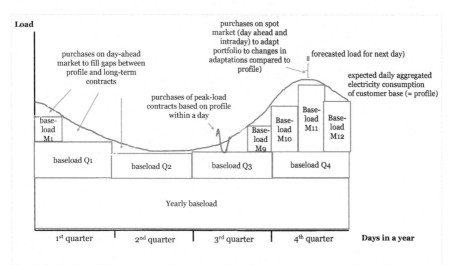

Fig. 7.6 Load profile and composition of portfolio of wholesale contracts of a retailer

From this example follows that the balancing scheme and pricing of imbalance energy plays a key role in electricity markets. One consequence of this is that the prices in the forward markets are correlated with market participant expectations related to the market situation in the following forward markets, including the situation at the time of physical delivery, which is the balancing market. The balancing market can, therefore, be seen as the last in a sequence of forward electricity markets (van der Veen and Hakfoort 2016). In other words, the price formation in any forward market is based on the expected imbalance (real-time) price. Therefore, it is crucial that the imbalance price is correctly set.

Market participants need to inform the system operator about their net positions before real time. This is called the energy schedule or energy programme. Typically, market participants need to inform the system operator on their schedule for each moment of time of a specific day immediately after the closure of the day-ahead market the day before. The period for which this information has to be submitted is called imbalance settlement period (ISP).

Besides the energy schedules, the market parties also need to inform the network operator how much will be transported through which connection with the grid. This information is called the transport schedule of physical nomination. The network operator uses this information to schedule how much the various units connected to different parts of the system will generate or use for every hour the next day. Hence, this information is important to manage the available transmission capacity.

After the submission of this information, market participants can go to the intraday market if they need to adapt their portfolio and schedules based on the latest information on availability of their power plants, expected weather circumstances affecting the production by renewable sources or the expected load. Just before real time, the intraday market is closed and market participants do not have the option anymore to adapt their portfolio.

If every market participant precisely acts according to its schedule, the system is in balance in real time as the sum of the energy schedules is also in balance. Most of the time, however, the actual levels of generation and load deviate from the commitments done in the wholesale market because of unforeseen circumstances. A power plant may have a sudden outage, the wind speed may be a bit higher or lower than expected an hour before while the temperatures may be a bit higher or lower affecting the demand for electricity. Because of such deviations between realizations and commitments, individual market participants cannot precisely act according to their commitments in the wholesale market.

From this follows that the definitions of imbalance on individual level differ from the one on system level. The definition of balance on individual level is defined as the equalization of net realized position (generation or load) (POS_{real}) and the net committed position (i.e. the allocated volume) (POS_{com}):

$$Balance \ of \ market \ participant:$$
$$POS_{real} = POS_{com} \tag{7.3}$$

In order to give market participants incentives to be in balance in real time, they can be given a responsibility to meet this balance requirement. In such a system, as it is for instance applied in the Netherlands, the system operator with its legal obligation to keep the grid in balance has given responsibilities to all market participants to ensure that their actual net position of power injection and withdrawal is in line with what was committed in the energy schedules which was submitted before real time. Because of this responsibility, these market participants are called Balancing Responsible Parties (BRPs).[8]

As a consequence, all the parties with programme responsibility are running an imbalance risk, which means that they run the risk that the actual net position of injections and withdrawals deviates from what they had committed to in the forward and spot markets and which deviation has financial consequences for them. Power producers, for instance, may be confronted with an unanticipated power plant failure. Retailers are uncertain about the actual power consumption by their customers. They sell electricity to consumers under the premise of a profile, i.e. standardized time patterns in consumer power consumption. Because they take over consumer programme responsibility, retailers need to ensure that their customer group's actual consumption is in balance with what they bought in the wholesale market and/or generate themselves and inject into the system. As a result, the consequences of changes in power consumption by consumers, for instance, when they are generating electric power themselves, are initially borne by their retailers.

Although the balance requirement on system level holds every second, the responsibility for the market participants refers to the imbalance settlement period (ISP) which is typically a period of 15 min. The use of an ISP means that market participants are only responsible for their net energy balance position over this ISP. Because the system has to be in balance every second, the system operator is responsible for this balance from second to second within the ISP.

7.4.4 Determining the Balancing Energy Price

When the system is imbalance, system balance has to be restored. An important aspect of the design of the balancing scheme is the determination of the price for balancing energy and the rewards for those who help the system operator to restore balance. When the prices of balancing energy, i.e. the real-time prices, are just based on the costs of the system operator, independent on the real (marginal) costs of restoring a particular imbalance, it does not give appropriate incentives to the BRPs to remain in balance. In such cases, these prices can be seen as kind of

---

[8]Strictly speaking, this programme responsibility task applies to all those who are connected with the network, including residential consumers. Retailers are obliged to take over this programme responsibility from the consumers, though.

regulated tariffs, not based on market conditions. A more efficient way of determining the balancing energy price is to make use of a market. Having real-time prices which reflect the marginal conditions in the systems is important as these prices are the basis for the prices of all forward products (see Box 7.5).

The demand side in the imbalance market is determined by the system operator who wants to maintain system balance. The magnitude of this demand is just equal to the deviation between injection and withdrawal of power on system level. On the supply side two types of suppliers can be distinguished: on the one hand, market participants, called Balance Service Providers (BSP), who have concluded contracts with the system operator to supply balancing power when needed, and on the other hand, the other hand market participants, called Balancing Responsible Parties (BRP), who respond to the actual system imbalance. The BSPs consist of major grid users who can guarantee the availability of capacity for balancing purposes.

An example of a market-based balancing scheme is the one implemented by the Dutch TSO (TenneT). In this mechanism, the price for balancing energy is based on a market including a merit order of all bids of parties who can help by offering reserves to restore an imbalance, while the price can be both positive and negative, depending on the market situation. The supply curve is constructed by ranking all bids from the perspective of the TSO with the objective to minimize the costs of the TSO for balancing. The sign and magnitude of the regulation needed is determined by the actual imbalance on system level. Figure 7.7 shows the sign and size of the system balance on the horizontal axis and the sign and size of the imbalance price on the vertical axis. In the quadrants on the right side, the system is short (i.e. there is excess load), and in those on the left, it is long (i.e. there is excess generation), while in the top quadrants, the imbalance price is positive and in the lower negative.

The supply curve in the imbalance market is based on bids by the BSPs to make a specific type and size of capacity available against their minimum required price (i.e. their marginal costs) for each ISP the next day. These bids may refer to the marginal costs of ramping up or down a power plant and decreasing or increasing load. The marginal costs curve in the right-top quadrant of Fig. 7.7 refer to the compensation BSPs require in order to provide upward regulation. This is a positive price as these BSPs are making costs (e.g. for ramping up a gas turbine). These positive price for upward regulation means that the system operator pays the BSP, while negative prices for upward regulation would mean that the BSP has to pay the system operator.

The marginal costs curve in the two left quadrants of Fig. 7.7 refers to the minimum required price for supplying downward regulation. This price can be both positive and negative. Positive prices for downward regulation mean that the BSP has to pay the system operator, while negative prices here mean that the system operator pays the BSP.[9]

---

[9]As an example, see the website of the Dutch TSO with real-time information on the imbalance volumes and prices: https://www.tennet.org/english/operational_management/System_data_relating_implementation/system_balance_information/BalansDeltawithPrices.aspx.

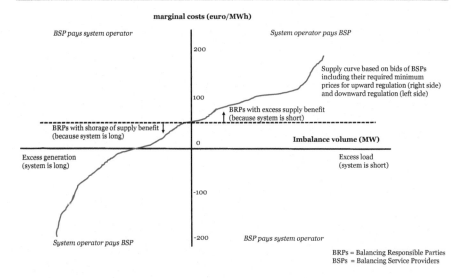

Fig. 7.7 Supply curve of energy balancing bids of Balancing Service Providers (BSPs) (*Source* derived from Tanrisever, et al. 2015)

The maximum price a BSP is prepared to pay for offering downward regulation by ramping down a gas turbine, for instance, is based on the marginal revenues of producing less electricity by this plant. These marginal revenues consist of the foregone marginal costs. As this BSP already has sold the electricity in a forward market, downward regulation of such a plant means that it has lower generation costs, while the revenues of selling electricity are not affected by this decision. Hence, in such cases, a BSP is willing-to-pay a (positive) price in order to produce less electricity (see also the example of redispatch in a zonal market in Box 7.3).

Some consumers may also be prepared to consume more energy (than contracted) against a positive price, as long as this price is lower than the price which should have been paid in the forward market. Other consumers may require a compensation by the system operator in order to increase their consumption. These latter bids are depicted in the supply curve in the left-bottom quadrant of the figure.

The system operator uses the above merit order to determine which reserves are used for system balancing over the course of an ISP. When there is a shortage of power at one second, the bid of an BSP with the lowest marginal costs for upward regulation will be asked to produce more or consume less. This results in an imbalance price that second. When in the next second, the shortage increases, more extra upward regulation is required and, therefore, the system operator asks the next bid in the merit order to determine which offer is honoured to provide this regulation. This results in an increase in the imbalance price. This continues until the end of the ISP. The last price in this period is given as the imbalance price. It is, however, also possible that during an ISP the sign of the imbalance changes, which means that a shortage of power changes into an excess of power. As a result, such an ISP will have two different prices.

Box 7.6 Energy transition and reliability of supply

Energy transition may have large consequences for energy networks. The traditional one-way transmission of power, from producer to consumer, changes partly into a two-way transmission system in which consumers also generate electricity. Furthermore, energy transition results in electrification and an increase in residential electricity consumption, among other things, for charging electric vehicles and heat pumps. The consequences of the greater variation in the flows and the higher peaks in grid load are that network operators must do more to keep the voltage level in the grids up and to prevent grid parts from becoming overloaded.

Network operators have different options available to deal with these consequences. With the help of information technology, for instance, generation by solar panels can be temporarily adapted when overloading in local network parts is going to happen (Martinot 2015). Another option to make the grid smarter is to integrate production and demand locally in order to relieve higher-level network parts. Investments in increasing the local network capacity are yet another way to solve congestions. Technically, network operators are capable of temporarily storing electricity for the purpose of preventing network parts from overloading; however, in doing so network operators would interfere with market participants. The more common method to address congestions is to apply redispatch (see Sect. 7.3). Another efficient solution would be if network operators make use of incentives to stimulate network users to consider the grid situation when determining the timing their electricity generation and demand. They can do this by translating network usage into dynamic network prices, as was discussed in Sect. 7.3. In Sect. 11.4, the consequences of dynamic tariffs for fairness will be discussed.

Another issue is whether a strong increase in weather-dependent renewable electricity generation will affect the balancing of the networks. In this respect, it is important to distinguish the concepts of volatility and uncertainty. Renewable electricity is highly volatile because of the weather dependency, but changes in weather circumstances in the near future can generally be well foreseen. Because weather conditions can be forecasted increasingly accurately, participants in the day-ahead or the intraday markets can adjust their portfolios to changing expected conditions. In some markets, participants can trade up to 5 or even 0 min before physical delivery, which enables them to adapt their own portfolio up to the very last moment.

In return for making reserves available for balancing, the so-called Balance Service Providers (BSPs) receive a compensation per unit of capacity (per MW) while in addition they also receive a price for each unit of balancing energy they deliver over an ISP (i.e. this is per MWh). In some countries, the regulator prescribes which type of market participants are obliged to offer capacity for balancing services to the system operator.

7.4.5 Financial Settlement

The market-clearing prices in the imbalance market can be used for financial settlement of the imbalances of the BRPs, which means that the imbalance prices are used to determine the financial value of being in imbalance. As during an ISP various bids within the merit order may be called upon, depending on the development of the imbalance during that period, there are several market-clearing prices during that period. In order to financially settle the imbalances, one option is to use the price of the last unit required to restore balance within the ISP, while an alternative option is to use the actual prices for each minute within the ISP.

As is shown in Fig. 7.7, the imbalance price can be positive or negative. When the system is long, while an individual BRP is short (e.g. it produces less than what has been sold in the forward and spot markets), then the shortage of supply by this market player has as a effect that the system oversupply is smaller, and the other way round. In such a situation, the economic value of the individual imbalance (V^{im}) is positive as it contributes to minimizing system imbalance. This value is equal to the product of the individual imbalance (Q_i^{im}), defined as the difference between its actual production and its commitments, and the price of balancing energy (p^{im}):

$$V^{im} = Q_i^{im} * p^{im} \tag{7.4}$$

When, for instance, the individual imbalance is negative (i.e. BRP is short) and imbalance price is negative (i.e. system is long), the economic value of being in imbalance is positive (see Table 7.3).

By such a mechanism, the imbalance market gives a reward to market participants with imbalance positions which are in the opposite direction as the imbalance position of the system, and v.v. As a result, such a balancing scheme gives extra incentives to market participants to restore system balance (Nobel 2016). These participants may respond to the imbalance price by regulating their production or demand, resp., up or down. Because of the above-mentioned mechanism, all participants have an incentive to minimize their imbalance or to be opposite to the system balance. This characteristic makes a balancing system more efficient

Table 7.3 Financial consequences of individual imbalances depending on sign of system imbalance

Individual balance	System balance	
	Short (excess load; shortage of production)	Long (excess production: shortage of load)
Short (realized production (load) is less (more) than committed production (load)	Negative	Positive
Long (realized production (load) is more (less) than committed production (load)	Positive	Negative

Note During an ISP, the system balance can switch from being short to being long and v.v

compared to systems in which BRPs are always financially punished for being in imbalance, irrespective of the sign of the system imbalance.

The financial settlement of imbalances during an ISP is based on the net imbalance over that period, which means that imbalances from second to second within this period are for the responsibility of the system operator. This implies that the system operator bears the costs of these imbalances within the ISP, but these costs can be financed by the payments for imbalances by the BRPs.

7.5   Pricing of Generation Adequacy

7.5.1   Introduction

In well-functioning markets, demand and supply can find an equilibrium. This means that the interaction between suppliers and buyers, for instance on an exchange, results in a commodity price at which all consumers having a WTP of that level or higher can buy the commodity from suppliers having a minimum required price below or equal to that price. All other consumers and producers are not successful in their bids, as their bids are too low or too high, respectively, compared to the market price. Hence, this process of market clearing not only results in an allocatively efficient outcome but also gives appropriate incentives for investments in production capacity. When the profits which can be realized by the inframarginal suppliers exceed the costs of extending capacity, firms have an incentive to do so. In this section, we first discuss how this mechanism of providing incentives for investments in generation capacity works in electricity markets, then we show that in some cases electricity markets fail to give the proper incentives for investments and finally, we discuss regulatory measures to address this market failure.

7.5.2   Investment Incentives in Electricity Markets

The above mentioned basic mechanism to take care of allocative efficiency and incentives for investments is also present in electricity markets which are based on the so-called energy-only mechanism. In energy-only markets, electricity prices are determined in two different ways, depending on the extent generation capacity is available. When there is sufficient generating capacity given the demand level, the electricity prices are based on the marginal generation costs, i.e. the costs for producing one unit of electricity, of the marginal supplier (see Sect. 3.3.4). These marginal costs include the direct operational costs, in particular the fuel costs for a conventional fossil-fuel plant, and the dynamic generation costs, such as the costs of ramping up and down a power plant. In addition, the marginal generation costs of selling electricity in a particular forward market (e.g. the day-ahead market) include the so-called opportunity costs of not being able to sell a specific amount of electricity in a following forward market (e.g. the intraday market). Moreover, the marginal generation costs also include the costs which are associated with the risk of not being able to meet real-time delivery (balancing) commitments due to, for example, technical problems with a power plant. All these costs constitute the marginal generation costs which determine the electricity prices when demand is below the totally available generation capacity.

When the demand for electricity, however, exceeds the available capacity, scarcity prices emerge. Scarcity prices are prices which are required in order to reduce demand until it equals the generating capacity. In these situations, therefore, prices are not based on the marginal costs, but on buyers' willingness to pay. Scarcity prices are not only meant to bring demand down to the level of the capacity constraint in order to clear the market but they also form incentives to invest in generating capacity. As scarcity prices are above the level of the marginal generation costs, they result in extra profits for all producers, so-called scarcity rents (see Fig. 7.8). Scarcity rents are profits which emerge because of constrains in production capacity.

Generally, an investment (I) in a power plant is profitable when the present value (with the discount rate r) of the expected aggregated operational revenues, based on the difference between the electricity price (p) and marginal costs (mc), of the production (G) in all future operating hours (h) are at least equal to the investment sum:

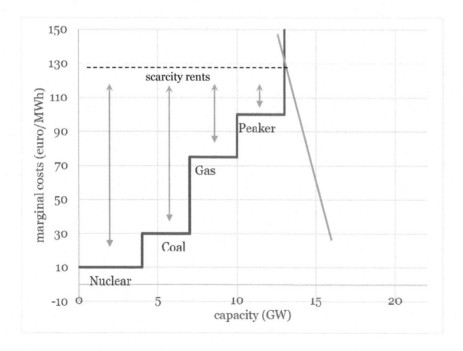

Fig. 7.8 The formation of scarcity (or: inframarginal) rents in an electricity market

$$I \leq \sum_h \frac{(p_h - mc_h) * G_h}{(1+r)^t} \qquad (7.5)$$

This pricing mechanism with scarcity prices ensures that the fixed costs of investments in generating capacity can be covered. This pricing mechanism is also called peak-load pricing (see also Sect. 6.6.1). A consequence of this mechanism is that, in theory, the market prices gives incentives for the optimal investments in electricity generation portfolio. This portfolio consists of different types of plants with different variable costs per MWh and fixed costs per MW.

The optimal generation portfolio in a market depends on the volatility in demand levels and prices. As the demand for electricity fluctuates from hour to hour (see the load-duration curve in Sect. 3.3.4), the utilization of generation capacity fluctuates as well. As a consequence, not all installed capacity can be utilized all the time. The revenues of every type of plant need, however, to be sufficient for cost recovery; otherwise, investors will not invest in them.

The relationship between demand volatility and investments in different types of power plants can be analyzed through so-called screening curves. These curves show how the total costs per MW per year of running a specific type of plant depend on the fixed costs per MW, the variable costs per MWh and the number of operating hours in a year. To clarify this, let's look at a stylized example shown by Fig. 7.9. Here, the horizontal axis of the figure gives the number of hours in a year and the vertical axis on the left side the total costs per MW per year. The fixed costs of a plant are depicted by the intersection with this vertical axis, while the variable costs are depicted by the slope of the line. The total costs per MW are also called the annual revenue requirement per MW, as this sum should be earned in order to compensate for all costs (see e.g. Stern 2002).

Suppose there are three plant types in an electricity system: A, B and C. Plant type A has low fixed costs and relatively high variable costs, while plant type C has the opposite cost structure, with high fixed costs and low variable costs and plant type B has intermediate fixed and variable costs. In this example, we assume that the investments per MW for plant types A, B and C are 1, 2.5 and 6 million euro, respectively, while it is assumed that the expected life time of these plants are 30, 40 and 50 years, respectively. For the sake of simplicity, we ignore the discount rate, so we work with the nominal values. Hence, the fixed costs per MW/year are defined by the investment sum per MW (I_{MW}) and the expected life time of the plant in number of years (L):

$$FC = \frac{I_{MW}}{L} \qquad (7.6)$$

Applying this formula on this example results in an annual fixed cost of 33,330 euro/MW/year for plant type A and 62,500 and 120,000 euro/MW/year, for plant types B and C, respectively. In addition, we assume constant variable costs in all cases, albeit at different levels: 60, 40 and 20 euro/MWh for types A, B and C,

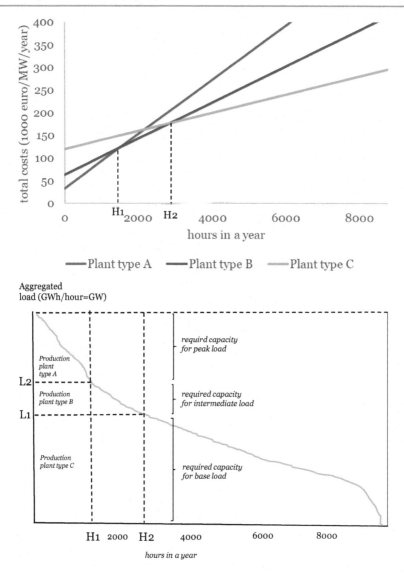

Fig. 7.9 Screening curves in relation to load-duration curve

respectively.[10] This results in an annual variable costs per MW of installed capacity (VC) which depend on these costs per MWh and the annual number of operating hours (H)[11]:

$$VC(H) = VC_{MWh} * H \tag{7.7}$$

The total number of operating hours can also be expressed as the capacity factor (cf) (see Sect. 2.3.4) times the total number of hours in a year:

$$H = cf * 8760 \tag{7.8}$$

Using the above two formulas, we can calculate the total costs per MW per year as a function of the number of operating hours:

$$TC(H) = FC + VC(H) \tag{7.9}$$

In this example, plant type A has the lowest total costs per MW per year when the annual number of operating hours of production is low because it has relatively low fixed costs. When the annual number of hours of production is higher, then plant type B has lower total costs per MW per year. The break-even number of hours (H1 in the figure) can be calculated by equating the total costs of plant type A with those of plant type B, where the total costs consist of the fixed costs plus the variable costs times the number of operation hours:

$$FC_A + VC_A * H = FC_B + VC_B * H \xrightarrow{yields} H = \frac{FC_A - FC_B}{VC_B - VC_A}$$
$$= \frac{33{,}000 - 62{,}500}{40 - 60} = 1475 \, hour \tag{7.10}$$

In a similar way, the break-even number of hours for plant types B and C can be calculated to be 2875 h (H2 in the figure).[12] So, plant type C has the lowest total costs per MW per year when it can produce more hours than this number of hours per year, which is due to its relatively low variable costs.

The results of this analysis of the screening curves can be used to determine the optimal generation mix. This optimal mix does not only depend on the screening curves of the various generation technologies, but also on the distribution of the

---

[10]Note that in reality, the variable costs are generally not constant due to the so-called dynamic costs which are related to the ramping up and down of a power plant. The magnitude of these dynamic costs influences the optimal generation portfolio. By using the load-duration curve, it is implicitly assumed that these dynamic costs do not play a role in the investment decisions.

[11]After all, variable costs in euro/MWh times the number of hours gives: $\frac{euro}{MWh} * h = \frac{euro}{MW}$.

[12]At the break-even number of operating hours, the average costs per unit of electricity (MWh) are also equal. These average costs can be found by dividing the total costs per MW per year by the annual number of operating hours. For instance, when the number of operating hours is 2875, the average costs per MWh for both plant type B and C are 61.73 euro/MWh, while the average costs for plant type A are much higher at that level of operating hours (i.e. 71.60 euro/MWh).

demand levels over time and the market prices. The volatility of demand is measured through the load-duration curve which shows all hours in a year ranked from the highest level of aggregated load to the lowest level (see also Sect. 3.3.4).

The bottom panel of Fig. 7.9 also shows a load-duration curve. These load levels are expressed in GWh/hour, which is equal to capacity (GW). Hence, from this load curve one can infer the required generation capacity to meet the electricity consumption.

In this example, the load levels above the level of L2 which occur less than H1 hours per year can be most efficiently served by plant type A. The load levels between L1 and L2 GW, which only occur between H1 and H2 hours per year, can be most efficiently served by plant type B. All other (lower) load levels (below L1 GW) occur at least during H2 hours, which implies that these levels can be most efficiently served by plant type C.

Note that plant type C also produces when plant type B serves the load between L1 and L2 GW and that both plant types C and B also produce when plant type A serves the extra load above L2 GW. Because plant type C is producing all time, this plant type is called a baseload plant, while plant type A is a peak-load plant as it only produces during peak hours. Plant type B produces intermediate load. Examples of plant type A are gas turbines; examples of plant type B are combined cycle gas plants and examples of plant type C are nuclear power plants and coal-fired power plants.

The above determination of the optimal mix of generation plants is based on total costs per MW per year and the load-duration curve. Such a mix can be determined through central planning, as it was done before electricity markets were liberalized, but this requires a lot of information for the central planner. Electricity markets based on the energy-only principle can also give incentives to market participants to invest in the optimal mix of generation plants. In such a market, as we have seen above, the electricity price is determined by the marginal costs in case of (expected) abundant capacity and by the willingness to pay of consumers in case of constraints in generation capacity. When we translate the results of the above analysis to a market situation, we see how the fluctuation in the electricity demand in combination with the differences in the marginal costs per type of plant results in fluctuating electricity prices (see Fig. 7.10). The supply curve includes the marginal costs of plant type C on the left side of the merit order (as this plant type has the lowest marginal costs), the costs of plant type B in the middle and the costs of plant type A at the right side.

Plant type C produces for all levels of demand, albeit that not all installed capacity will be used permanently. When this plant type is the marginal producer, the electricity price is equal to its marginal costs (20 euro/MWh, in this example). This implies that in these hours, this plant type does not make any profits to cover the fixed costs. When, however, plant type B is the marginal producer, the electricity price rises to 40 euro/MWh. As a consequence, plant type C makes an hourly profit equal to (40–20=) 20 euro/MWh times its production. During peak demand, plant type A is also needed and then the electricity price will rise to 70 euro/MWh or even way above in case all capacity of this plant type is utilized. In the latter

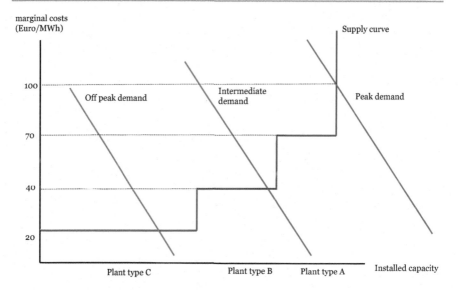

Fig. 7.10 Marginal costs per plant type and electricity price

situation, also plant type A makes a profit to recoup its investment costs equal to for instance (100–60=) 40 euro/MWh, while plant types B and C make an even larger profit (60 and 80 euro/MWh, respectively).

Looking at Fig. 7.9, one can also easily see the impact of change in the load pattern on the generation portfolio. When the load is more stable, less installed capacity of peak plants is required and the other way around. This analysis of screening curves in relation to the load-duration curves gives information on which type of plant should be used for which type of demand, but it does not answer the question how much should be invested in each type. The answer to this question depends on the prices for electricity, and in particular the prices during peak hours. In the above example, we have just assumed a peak-load price of 100 euro, but this price depends on the scarcity in the market, i.e. it depends on the magnitude of installed capacity. This magnitude in turn depends on how much firms have invested. Firms will only invest in new capacity if they expect that the future revenues will be sufficient to recover the investments. In this example, an investor will, for instance, only invest in extension of the capacity of plant type C, when the present value of the expected revenues which can be realized when plant type A and B are the marginal producers is higher than 6 million euro/MW. When the expected scarcity prices are not sufficient for investments in new power plants, the capacity will remain the same. This may change when at some future point in time the scarcity prices rise because of an increase in demand or a closure of existing power plants.

This mechanism implies that the profitability of investments of all plant types, including the base load and intermediate-load plants, depends on the presence of scarcity prices. Hence, the presence of scarcity prices is a key condition of

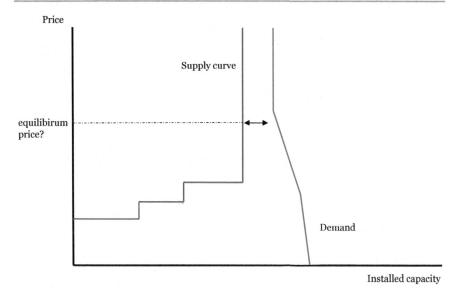

Fig. 7.11 Market failure: electricity market cannot find an equilibrium. *Source* Based on Cramton et al. (2013)

electricity markets to ensure that there is sufficient generating capacity at all times. In addition, for a particular type of investment it is relevant which prices it will be able to capture (see Box 7.7).

7.5.3 Market Failure: Insufficient Generation Capacity

In some situations, however, electricity markets are not able to generate scarcity prices. This occurs when these markets cannot find an equilibrium. In the event of a fully utilized generating capacity and a high, but inelastic demand, the market may not be able to clear (see Fig. 7.11). Such situations may happen at those moments when not all the installed generation capacity is available for generation, which can be due to, for instance, a sudden, unexpected technical outages of some plants, while, at the same time, the demand for electricity is high because of, for instance, a high demand for cooling during a hot summer day.

Box 7.7 Investments in renewable energy, merit-order effect and market-value effect

To determine the expected profitability of an investment in generation capacity, one has to assess the future power prices. These prices depend on several factors. These factors can be distinguished in factors that affect the demand curve and factors that affect the supply curve (merit order). Higher average income levels or higher temperatures may shift the demand curve to the right and increasing the likelihood of scarcity prices. Higher fuel prices do the supply curve shift upwards, which may result in higher electricity prices if the marginal costs of the price-setting plant also increase, while investments in generation capacity make the supply curve to shift to the right, which results in lower electricity prices.

The latter effect is in particular relevant for investments in renewable energy. Because of the low marginal costs of wind turbines and solar panels, these technologies are on the left side of the merit order (see also Sect. 3.3.4). When the capacity of these technologies increase, the supply curve shifts to the right resulting in lower electricity prices. This is the so-called merit-order effect of renewable energy (see e.g. Mulder and Scholtens 2013). The precise size of this effect depends on the shape of the remaining parts of the supply curve and the position of the demand curve, but in general one can say that an extra unit of renewable energy production (G^R) reduces the average electricity price (P) in a market. This effect can be expressed as:

$$merit\, order\, effect = \frac{dP/p}{dG^R/G^R} \qquad (7.11)$$

This merit order effect of renewable energy is relevant for all types of producers as it affects the average electricity price everyone is receiving. For the renewable energy producers, there is an extra effect which needs to be taken into account, which is the so-called market-value effect. This effect entails that the higher the share of renewable energy in a market, the lower the average price the producers of renewable energy receive (P^R) compared to the time-weighted average electricity price (P):

$$market - value\, effect = \frac{\sum_h P_h^R}{\sum_h P_h} \qquad (7.12)$$

This effect results from the fact that the production of renewable energy by the various installed units is correlated (e.g. most wind turbines produce electricity when the wind blows, while no one produces when there is no wind). As a result of this correlation effect, producers of renewable energy do not benefit from the higher electricity prices when there is no renewable energy, i.e. when the price is higher due to the absence of the merit-order effect.

When the electricity market is not able to clear, there is no market price and, hence, suppliers cannot be remunerated for their production. In addition, because of the lack of supply to meet total demand, the network balance is in danger. Because of the network requirement of balance, such a situation would result in a physical imbalance, implying that no producer is able to deliver its product to consumers and

Table 7.4 Regulatory measures to deal with risk that electricity market cannot find equilibrium

Regulatory measure	Objective	Risk
Setting a price	Determining a remuneration for power in case the market does not clear	Price cap is lower than Value of Lost Load (VoLL), resulting in suboptimal investments in generation capacity
Organizing extra capacity	Preventing system imbalance in case the market has insufficient capacity	- Moral hazard (i.e. reducing incentives for market parties to invest in capacity) - Windfall profits (i.e. payments for capacity are given for already realized capacity) - Over insurance (i.e. payments and obligations are more expensive than benefits of having lower risk of shortage)

no consumer is able to use electricity. While in most other markets, the infra-marginal producers would still be able to deliver its products to the consumers, in electricity markets no producer is able to this if the system not able to restore the balance.

Hence, the inability of the market to find an equilibrium price results in two problems: (a) no price is set for the producers which would be able to produce and (b) the shortage of supply results in physical imbalance and malfunctioning of the system. Without any specific regulatory intervention, electricity markers are not able to solve this. The regulatory challenge to solve this market failure is, hence, to (1) determine a market price for the supply that can be produced by market parties and (2) to get additional generation capacity to prevent system imbalances (see Table 7.4).

Box 7.8 Generation portfolios in number of OECD countries

Generation portfolios differ strongly among countries. These differences are partly due to differences in national circumstances, such as the possibility to utilize differences in altitude to generate electricity by hydropower plants. Figure 7.12 shows that in particular Norway, but also France, Italy and Japan produce a significant amount of hydropower. Another factor contributing to differences in national generation portfolios is the differences in national energy policies. In France, for instance, the government strongly supported investments in nuclear power, while in the Netherlands the presence of a domestic gas resources made it more attractive to build and utilize gas-fired power plants.

Despite these differences in generation portfolios among countries, a general principle appears to hold in most countries. This principle is that power plants with low variable costs are utilized more often than plants with lower fixed costs, as is explained in this chapter. To the former group of plants belong coal-fired and nuclear plants, while gas-fired power plants belong to the latter.

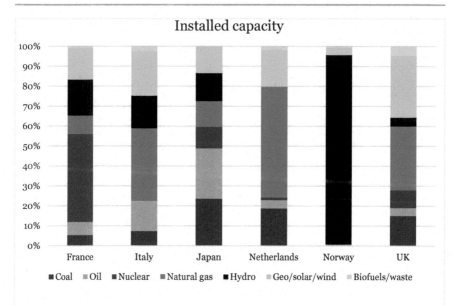

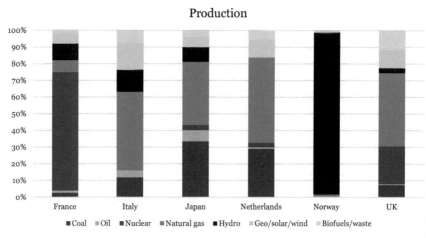

Fig. 7.12 Share of generation techniques in installed capacity and annual production in a number of European countries, 2018. *Source* IEA, Electricity Information 2019

From the figures, it appears that the share of coal-fired and nuclear power plants in total production is higher than in installed capacity, while the share of the gas-fired power plants in production is lower than in installed capacity. This indicates that the latter type of capacity is less often used.

7.5.4 Regulation: Setting the Price

The theoretical basis for setting a market price in case the market itself is not able to do this is to define the maximum price users are willing-to-pay in order to have electricity. This electricity price is equal to the marginal value consumers attach to having electricity. This willingness to pay for electricity is called the Value of Lost Load (VoLL), which is the value consumers are prepared to pay to avoid the situation of disruption (see Box 7.9). The VoLL is related to the costs consumers have to make when such a situation occurs. In theory, when the electricity price consumers have to pay for electricity exceeds these opportunity costs, then it is more efficient for them not to use electricity and to accept the costs of having no electricity. Hence, if electricity prices exceed this value, consumers prefer having no electricity above paying a higher price.

Although theoretically, the best solution is to set the prices in shortage situations at the level of the VoLL, in practice regulators often choose lower levels of the maximum price. The EPEX day-ahead market, for instance, has set its maximum price at 3000 euros/MWh, which is increased when the actual price is close to this level, while the maximum price at the intraday market is 9999 euro/MWh, and the imbalance market does not have a price cap.[13]

The price cap set by regulators is not only relevant for the hours in which this market problem occurs, but also for investment decisions. After all, if the regulator informs market parties that the price in such situations will be set at a specific level (i.e. the price cap), these parties can formulate their expectation regarding the future revenues and profits. The higher the price cap, the more profits can be realized during the hours of lack of capacity and the less profits are required during other hours to recoup the investments costs. The level of the price cap is, therefore, relevant for all investments in power plants. A price cap that is below the VoLL has as a result that market parties invest less in generation capacity than what would be socially efficient.

Box 7.9 Estimating the Value of Lost Load (VoLL)

Regulators can determine the VoLL in different ways. One approach is to estimate the damage the various types of electricity consumers experience when electricity supply is interrupted. For industrial users, this damage can be estimated on the basis of the value added per unit of time. This is the so-called macroeconomic approach. If electricity is not available, these users will not be able to realize this value added during the period of interruption. Another approach is to estimate the willingness to pay of consumers through stated-preference methods, like conjoint experiments (see Sect. 4.3.3). In this approach, respondents are asked to make a choice among various options regarding the availability of electricity. Based on the choices they make, their (implicit) willingness to pay for electricity can be estimated.

---

[13]See https://static.epexspot.com/document/38579/Epex_TradingBrochure_180129_Web.pdf

The results about the size of the VoLL vary strongly among the various studies, but in general, the outcomes of the macroeconomic studies are in the range of 10,000–25,000 per MWh, while the stated-preference methods give as result a VoLL of about 10,000 euro per MWh (Schröder and Kuckshinrichs 2015).

From this follows that, when formulating the expectations regarding the future revenues, market participants also have to assess the likelihood that hours with the regulated prices (based on VoLL or at a lower level) will occur. This likelihood is called the Loss of Load Probability (LoLP), which is the defined as the probability that in a given period the load has to be shed. Based on this probability, the Loss of Load Expectation (LoLE) can be calculated, which is the number of time units (e.g. hours) in that period in which the load has to be shed because of shortage of production capacity (Creti and Fontini 2019). The product of the LOLE and the VoLL determines the value of power (V) during hours of lost load.

$$V = VoLL * LOLE \qquad (7.13)$$

An investment in a power plant that is meant to use during hours of scarcity breaks even when this value exceeds the investment costs. To illustrate this, let's discuss the following example.

Suppose that the regulator has set the electricity price in shortage circumstances (i.e. when the demand exceeds the available production capacity) equal to the VoLL of 10,000 euro/MWh. Suppose further that the costs of a peaker plant are equal to those of plant type A above (so, fixed costs of 33,333 euro/MW/year and variable costs of 60 euro/MWh). Both the VoLL and the costs of having and using the peaker plant can be related to the number of hours of shortage (see Fig. 7.13). When there are no shortages at all, the the value of shortage is equal to zero, while the costs of having the peaker consist of the fixed costs. When the number of hours of shortage per year increases, the value of shortages increases linearly. The costs of operating the peaker only increase with the variable costs, which are negligible compared to the fixed costs (at such low numbers of operation). From the figure follows that after a number of hours of shortage, the investment in the peaker is efficient as the costs of the peaker are lower that the costs of having no electricity. This number of hours of shortage (n) after which the investment in peaker become profitable can be estimated by dividing the fixed costs per MW per year (so ignoring the variable costs) of the plant by the VoLL per hour:

$$n = \frac{FC}{VoLL} \xrightarrow{yields} \frac{33,333}{10,000} = 3.33 \, hours \qquad (7.14)$$

When, however, the regulator has set the regulated price during shortages at a lower level (e.g. 5,000 euro/MWh), then firms will only invest in a peaker capacity

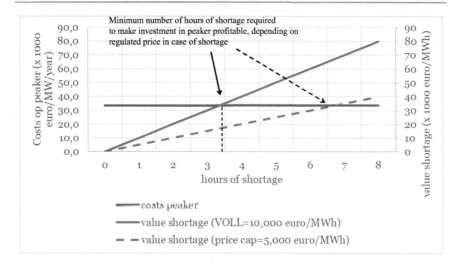

Fig. 7.13 The required number of shortages to make an investment in peak capacity profitable, depending on the regulated price during shortages

when they expect that the number of hours of shortage is twice as high (i.e. 6.66 h per year). Because the regulated price cap is below the VoLL, market parties are investing too less from a social-welfare point of view.

7.5.5 Regulation: Organizing Reserve Capacity

When the actual LOLE is below the required number of hours to make an investment profitable (in this example, this number is 6.66 h when the price cap is 5,000 euro/MWh), no firm is going to invest in a peaker plant. As a consequence, additional regulatory measures are required to reassure that there is sufficient generation capacity to keep the system in balance. These measures consist of organizing extra generation capacity or demand response capabilities which is only used in these circumstances and organizing involuntary load shedding (i.e. reduction in electricity consumption). Organizing extra capacity can be done in several ways. These options can be distinguished in (a) reserves owned by the network operator, (b) capacity payments, (c) capacity obligations and d) market-wide capacity auctions.

In option (a), the network operator purchases a number of generation plants, which can be used in times of scarcity. This reserve capacity is only used when all other capacity in the market is operating on full capacity.

In option (b), the network operator does not own and operate the reserve capacity itself, but it pays market participants to have reserve capacity available which should be made available to the network operator in times of shortages.

In option (c), market participants are obliged to hold reserve capacities, which implies that they are not allowed to fully sell their capacity in the wholesale market, but that a minimum part should be kept available for the network operator in case of shortage.

In option (d), finally, the network operator organizes a market in which market participants can offer capacity. The minimum required prices of the bidders in this market are related to the opportunity costs of keeping capacity available. In addition to this price, the suppliers also need compensation for the variable costs when the capacity is actually used. Hence, a capacity market has two different pricing mechanisms: a price for capacity (MW) and a price for producing electricity (MWh).

The organization of extra reserve capacity, however, does not necessarily increase the reliability of supply because of the risk of adverse effects. When market participants know that the network operator has implemented a system for reserve capacity, they may expect that this capacity will also be used when there is not a real shortage. If that happens, the scarcity prices in the wholesale market are reduced, which means that firms have less incentives to invest in generation capacity. Hence, the presence of a system of reserve capacity may reduce the incentives for investments. This is the so-called moral hazard effect of reserve capacity, comparable to the moral hazard effect of an insurance on the behaviour of insured persons.

Besides this risk of the implementation of strategic reserves, there is the risk that a capacity system results in windfall profits for incumbent participants. For instance, when the regulator introduces a system of capacity payments, firms that already have invested in power plants based on a power system without capacity payments may also receive these payments, while it does not lead to more generation capacity.

Another risk is that the capacity payments or obligations are too high from a social-welfare point of view. As a result, society is making too much costs compared to the benefits of having a lower risk of shortage. Hence, this is the risk over insurance.

When the utilization of the reserve capacities is not sufficient, the option of last resort for network operators in order to prevent a total electricity blackout from happening is involuntary load shedding. Load shedding can be done automatically and in stages as soon as frequency values become very low (less than 49 Hz in the European system where the grid frequency is 50 Hz.). This measure is only used when all other measures are utilized and not sufficient to keep the system in balance.

All these types of measures, which are meant to guarantee sufficient generation capacity, are only needed when the market itself is not able to find an equilibrium because of capacity constraints on the supply side and inflexibility on the demand side. The first best option, however, to address this risk, is to increase the short-term price sensitivity of consumers as that would help the market to find an equilibrium price itself.

7.6 Exercises

7.1 What may go wrong in absence of regulation of network capacity?

7.2 Explain how the regulator can address the risk of limited available network capacity?

7.3 What may go wrong in absence of regulation of the balance of energy systems?

7.4 How can the need to balance energy systems be realized?

7.5 What may go wrong in absence of regulation of generation adequacy?

7.6 How can the need to assure the adequacy of generation capacity be addressed?

References

Cramton, P., Ockenfels, A., & Stoft, S. (2013). Capacity market fundamentals.

Creti, A., & Fontini, F. (2019). *Economics of electricity: Markets, competition and rules.* Cambridge University Press.

Frontier (2009), International transmission pricing review, A report prepared for the New Zealand Electricity Commission. July.

Jafarian, M., Scherpen, J. M. A., Loeff, K., Mulder, M., & Aiello, M. (2020). A combined nodal and uniform pricing mechanism for congestion management in distribution power networks. *Electric power systems research, 180* (106088).

Martinot, E. (2015). How is Germany integrating and balancing renewable energy today? Education Article, January.

Mulder, M., & Scholtens, B. (2013). The impact of renewable energy on electricity prices in the Netherlands. *Renewable Energy, 57,* 94–100.

Neuteleers, S., Mulder, M., & Hindriks, F. (2017). Assessing fairness of dynamic grid tariffs. *Energy Policy, 108,* 111–120.

Nobel, F. (2016). On balancing market design; Eindhoven University of Technology.

Schröder, T., & Kuckshinrichs, W. (2015). Value of lost load: An efficient economic indicator for power supply security? A literature review. *Frontiers in Energy Research,* 24 December.

Stoft, S. (2002). *Power System Economics; designing markets for electricity.* The Institute of Electrical and Electronics Engineers, IEEE Press.

Tanrisever, F., Derinkuyu, K., & Jongen, G. (2015). Organization and functioning of liberalized electricity markets: An overview of the Dutch market. *Renewable and Sustainable Energy Reviews, 51,* 1363–1374.

TenneT. (2019). Onbalansprijssystematiek; hoe komen de geldstromen tot stand? 13 februari.

van der Veen, R. A. C., & Hakfoort, R. A. (2016). The electricity balancing market: Exploring the design challenge. *Utilities Policy, 43,* 186–194.

van Dinther, A., & Mulder, M. (2013). The allocative efficiency of the Dutch gas-balancing market. *Competition and Regulation in Network Industries, 14*(1).

Externalities in Production and Consumption in Energy Markets

8

8.1 Introduction

In well-functioning markets, economic agents are incentivized to take into account all effects of their decisions. Sometimes this is not possible because of externalities. Examples of externalities in energy markers are the environmental effects of using fossil energy and the impact of importing fossil energy on security of supply. This chapter first discusses how the various types of externalities in energy markets distort market outcomes (Sect. 8.2). Then, the chapter goes into the design of environmental taxes (Sect. 8.3), support schemes (Sect. 8.4) and emissions trading schemes (Sect. 8.5), before discussing how these three types of environmental regulation affect electricity markets and how they interact with each other (Sect. 8.6). The chapter concludes by discussing the regulation of security of supply externalities (Sect. 8.7).

8.2 Externalities in Energy Markets

8.2.1 Theory

The presence of negative externalities means that the supply curve in a market based on the private bids of economic agents is below the so-called social supply curve (see Fig. 8.1). The private supply curve only reflects the costs which are considered by the agents and do not include the costs which are not relevant to them, because they don't see the financial consequences of them. The social supply curve includes both the private costs and the costs which are ignored by the agents. Because the social supply curve includes more costs, it intersects the demand curve at a higher price (P2) and lower volume (Q2) than the private supply curve does (P1 and Q1, respectively). Hence, the market outcome based on the private supply curve is distorted because the equilibrium volume is too high. This welfare loss is

© Springer Nature Switzerland AG 2021
M. Mulder, *Regulation of Energy Markets*, Lecture Notes in Energy 80,
https://doi.org/10.1007/978-3-030-58319-4_8

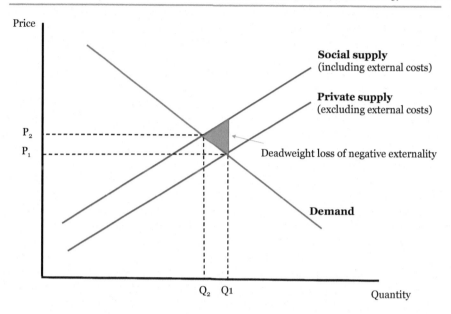

Fig. 8.1 Effects of negative externality on market and welfare

given by the triangle in the figure. This is the so-called deadweight loss of a negative externality. It is a loss to society, because the extra production on top of what would be socially optimal has marginal costs which exceed the marginal willingness-to-pay. Hence, these marginal consumers are consuming the volume between Q2 and Q1, although the value they attach to these commodities is below the costs that are caused by their consumption.

Positive externalities occur when economic agents do not take into account all benefits of their activities. In this case, the demand curve that the suppliers face (the so-called private demand curve) only reflects the marginal benefits which can be captured by the suppliers, excluding other benefits for society which do not result in extra revenues for the suppliers. The social demand curve includes both the private benefits and the benefits which cannot be captured by the suppliers. Because the social benefits curve includes more benefits, it intersects the supply curve at a both higher price (P2) and higher volume (Q2) than the private demand curve does. Hence, the market outcome based on the private demand curve is distorted because the volume of consumption is too low from a social-welfare point of view. This welfare loss is given by the triangle in Fig. 8.2. This is the so-called deadweight loss of a positive externality. It is a loss to society, because the extra consumption on top of what is privately optimal (i.e. the difference between Q2 and Q1) has marginal benefits which exceed the marginal costs of that extra supply.

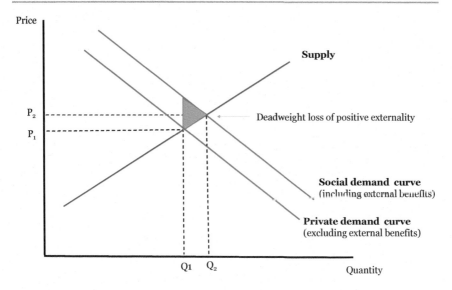

Fig. 8.2 Effects of positive externality on market and welfare

8.2.2 Examples of Externalities in Energy Markets

There are many examples of negative externalities in energy markets (Table 8.1). The use of fossil energy results in emissions that contribute to climate change for which effect no market exists (without regulation), and consequently, economic agents do not take these costs into account. The is mainly due to the absence of a price for these effects in case of unregulated markets. In addition, agents do not have an incentive to include these emissions in their economic decisions as the direct negative environmental effect on their utility or profit is generally very small compared to the costs they have to make to reduce that effect. This holds in particular for carbon emissions as the negative effects of these emissions on climate are spread over the globe and a long period of time. Hence, the social marginal benefits of a reduction of carbon emissions have to be shared with all global inhabitants over a significant number of years. Because of the global and long-term nature of climate change caused by emissions of carbon (and other greenhouse gases), climate policy requires global cooperation in order to be effective. Without such an approach, individual agents (e.g. firms, countries) may be hesitant to contribute to emissions reduction as they may be uncertain about what others will do (Barrett 2007).

Some environmental effects of energy use occur on a more regional or even local scale, such as the emissions of small particles resulting from fossil-fuel use as these emissions directly affect the air quality in the direct neighbourhood of the emitter. This limited spatial scale of environmental effects makes it less complicated to implement effective collective measures as individual members can be less afraid of free-rider behaviour by others.

Table 8.1 Examples of externalities in energy markets

Type of externality	Examples
Negative, environment	• Impact of emissions of CO_2 through burning of fossil fuels on climate • Impact of emissions of NO_x, SO_x and ultrafine particles through burning of fossil fuels on health • Noise produced by heat pumps • Risk of nuclear radiation • Impact of biomass production on biodiversity • Tragedy of the commons
Negative, security of supply	• Increased political dependence resulting from import of energy • Reduced flexibility to respond to shocks resulting from lack of investments in infrastructure
Positive	• Expenditures in research and development • Reliability of energy networks as public good

Other examples of negative environmental effects in energy markets are the risk of nuclear radiation, the impact of the use of biomass on biodiversity and the noise produced by heat pumps. Regarding the former, this risk cannot be priced by market participants because of the potential huge magnitude. Without regulation this risk is not priced in a market, and as a result, unregulated market parties would tend to invest too much in nuclear energy. As using biomass for energy production, for instance in the generation of electricity, may have negative effects on biodiversity in the regions where the biomass is taken from, this use is another example of a negative environmental externality. Moreover, the noise produced by heat pumps can also be considered as an externality insofar other agents (e.g. neighbours) are confronted with this more strongly than the agent himself.

A special type of negative externalities is called the tragedy of the commons. This market failure refers to situations in which individual agents do not have the incentive to take care of an asset which is also used by other agents, while it would be a joint interest when all agents would carefully deal with that asset. An example in energy markets is the exploitation of energy resources, like an oil or gas field. If the production of oil or gas occurs in a too high pace, the total aggregated production can be less than what would be possible in case of a more modest pace of exploitation. When this field is operated by many different agents, however, every individual agent faces the risks that others will not produce according the optimal path, and hence, every agent will produce as much as possible. This market failure can only be solved through coordination.

Another category of negative externality refers to the security of energy supply. Consumers of energy may not take into account the geopolitical consequences of consumption on import of fossil energy and the resulting dependency on a few exporting countries. When consumers use energy that has to be imported from a limited number of countries, as is generally the case for gas and oil (see Sect. 3.3.2), an increase in fossil energy use implies a higher import dependency. A higher dependency on import in itself is not a market failure, but it may make a country more vulnerable to international political risks.

The security of energy supply may also be negatively affected if network operators do not invest sufficiently in the infrastructure, which reduces the option of energy users to respond in case of a shock, such as a disruption in supply. If network operators do not take into account the costs for network users if they cannot respond quickly to such a shock, then this is also an example of a negative externality in the energy sector. In Sect. 10.3.4, we will see that international integration of gas markets which is facilitated by investments in infrastructure has indeed fostered the ability of energy users to respond to unexpected shocks in demand or supply.

Box 8.1 District heating systems as examples of network externalities in energy markets

Network externalities occur when the benefit of consuming a commodity depends on the number of other agents also consuming that commodity. An example of such a market failure is a local energy system, such as a district heating system, where local producers and consumers interact. The value of such a system increases the more local agents participate. After all, a local energy system can only exist when there are sufficient number of participants who can supply the energy that is needed by the local community, while the local producers need a number of consumers in order to make the production profitable.

The occurrence of a network externality may hinder the initial development of such a system. If such network externalities exist in a market, market parties could coordinate how they want to organize the market or, alternatively, a regulator could impose regulations on market participants.

The typical example of positive externalities refers to expenditures on research and development. If firms cannot capture all the benefits of this activity, they will innovate less than what would be socially optimal. In energy markets, this refers to innovation in generation technologies (e.g. technologies that increase the capacity factors of solar power), other types of energy conversion (e.g. innovation in electrolysis to produce hydrogen), energy storage (e.g. increasing efficiency of batteries), energy transport (e.g. making distribution grids more smart) or energy consumption (e.g. increasing energetic efficiency of gas boilers).

Some commodities with positive externalities are called *public goods*, which are commodities characterized by two aspects: non-excludability and non-rivalry. The first aspect means that no agent can be excluded from enjoying the commodity and, as a result, no one has an incentive to pay for it because it is more efficient for everyone to freeride on others for the supply of the commodity. Non-rivalry means that the consumption of a commodity by one consumer does not affect the consumption of others. The classical example of public goods is dike to protect against sea water. Once a dike has been built, everyone behind the dike is protected while consuming this protection by one agent does not affect the availability of it for others. As a result, dikes will only be built when there is collective arrangement how to share the costs. Examples of public goods in energy markets are the reliability of energy networks as discussed in Chap. 7.

8.2.3 Regulatory Measures

The various types of externalities can be addressed by various types of regulatory measures. In general, these measures can be distinguished in three groups: price-based measures, command-and-control measures and combinations of these two. Price-based regulation refers to measures which are meant to directly correct the financial incentives of economic agents, while command-and-control measures are directed at the volume of the inputs used or outputs produced by these agents. Examples of price regulation are subsidies for renewable energy and energy-conservation technologies, and taxes on the consumption of (fossil) energy. Examples of command-and-control regulation are environmental standards and quantitative constraints on the amount of production by specific technologies or plants. Examples of regulatory measures that have elements of both are emissions trading schemes based on a cap-and-trade design and renewable energy obligations.

Both subsidies for renewable energy and energy-conservation technologies, and taxes on the consumption of fossil energy can be effective instruments to change the decisions of economic agents in order to reduce the emissions of carbon. A common characteristic of these instruments is that they both consist of transfers of financial funds within a society. In principle, the transfer itself is not a cost, but only a redistribution of wealth among economic agents. The direct costs of these policy instruments are related to the transaction costs of implementation. In case of energy taxes, the transaction costs refer to the process of collecting taxes, while in the case of subsidies the transaction costs not only refer to the process of distributing the subsidies, but also to the process of organizing the financing of the subsidies.

Box 8.2 Regulatory measures to address positive externalities

Regulatory measures to address the problem of positive externalities can be distinguished in measures which raise the benefits for the producers and measures which reduce their costs. Both measures will result in a higher equilibrium output and, hence, reduce the deadweight loss of the positive externality.

The key example of a regulatory measure to raise the revenues for innovation activities is patent policy. This policy gives economic agents the sole right to exploit a specific innovation, which overcomes the risk that an innovator has that other firms will free ride on its innovation. Hence, the private demand curve shifts towards the social demand curve as the firm with a patent is able to serve the full market and, hence, realize higher volumes of sales and/or higher prices.

Another type of regulatory measure to address the presence of positive externality is to reduce the costs of producers. An example of such a measure is giving subsidies for renewable energy production which reduces their minimum required prices. Consequently, the supply curve shifts downwards and the equilibrium market volume is raised, which is intended to remove the deadweight loss of the positive externality.

A fundamental difference in the functioning of subsidy schemes and tax schemes, however, is how they affect demand. Subsidies for renewable energy make energy cheaper, which may result in a higher demand for energy (depending on the price elasticity of demand, see Sect. 2.4.7). A clear example of this mechanism is subsidies for electric cars. Besides stimulating the substitution of conventional cars by electric cars, they also stimulate people to drive more because driving by the latter type of cars has become cheaper (or: less expensive), while the costs of the former remain the same. Hence, economic agents are stimulated to drive more. This is the so-called rebound effect of subsidies (see Box 8.5). Another example of this effect is provided by subsidies for energy saving (e.g. insulation of houses) which may trigger people to increase the in-house temperature as energy has become cheaper because of the energy-saving measures. Such rebound effects do not exist in the case of taxes on fossil energy. On the contrary, besides stimulating economic agents to substitute away from fossil energy, it also has a negative effect on energy demand as energy has become more expensive.

Another adverse effect of subsidies is that subsidy programmes easily result in windfall profits, which is that subsidies exceed the amount which is needed to let agents make another decision. Hence, in such a situation, agents receive more financial compensation than economically required. These two effects plus the above-mentioned extra transaction costs of funding subsidy schemes make that economists generally have a strong preference for taxes above subsidies.

Command-and-control regulatory measures can be distinguished in constraints with and without the use of market instruments. Examples of the latter are environmental standards for production processes (e.g. in terms of maximum noise or emissions levels), energy-efficiency standards for the construction of houses, constraints on the use of inputs (e.g. a minimum level of biofuels in the production of gasoline) and obligations on retailers to purchase a minimum level of renewable energy (so-called renewable energy obligations). Such regulatory measures can be effective to reach a specific objective, but the risk is that the realization of these objectives is not efficient. This is related to the presence of information asymmetry between regulator and regulated agents, which makes it difficult to determine the most efficient technologies and behaviour to reach the regulatory objectives.

In order to increase the efficiency of command-and-control regulatory measures, it may help to allow the agents to trade with each other, i.e. by adding a market instrument. The most obvious example of such a measure is emissions trading, since a quantitative constraint is combined with the possibility to trade. Another example is to give retailers with a renewable energy obligation the option to buy and sell certificates.

Below we will further discuss the design of various types of regulatory measures to address externalities and how they may interact with each other.

8.3 Environmental Taxes

8.3.1 Pigouvian Tax

The theoretical solution to address negative externalities is to make use of so-called Pigouvian taxes (Sandmo 2000).[1] The fundamental idea behind this tax is to impose a tax on the agent who generates a negative externality, which makes that the supply curve shifts upwards and, consequently, the equilibrium output becomes similar to the one that would occur when the negative externality does not exist. As a result, the deadweight loss of the negative externality disappears because the external effect has been internalized. In terms of market outcomes, this means that the equilibrium output is lower and the prices are higher compared to the situation without this tax.

In theory, the tax imposed on the polluting agent should be equal to the marginal social benefits of a reduction of pollution for other agents. These benefits depend on the damage that is caused by pollution. In the case of carbon emissions, the benefits occur on a global and long-term scale as carbon emissions affect the global climate on a long period of time. In order to estimate the marginal benefits of emissions reduction, therefore, one has to discount the future effects, which makes this calculation highly sensitive to the value of the discount rate used (see Sect. 4.3.4). In addition to this, the marginal valuation of these environmental effects is also highly complicated and uncertain as the preferences of economic agents are often not explicit, while there is also often a large variation among the group of agents that are involved. Nevertheless, methods like Choice Experiments which were discussed in Sect. 4.3.3 can be applied to form an estimate of the (average) marginal benefit of changes in non-priced environmental qualities.

Because of the difficulties to set the environmental tax properly, in practice the tax is often indirectly related to the externality. In this approach, the tax level is chosen such that a predefined environmental objective can be realized. This more pragmatic view can be attributed to the economists Baumol and Oates (Speck

---

[1]This theory was first formulated by the English economist Pigou.

2013). A common way to do this is to set the tax on the use of fossil energy. If such taxes are not properly differentiated according the carbon content of energy, however, inefficiencies result as energy with a relatively low carbon content (e.g. natural gas) may be taxed to the same extent as an energy carrier with a higher carbon content (e.g. coal).

No matter how the environmental tax is determined, the primary objective of the tax is to influence the decisions of economic agents, not to raise the financial burden on polluting firms as is often assumed and misunderstood. It is often said that a tax on fossil energy can only be effective to reduce, for instance, carbon emissions when the revenues from the taxes are used to stimulate renewable energy or energy efficiency. This is, however, not correct. The key objective is to get the incentives right, the spending of net revenues from taxation is another topic, at least seen from a social-welfare perspective.

Box 8.3 Energy tax based on a relative benchmark

Energy taxes can be based on the amount of fossil energy, but this may have negative financial effects for the companies involved which cannot easily be repaired through a reduction of, for instance, corporate income taxes. As an alternative design, the energy tax can be related to the efficiency of companies, i.e. to let the companies only pay a tax on the use of energy which exceeds an efficiency benchmark. Such a measure gives incentives to the companies to become at least as efficient as the benchmark, without imposing a financial burden on the company.

A caveat of relative benchmarks as basis for energy taxes is, however, that they do not give incentives to companies to become more efficient than the benchmark. After all, when the efficiency level of the benchmark is achieved, the benefit of further increasing efficiency becomes zero.

Another caveat of a relative benchmark as basis for energy taxes is that it is only directed at productive efficiency, not at allocative efficiency. This follows from the fact that, in this design, companies can increase their use of energy without facing any higher tax burden as long as the energy use per unit of production is not higher than the benchmark. Hence, this tax design does not fully address the negative externality as this externality is generally related to the absolute levels of emissions, not to the relative levels per unit of output.

Another practical caveat of this tax design is related to the definition of the benchmark. If the benchmark is defined as the average level within the industry, about half of the firms (or: production) does not face any incentive as they are already operating on or below the benchmark. Moreover, the average can be calculated on various ways. When larger firms are more energy efficient than smaller firms, the industry will lobby for using the

Table 8.2 Alternative ways to determine relative benchmarks as basis for energy tax

Firm	Output	Energy use	Energy efficiency (energy use/output)
A	100	50	0.5
B	200	70	0.35
C	300	80	0.27

Unweighted average energy efficiency = 0.37
Weighted average energy efficiency = 0.33

unweighted average energy efficiency, as that sets the target at a less ambitious level. In the example of Table 8.2, only firm A has to pay energy taxes when the efficiency benchmark is based on the unweighted average energy efficiency.

During the process of choosing the benchmark, the industry will always lobby for a low (less ambitious) benchmark, such as the average benchmark in the global market if they know that the domestic industry has a relative high efficiency. Hence, an environmental tax based on a relative benchmark is vulnerable to rent-seeking behaviour of interest groups.

Governments may, however, have as an additional objective to change the tax base from labour or capital income to energy. By implementing such a change, a so-called double dividend may be realized. The idea behind this policy is that implementing a tax on the energy which economic agents use, government controls for the negative externality of fossil energy use, while lowering the tax on labour or capital income reduces the deadweight loss of taxation (Speck 2013). This deadweight loss of taxation results from the increase in the marginal costs of labour or capital resulting from taxation, which shifts the respective supply curves upwards, resulting in a deadweight loss in the labour or capital market, respectively. Hence, by replacing taxes on labour and capital income by taxes on polluting activities, like use of fossil energy, two benefits may be realized without raising the net tax burden on economic agents.

Hence, the challenge of designing a tax on fossil energy is to influence the marginal decisions of economic agents by raising their marginal costs without harming their financial position in general. A solution to this challenge is to redistribute the revenues from the energy taxes back to the producers and consumers who have to pay the energy taxes, but in a different way, for instance by lowering the tariffs in corporate or income tax. As a result, the economic agents do not pay net taxes to the government and the latter does not receive net income (i.e. the tax scheme is budget neutral), while the economic agents receive appropriate incentives to make other choices regarding the use of, for instance, fossil energy. In practice, however, this measure is more complicated as the firms which use a lot of energy do not necessarily pay a lot of corporate income tax, and the other way around. An alternative measure to reduce the financial burden on companies is using a relative benchmark for determining the tax base (see Box 8.3).

8.3.2 Differentiation in Tariffs

Because of the concerns on adverse effects of domestic environmental taxes on the competitive position of companies operating on international markets (the so-called exposed industries), governments usually give these companies exemptions or much lower tariffs than companies operating on domestic markets and households. This sometimes results in strong regressive tariff structures in which the latter types of users pay relatively high tariffs per unit and the former type of users much lower tariffs. Although such designs are meant to prevent adverse effects for the energy-intensive industry, they also result in inefficiencies in how emissions are reduced.

Regressive environmental tax tariffs result in productive inefficiencies as they favour some more expensive reduction options above less expensive options. This can be illustrated by looking at the marginal abatement cost curves. These curves rank all options to abate an environmental effect (e.g. technologies to reduce carbon emissions) from the lowest levelized costs (LCOE) to the highest. Such curves can be constructed for every industry. Sometimes, these curves include a number of options with negative marginal costs, which means that these options with a positive environmental effect could be taken with a positive net revenue. Examples of such options are some energy-saving measures. Figure 8.3 gives examples of the marginal abatement costs curves of two industries, A and B. In this example, industry A has cheaper options than industry B to reduce carbon emissions.

Suppose the government has imposed a tax on carbon emissions of 100 euro/ton for industry B and of 50 euro/ton for industry A, the latter industry may be viewed to be an exposed industry and has to be protected against adverse competitive

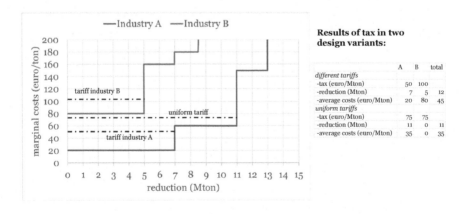

Fig. 8.3 Inefficiencies of different tax tariffs illustrated through marginal abatement costs curves

effects of environmental taxes. This differentiation in tariffs may be the result of intensive lobbying by industry A or just governmental concerns about the international competitive position of this industry. Anyway, because of this design of the environmental taxes, industry A will only take measure that costs less than 50 euro/ton, resulting in an emission reduction by this industry of 7 Mton. Because industry B faces a twice as high tax tariff, this industry takes all measures which costs less than 100 euro/ton. Hence, industry B is taking relatively expensive measures, in this example of 80 euro/ton. In total, this industry realizes 5 Mton emission reduction. These reductions by this industry occur while less expensive measures in industry A remain unused.

If both industries would be subject to the same tax level of, for instance, 75 euro/ton, then industry B would not take its relatively expensive measures, while industry A would take more measures. As a result, in this example, industry A would reduce 4 Mton more, while industry B reduces 5 less and would not do anything. The benefit for productive efficiency, hence, is that more expensive reduction options (costing more than 75 euro/ton, which is the uniform tariff) are replaced by options that costs less. So, the average costs of reducing emissions go down, in this example, by 10 euro/ton (from 45 to 35 euro/ton). Hence, the change in productivity efficiency (PE) due to a change in the design of tax tariffs can be calculated as the size of emission reduction (Q^e) times the difference in average costs (C^e) in both tariff systems:

$$\Delta PE = Q^e * \left(C^e_{different\,tariffs} - C^e_{uniform\,tariffs} \right) \tag{8.1}$$

The overall level of emission reduction in the case of uniform tariffs is, however, slightly lower, which illustrates that the quantitative effect of an environmental tax is uncertain as it depends on the characteristics (i.e. the marginal costs) of the agents subject to the taxes. As the overall emission reduction is still 11 Mton, while the decrease in average costs per unit is 10 euro/ton, the overall societal savings are 110 million euro (i.e. 11 Mton × 10 euro/ton).

In order to fully assess the different tariff schemes, one has to compare this loss of productive efficiency in reducing carbon emissions with the welfare loss due to reduced international competitiveness of the exposed industry. Only when the latter exceeds the productive efficiency loss, such a differentiation in tax tariffs is welfare improving.

Box 8.4 Internal carbon prices

Not only governments (public agencies) can incentivize economic agents to internalize negative externalities, but they can also do it themselves. An increasing number of firms is doing this by applying so-called internal carbon prices (ICP). By these prices, they use an internally determined price for carbon emissions in their economic analysis of investment projects. A motivation to do this can be that these firms anticipate future environmental regulation and that they want to reduce the risk of being locked-in in carbon-intensive production process. Hence, by assuming that there will be a higher costs of emitting carbon in the future through an ICP, firms give more weight to those activities with less carbon emissions.

8.4 Support Schemes

8.4.1 Multiple Objectives of Support Schemes

The opposite regulatory measure of taxing a negative environmental effect is giving support to a positive alternative. In case of energy this means, for instance, giving subsidies to alternative energy carriers (e.g. renewable energy) which do not have a negative externality or to promote energy conservation. The optimal level of subsidies for renewables or energy conservation is, in theory, also related to the value of (reducing) the negative externality (Perman et al. 1999). If the subsidy for renewable energy or energy-conservation technologies would exceed this value, it may result in a welfare loss for society. This will happen if the subsidy stimulates technologies which costs more than the size of the marginal benefits for the environment. Hence, one condition for setting the value of the subsidy (S) is that it should not exceed the marginal social benefit (MSB)

$$S \leq MSB \tag{8.2}$$

Another condition for the level of subsidies is that it should not result in windfall profits for the recipients of the subsidy. Hence, the subsidy should not be higher than the amount actually needed to make a project to become break even. This amount can be determined as the difference between the LCOE and the marginal private benefit (MPB) that can be realized in the market:

$$S \leq (LCOE - MPB) \tag{8.3}$$

The marginal private benefit can be seen as the average price per unit that can be realized in the future. This price depends on the distribution of the future production (Q) and the discounted value of the prices that can be realized for the various parts of the future production:

$$MPB = \frac{\sum_t \frac{P_t}{(1+r)^t} * Q_t}{\sum_t Q_t} \tag{8.4}$$

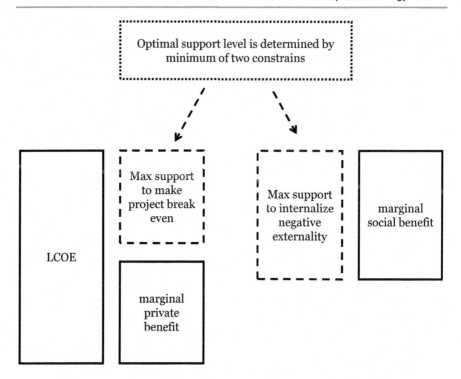

Fig. 8.4 Optimal support level depends on value externality and break-even constraint

Note that in the case of renewable electricity, the marginal private benefits of selling electricity are not equal to the electricity price, as the average price that can be realized by wind turbines and solar panels are generally lower than the average price in the market. This effect is called the market-value effect (see Sect. 7.5.3).

From the above follows, that the subsidy for measures that reduce the negative externality, such as renewable energy projects, should be set at such a level that it is not higher than the marginal social benefit and also not higher than the amount really needed to make a project break even (see Fig. 8.4). If for a project, the maximum support to make the project break even exceeds the value of the negative externality, then subsidizing such a project would result in a loss of welfare. On the other hand, when a subsidy is set at a higher level than needed to make a technology break even, it results in a windfall profit.

When subsidy levels are differentiated for different types of projects, it may also result in productive inefficiencies. These two effects can be illustrated through the marginal abatement cost curve on industry level (see Fig. 8.5). This abatement cost curve ranks all options in an economy from lowest marginal costs to the highest. In the example of this figure, the options A and B can be taken with a positive net private benefit, while the options C until F are costly to the economic agents. Without any additional compensation, these latter options are not taken by rational

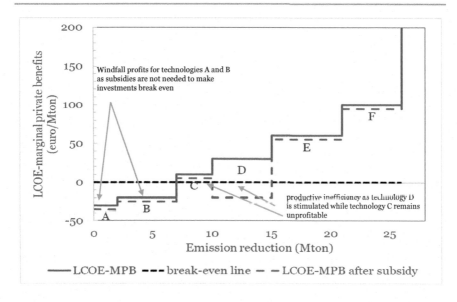

Fig. 8.5 Inefficiencies and windfall profits due to subsidies scheme illustrated with a marginal abatement cost curve on economy-wide level

economic agents, while the former options would not require any support when the agents behave rationally.

Suppose that the government introduces a subsidy scheme which offers all options 5 euro subsidy per ton of carbon reduction, but that, for some reason, such as societal preferences, the government decides that technology D receives ten times as much: 50 euro/ton. Such a design of the subsidy scheme results in two effects: a productive inefficiency and windfall profits. The first effect occurs because now technology D will be implemented, because it has become profitable, while technology C is still unprofitable, despite its much lower LCOE. The windfall-profit effect occurs as a result of the subsidies given to technologies A and B which were already profitable without any subsidy, and also because the subsidy level for technology D is (much) higher than required to make it break even.

Summarizing, when looking for the optimal design of support schemes, regulators have to assess a number of trade-offs (Table 8.3). In general, support schemes are meant to trigger investments in renewable energy at the lowest costs for society. Hence, the scheme should not only make it attractive for investors to invest in renewable energy projects, but also stimulate them to innovate and to improve the productivity of their technologies. In addition, support schemes should trigger the cheapest options before other, more expensive options are chosen. Moreover, support schemes should not result in supra-normal profits for some economic agents, which means that the support should not be higher than the actual costs of the investments. Finally, support schemes should not trigger inefficient behaviour, such as producing energy when there is no actual demand for it.

Table 8.3 Criteria for the design of support schemes

Criteria	Measures
Triggering most efficient investment options	Choose options with the lowest LCOE
Stimulating productive efficiency	Subsidies should be based on exogenous information
Preventing windfall profits	Subsidies should not be higher than the LCOE
Preventing allocative inefficiencies	No subsidies if market/economic value is negative + subsidy per unit should not be higher than (absolute) value of negative externality

8.4.2 Design Choices of Support Schemes

Depending on their objective, regulators can choose from several types of support designs (see Table 8.4). When effectiveness is the key priority, the support scheme should make the investment in renewable energy attractive by removing the risk from the investor and guaranteeing a return on investment which is at least equal to the return that can be realized through an alternative investment. A support scheme able to do this is the so-called Feed-in-Tariff scheme (FiT). In such a scheme, that was used, for instance, in Germany in the period until 2012, guaranteed prices are paid for each unit of renewably generated power during the lifetime of the investment. These subsidies are only related to the LCOE of the generation plants,

Table 8.4 Financial support schemes for renewable energy

Type of support scheme	Characteristics
Price schemes	
• Feed-in-Tariff (FiT)	Fixed subsidy per unit of production; producers sell their energy to the network operator
• Feed-in-Premium (FiP)	Fixed subsidy per unit of production, based on Levelized Costs of Energy (LCOE) and expected or realized energy prices (including several variants, such as a cap and floor on subsidy, and conditions regarding level of electricity price); producers sell their energy on the market
• Netting of consumption and production of so-called prosumers	Consumers can subtract their own production during a period (e.g. a year) from their consumption in that period; subsidy results from the taxes on consumption which don't have to be paid anymore
Combination of price-based schemes and command-and-control measures	
• Auction for support	Tendering of a limited budget; support can be based on FiT, FiP or sFiP; or tendering of locations for renewable energy projects
• Renewable energy obligations	Obligation on energy users or retailers to comply with the rule to have a minimum percentage of renewable energy

without taking into account the market price of energy that will be produced. This type of support is highly effective as it gives a strong incentive to invest in renewable energy, provided that the tariffs are sufficiently high to compensate for the costs of capital of that investment. The downside of this measure, however, is that renewable energy producers receive no financial incentive to work as efficiently as possible, but only to produce as much as possible.

Investors in renewable energy can be given incentives to operate efficiently by using a Feed-in-Premium (FiP) support scheme. In such a scheme, producers are supposed to sell their power on the market themselves and receive a subsidy for the difference between the average annual power price and LCOE. This scheme reduces the financial burden of paying subsidies for the government, as now only the difference between costs and market prices is subsidized.

Box 8.5 Rebound effect of subsidies for energy efficiency

As the level of energy demand is a major factor behind the overall level of carbon emissions (see Sect. 2.4.7), increasing the efficiency of energy use can be an effective instrument to curb these emissions. Therefore, governments give subsidies for measures which reduce the energy use, for instance for heating or cooling a house. These subsidies can, for instance, be provided as a contribution to the expenditures for insulation of houses. Such subsidies, however, may be partly offset by the so-called rebound effect. As insulation a house implies that less energy is needed to heat or to cool it, heating and cooling become cheaper. Generally, when products become cheaper, consumers increase their consumption, which is indicated through the negative value of the price elasticity of demand (see Sect. 2.4.7). Hence, the higher the price elasticity of demand, the stronger the rebound effect of measures like subsidies for insulation. This is the so-called direct rebound effect (Sorrel 2009).

Besides its direct effect on consumption of energy, the reduced costs of energy use may also result in an indirect rebound effect. This effect results from the increased net income of consumers because they have to spend less on, for instance, heating and cooling. When these cost savings are used for other consumption, such as making more long-distance journeys with a car, it will result in a higher use of other types of energy. In addition, when industries realize a higher energy efficiency, the overall productivity may increase and result in lower product prices and, hence, higher consumption of the products. The overall, economy-wide rebound effects are estimated at about 50% of the initial savings through the energy-efficiency measure (Sorrel 2009).

FiP schemes can be designed in more advanced ways to pursue more objectives. An advanced design is in which the premium is adapted frequently in relation to the

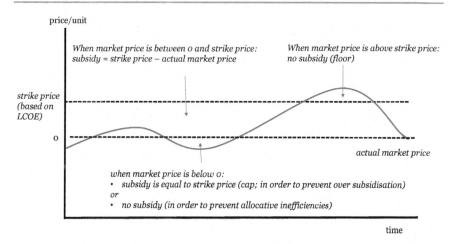

price/unit

When market price is between o and strike price:
subsidy = strike price – actual market price

When market price is above strike price:
no subsidy (floor)

strike price
(based on
LCOE)

o

actual market price

when market price is below o:
• subsidy is equal to strike price (cap; in order to prevent over subsidisation)
or
• no subsidy (in order to prevent allocative inefficiencies)

time

Fig. 8.6 Design of sliding (or: floating) Feed-in-Premium scheme with floor and cap

actual market price in order to reduce the risk of windfall profits (see Fig. 8.6). The basis for the support level is the so-called strike price which is an estimate of the actual costs of the renewable energy plant, generally measured through the LCOE. If the realized market price exceeds the strike price, the support level can be set to zero, as no support is needed as compensation for the costs. This is the floor in the subsidy scheme. If the realized market price is below zero, the subsidy can be set equal to the strike price, in order to prevent over subsidization. This is the cap in the system. Instead of this cap on the support, the support could also be set to zero during periods of negative prices, as the actual value of the energy is negative as well in those periods. The latter rule may be appropriate when oversupply has negative effects on the network causing local bottlenecks, as discussed in Sect. 7.3. When there is no appropriate congestion management scheme in place which gives incentives to producers of renewable energy to take congestion into account, such subsidies could results in inefficient network usage. In addition, if negative electricity prices are due to inflexibility of conventional production (such as coal-fired power plants), then subsidizing renewable energy during hours of oversupply does not result in less conventional fossil production (because of their inflexibility), but only stimulates electricity consumption through its negative effect on electricity prices.

In order to foster productive efficiency, two elements can be added to the support scheme: a limited budged per period and dividing the period in a number of phases in which the subsidy level increases over time. The limited budget gives an incentive to firms not to wait to long for claiming support, which means that they have an incentive to ask for the subsidy as soon as the amount is sufficient to make their investment break even.

Box 8.6 Auctioning of renewable energy subsidies or locations

Instead of offering a fixed subsidy for a specific renewable energy project, governments may also organize a beauty contest in which potential investors have to submit their plans and which are consequently assessed on a number of criteria. Another regulatory tool is to use an auction in which the bidders have to indicate which amount of subsidy they require or which price they are prepared to pay.

In these systems, the size of the project is determined by the requirements the governments have determined (such as the magnitude of the installed capacity on a particular place). In case of an auction, the price to be paid or received by the government is endogenous. Both the beauty contest and the auction promote competition among potential developers of such projects, which may foster productive efficiency.

This kind of support schemes is increasingly used for allocating slots for investments in offshore wind parks in the North Sea region. The UK, for instance, has organized auctions, while the Netherlands has organized beauty contests.

Participants interested in building an offshore wind farm compete for the lowest subsidy fee. When they formulate their bids, the investors not only have to assess the costs of building a wind farm, but also the future electricity prices that can be captured by the wind park and the future production.

Another support scheme with exogenous prices is the net-metering scheme. A net-metering scheme gives consumers the option to offset their power production (i.e. the feed-in of electricity into the network) with their electricity consumption on an annual basis. Hence, a net-metering scheme gives a reward for the production by renewable energy sources on an annual basis, while all prices in the electricity market are related to the (expected) real-time situations on a 15-min basis. As was shown in Sect. 3.3.4, a net position on an annual basis is not a relevant quantity in the wholesale market.

Because of this offset fiction, consumers do not receive the wholesale price for their electricity production, but the average annual consumer price in the retail market. The latter price is based on a number of components: the average wholesale price retailers have to pay, the costs and markup of retailers, plus any tax on energy use, such as an energy tax and a renewable energy surcharge, and the value-added tax on the retail energy bill. This means that the financial compensation on electricity production by residential households can be significantly higher than the price that regular electricity-producing firms receive in the wholesale markets.

This high payment is directly associated with the relatively high rates in the energy taxes for residential users. Since the LCOE of solar PV is gradually going down and the financial compensation in this support scheme is not based on those costs but, among other things, on the increasing taxes on energy consumption, making use of this scheme may become increasingly appealing to residential consumers, but it may also result in increasing windfall profits.

Such net-metering schemes have as an additional effect that consumers who generate electricity themselves are paying less energy tax and, consequently, are contributing less to the financing of public-sector spending, including expenditure to encourage the use of renewable energy. While the net-metering scheme may increase support for energy transition among residential users who make use of the scheme, it may decrease support among other consumers who do not benefit from it, but are faced with higher taxes.

In short: the net-metering scheme contains few incentives to make efficient investments in renewable energy, while the risk of supra-normal profits is positively related to the level of the tax on energy consumption. Besides that, consumers are not stimulated to take the actual conditions in the wholesale market into account when they generate power, because the revenues are based on the balance of their annual consumption and production.

Box 8.7 Support schemes in Germany and the Netherlands

Countries choose different types of support schemes for renewable energy. Germany, for instance, initially introduced a FiT scheme, later on replaced by a FiP scheme, while the Netherlands introduced several variants of FiP schemes.

In Germany, the FiT costs have significantly increased after 2012 due to the increase in subsidies granted for solar energy. While solar power represents only a quarter of the total renewable production in Germany, it was paid half of the total amount in subsidy granted in 2014. Although solar power premiums have come down considerably, 300 euros/MWh in premiums were still paid under the FiT scheme in 2014. This amount was about 75 euros for onshore wind energy and about 150 euros for offshore wind energy.

While the previous Dutch support scheme (MEP) made a subsidy amount available for each separate technology, the SDE(+) regime (started in 2013) requires the individual technologies to compete for the subsidy budget which was made available. As a result, the premiums per unit of generated renewable energy have gone down. In this system, the subsidy amount gradually increased over a subsidy period, so only the most efficient technologies will apply for a subsidy at the beginning of the subsidy period. In the course of the subsidy period, when the subsidy amount gets higher, the subsidy scheme starts to become appealing to economically less efficient technologies as well. Partly as a result of this change made to the scheme, the costs per unit of generated renewable power decreased compared to the old scheme (ECN 2016).

Another quantitative support scheme is the renewable energy obligation. Such a scheme was introduced in the United Kingdom in 2005. Energy retailers in this system were obliged to buy a minimum percentage of renewable energy. They must buy certificates from renewable energy producers, who receive these certificates for the electricity which they produce. In this system, the different technologies are treated similarly and, as a result, the less efficient technologies are hardly used and the average premium for renewable energy is low. Consequently, the share of solar energy in the UK energy mix was extraordinarily low (1%), whereas Germany had a 25% share of solar energy.

These examples show that the cost of stimulating green energy can be reduced when market participants are given more freedom of choice, as in a quota system, or when they are allowed to compete for subsidies. Where there is no competition for subsidies and the compensation level is more or less guaranteed, as in the net-metering scheme and the FiT scheme, there is a significant risk that less efficient technologies will be promoted and that society will pay more for renewable energy than what is necessary.

8.5 Emissions Trading

8.5.1 Primary and Secondary Allocation of Allowances

Emissions trading schemes are quantitative environmental policy measures with endogenous prices which means that this type of policy measure has precisely the opposite design as environmental taxes. In theory, however, both environmental policy measures are equally efficient in internalizing negative externalities (Tietenberg 2006). Emissions trading schemes can be distinguished in cap-and-trade schemes and baseline-and-credit schemes (Nentjes and Woerdman 2012). Here, we focus on cap-and-trade schemes, as this is the one introduced in the EU for greenhouse gas emissions.

The basic principle of a cap-and-trade emissions trading scheme is that the government sets a cap on the total amount of environmental effects (e.g. carbon emissions) and that this cap is translated into emission rights (or: allowances, permits), which are allocated among all those economic agents which have been made subject to the scheme (which is the so-called primary allocation) and that these agents can trade in these allowances (which is called the secondary allocation) (see Fig. 8.7).

In theory, the total cap on the environmental effect has to be set at that level at which the marginal costs of reaching it are equal to the marginal benefits. After all, a more ambitious objective (i.e. a lower cap) would result in higher marginal costs than the marginal benefits, which results in a welfare loss. The same would occur if the objective is less ambitious. In practice, however, the caps are based on policy objectives regarding future emission levels, which may be based on what is viewed to be sustainable levels, although information on what is seen as the optimal policy

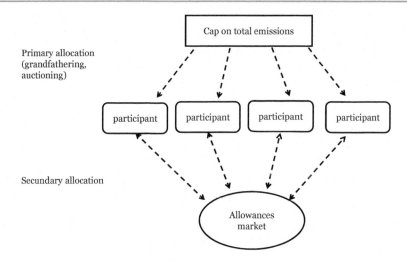

Fig. 8.7 Allocation of allowances in emissions trading scheme

from an economic point of view may also be used. In climate policies, these caps are related to objectives regarding the maximum concentration of greenhouse gases to keep global temperature increase beneath a specific percentage.

When the cap has been set, for each unit of emission (e.g. ton of carbon) an allowance can be created. The primary allocation of these allowances can be done in different ways. The basic choice options regarding the primary allocation method consist of either giving the allowances for free to the participants of the scheme or to auction them. Free allocation can be done through grandfathering, in which case the allocation is based on historical levels of emissions of the participants, or through an allocation scheme based on relative emissions levels (in comparison to a benchmark).

In both methods of free allocation, there is no immediate price for the allowances, but this price emerges when the participants start trading with each other in the allowances market. The allocation on this market is called the secondary allocation. In this market, various types of products are traded, such as long-term and spot contracts. As discussed in Sect. 3.2, economic agents may be prepared to pay a premium for a forward contract that mitigates the price risk. In this case, buying parties may be willing to reduce the risk of high spot prices, because they need to have sufficient allowances in order to be compliant with the regulation at the end of a year. By buying allowances in advance (i.e. physical forwards) or by hedging the price risk related to spot prices (i.e. financial forwards), they can mitigate the risk that they have to pay high prices for allowances at the end of the year.

8.5.2 Determination of Allowances Price

When the primary allocation is done through an auction, then this results immediately in a carbon price through the interaction of demand and supply. The demand
for allowances results from the legal obligations of participants to be compliant with
the imposed rule that the total level of annual emissions may not be larger than the
total level of allowances they possess at the end of the year. If the participants are
not compliant, they will face a financial penalty. In order to prevent such a penalty,
participants want to buy allowances, which results in a demand curve. This curve is
determined by the willingness-to-pay (WTP) for allowances of all participants. This
demand curve is the reciprocal of the marginal abatement cost curve. The latter
curve shows all abatement options ranked from the lowest to the highest costs per
unit (see Fig. 8.8).

To analyze this further, lets suppose that each of the technologies of the
abatement curve is operated by a different firm. Hence, each of the technologies
stands for a different firm. Firm D, for instance, can reduce its carbon emissions at
30 euro/ton and, as a result, for this firm it would be efficient if it could buy an
allowance for less than 30 euros. Hence, the maximum price this firm is prepared to
pay for an allowance is 30 euro/ton. In this example, firm F has the most expensive
options to reduce carbon emissions and, therefore, it has the highest WTP for an
allowance. So, by using the marginal abatement costs per firm, one can determine
the demand curve in the market for carbon allowances (see Fig. 8.9).

The supply curve in the market for the primary allocation of allowances is
determined by the regulator who sets the total amount of allowed emissions. From
the intersection of demand and supply curve, the equilibrium price for carbon
allowances results. In this example, the interaction between the demand curve based

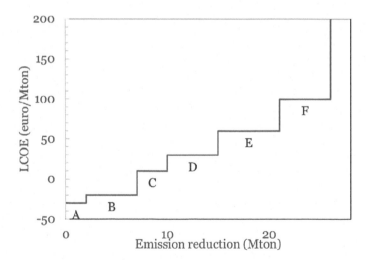

Fig. 8.8 Example of marginal abatement cost curve

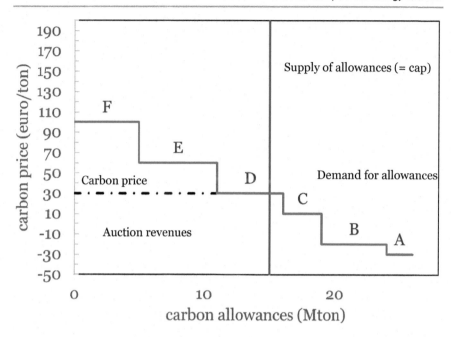

Fig. 8.9 Demand for and supply of allowances in market for primary allocation

on the marginal abatement cost curve and an emissions cap of 15 Mton results in an equilibrium carbon price of 30 euro/ton. This means that all firms having more expensive options to reduce emissions will buy allowances (i.e. firms F, E and D), while firms with less expensive options will not buy but choose to utilize their own reduction options (i.e. firms C, B and A). The firms F, E and D will buy allowances at a price of 30 euro/ton, which is more efficient for them than using their own reduction options. As a result, the total level of emissions is reduced to 15 Mton. In addition, because of the auctioning of the allowances, the government receives (15 Mton × 30 euro/ton=) 450 million euro.

Except for these auction revenues, the same result would result in case of grandfathering, at least in theory. Free allocation of emission allowances can be viewed as a lump sum subsidy, which means that the firms do not need to pay for the allowances and, hence, receive a subsidy equal to the value of the allowances. The incentive to curb emissions, however, is only determined by the marginal costs of emission reduction which is equal to the price of CO_2 allowances. This price is determined by the marginal costs to curb emissions up to the emissions ceiling, and these costs are the same in both forms of allocation, provided that the allowances market works efficiently.

After the primary allocation, the participants in this scheme can trade. This is called the secondary allocation. As an example, suppose Firm E wants to extend its activity which results in more emissions. Then, this firm has to choose from two options: either to buy more allowances or to utilize its own reductions options which have not been used yet. The marginal costs of the latter option remain, in this example, 60 euro. If this firm is able to buy allowances at a lower price, that would be more efficient. In this example, firm E can ask firm D to buy some of his allowances. Firm D is willing to do so if the price he can receive exceeds his marginal abatement costs, which are 30 euro/ton in this example. Hence, any price between 60 and 30 euro/ton can be an equilibrium price which is beneficial for both parties (see Fig. 8.10).

In this example, we assumed that there are only two players, which implies that a transaction is a bilateral one, based on negotiation. If firm D would represent a large number of similar firms with similar technologies, then the negotiation position of an individual firm is negligible, and consequently, the transaction price will be close to the marginal abatement costs of these firms. The negotiation position of firm D would be very strong, when it is the only firm able to supply the allowances at that price while there are many firms like firm E that want to buy them. In such a situation, the allowance price will be close to the maximum price the buyers are prepared to pay.

8.5.3 EU Emissions Trading System

The largest emissions trading scheme worldwide is the EU Emissions Trading System (EU ETS) which has been introduced in 2005. This cap-and-trade scheme is viewed to be one of the corner stone instruments of the EU to curb greenhouse gas

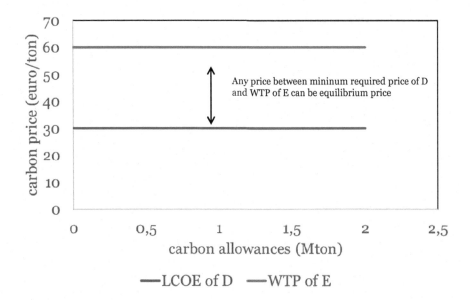

Fig. 8.10 Demand for and supply of allowances in market for secondary allocation

emissions. Since its start, the design has been further developed by extending the number of participating countries, industries and gases (see Table 8.5).[2]

The EU objectives regarding the future aggregated levels of greenhouse gas emissions were initially allocated to the sectors that are subject to the European emissions trading system (the ETS sectors) and the other sectors (the so-called non-ETS sectors). Subsequently, the objectives for both groups were allocated to the EU Member States in relation to the relative level of economic development. Since a number of years, the determination of the total amount of emissions is centrally done on European level. In the early stages of the ETS, the allowances were allocated for free, but they are increasingly sold by auction. The electricity sector is one of the ETS sectors, which means mandatory participation in ETS for all electricity generation in power plants above a certain lower limit.

Initially, the allocation was organized in a decentralized manner based on national allocation plans. This contributed to an oversupply of allowances. Other factors which resulted in the supply exceeding the demand for allowances were the economic crisis in 2008–2011, the high import of international credits and the promotion of renewable energy (see also Sect. 8.6).

In order to address this oversupply, a number of measures were taken in the 3rd phase (period 2013–2020). First of all, the allocation became centrally organized based on the EU wide objectives regarding emissions of greenhouse gases. In addition, the mechanism of backloading was introduced which enabled the EU to shift the auctioning of a part of the allowances to later years. This measure does not result in a structural lower level of allowances, but it only affects the distribution of the allowances over time. In order to structurally affect the number of allowances in case of an oversupply, the Market Stability Reserve (MSR) was introduced. This is a quantity-based adaptation of the system to solve the problem of low carbon prices (Hepburn and Teytelboym 2017). This reserve is meant to improve the resilience of the trading system to major shocks in the demand by adjusting the supply of allowances. As a result, the carbon prices are expected to become less volatile, which may give more certainty to investors.

The operation of MSR is fully based on predefined rules, which means that there is no political discretion. The withdrawal of allowances from the market or the release of allowances from the reserve is based on the so-called Total Number of Allowances in Circulation (TNAC). This number is equal to the supply of allowances in a year minus the demand for allowances and the number allowances in the MSR. The supply of allowances is based on the number of allowances auctioned or grandfathered, plus the allowances banked in the previous phase plus the allowances reserved for new entrants. The demand for allowances is equal to the level of verified emissions minus the cancelled allowances. When the TNAC exceeds the number 833 million, allowances will be withdrawn from the market by auctioning less, while allowances will be released from the reserve when the TNAC is below the number of 400 million. As from 2023, allowances in the MSR which exceed the volume of the auction volume of the previous year will no longer be valid.

---

[2]Source: https://ec.europa.eu/clima/policies/ets_en.

Table 8.5 Development of the EU Emissions Trading Scheme since its start in 2005

	Phase I (or trading period I) (2005–2007)	Phase II (2008–2012)	Phase III (2013–2020)	Phase IV (2021–2030
Cap for fixed installations	EU wide cap established in a decentralized way as sum of national allocation plans NAPs)	Overall cap was about 6.5% lower than in 2005	Introduction of centralized system to set the EU wide cap for stationary installations Cap reduced by annual linear reduction factor: 1.74%	EU wide cap annual linear reduction factor of 2.2%, based on the 2030 target of 40% reduction
Cap for aviation	–	–	Only for flights within the European Economic Area	Linear reduction factor of 2.2%
Industries	Electricity and energy-intensive industries (only firms with installations above a minimum size)	Aviation (flights within Europe) was added in 2012 Airlines can surrender general allowances or aviation airlines, other industries can only surrender general allowances		Other industries may also surrender aviation allowances
Countries	EU Member States	EU Member States + Iceland, Liechtenstein, and Norway		
Gases	CO_2	CO_2, N_2O	CO_2, N_2O and PFC's	
Penalty for non-compliance	40 euro/ton	100 euro/ton (in euro of 2013) (so, nominal value increases annually with rate of inflation) plus: the shortfall in compliance (i.e. difference between verified emissions and allowances) is added to the company's target for the next year		
International links		Business can buy, with restrictions, international credits based on Clean Development Mechanism (CDM) or Joint Implementation (JI) projects		Use of international credits will be replaced by linking EU ETS with other trading schemes
Allocation method	Almost all allowances were grandfathered	Free allocation about 90%	About 60% was auctioned	Free allocation will be continued for internationally competing industries
Auction platforms			EEX (European Energy Exchange), Leipzig and ICE Futures Europe in London	
Allocation of allowances over Member States	Based on national emission targets, translated into national allocation plans (NAPs)		88% based on verified emissions in 2005–2007	90% based on verified emissions 10% to less wealthy Member States

(continued)

Table 8.5 (continued)

	Phase I (or trading period I) (2005–2007)	Phase II (2008–2012)	Phase III (2013–2020)	Phase IV (2021–2030
			10% to less wealthy Member States 300 million allowances in New Entrants Reserve	
Use of auction revenues			At least 50% of auction revenues have to be used for energy and climate purposes	
Measures to address surplus	Overallocation resulted in very low carbon prices	Economic crisis resulted in surplus of emission allowances	2014: backloading to reduce surplus: postponing the primary allocation of 900 million allowances until the end of this phase 2019: Market Stability Reserve	2023: Allowances in the MSR above the previous' years auction volume will no longer be valid
		Possibility of **banking** between trading periods and to borrow within a trading period		

Source https://ec.europa.eu/clima/policies/ets_en

Finally, in order to realize the climate-policy objective to have lower emissions of greenhouse gases, the cap on the allowances is annually reduced. From 2021 onwards, the cap is reduced by 2.2% per year.

Box 8.8 Carbon price and effectiveness of emissions trading schemes

In discussions on ETS, its effectivity is often measured by the level of the CO_2 price. The fact that for many years, this price was fairly low and considerably lower than expected is sometimes wrongly seen as a sign of malfunctioning of the ETS (see Fig. 8.11). However, the performance of ETS must be judged by (a) the degree of emission reduction and (b) emission being realized as efficiently as possible.

Since the start of this system, emissions in ETS have been curbed by nearly 3% annually, more than necessary according to the ceiling (Fig. 8.12). Therefore, ETS has been effective in reducing the greenhouse gas emissions. The level of the allowances price must be seen as an indication for the marginal costs of reaching the emission target. The lower the price, the cheaper it is for society to achieve the intended emission reduction. The low prices in the past, as shown by this Figure, however, were partly due to overallocation, which is that Member States were too generous in their

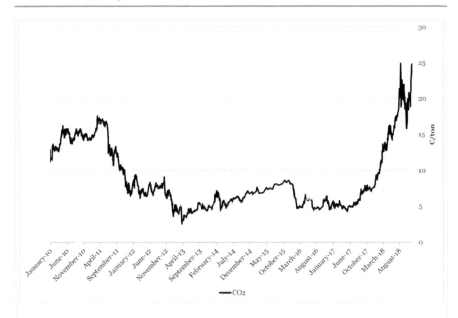

Fig. 8.11 Carbon price in the EU ETS, 2010–2019. *Source* Bloomberg

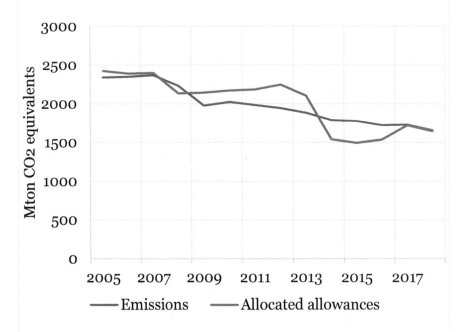

Fig. 8.12 Greenhouse gas emissions and allocated allowances in ETS in EU, 2004–2018. *Source* European Environmental Agency

allocation. From that perspective, these low prices can be seen as indirect indication that the system was not functioning well at that time.

8.6 Interaction Among Environmental Measures

All of the above regulatory measures are meant to intervene in energy markets and may, therefore, interact with each other when they are implemented at the same time. This interaction of these various types of environmental regulation may affect the overall effectiveness (see Table 8.6). In order to understand this, we need to analyze how each measure affects energy markets. Let's first look how an emissions trading scheme may affect the electricity market.

The participation of the electricity sector in an emissions trading scheme has consequences for the supply curve of electricity as the marginal costs of producing electricity increase by the marginal costs of carbon emissions (i.e. price of the allowances). This magnitude of this increase depends on two factors: the size of the carbon price and the carbon intensity of electricity generation per type of plant. As the carbon intensity of coal-fired power generation is about twice as high as the intensity of an CCGC plant (see Sect. 2.4), its marginal costs increase twice as much when the carbon price increases (see Fig. 8.13).

The effect on the electricity price of an increase in the carbon price is determined by the increase in the marginal costs of the system-marginal plant, i.e. the price-setting plant (see Sect. 3.3.4). If this plant is an CCGC plant, coal-fired power plants will face a higher increase in costs than an increase in revenues, while the CCGC plant is fully compensated for the increase in costs (hence, they are hedged against the risk of higher carbon prices). If the CCGC plant remains the

Table 8.6 Interaction between environmental policy measures

Effect of instrument below on the effects of instruments on the right	Emissions trading	Subsidies for renewable energy	Carbon tax
Emissions trading	x	higher price of electricity, less subsidy required	less emission reduction per unit of tax as relative effect is smaller
Subsidies for renewable energy	lower carbon price, no effect on emissions	x	higher emission reduction per unit of tax as alternative is cheaper
Carbon tax	lower carbon price, no effect on emissions	higher electricity price, less subsidy required	x

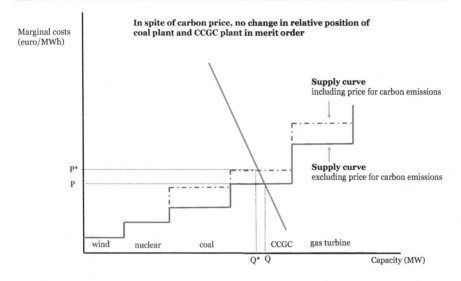

Fig. 8.13 Impact of a carbon price on electricity market (variant 1)

system-marginal plant, the generation mix is hardly affected in the short term as the coal-fired power plant keep producing, although their (inframarginal) profits are reduced. The latter implies that the incentives for investments in such plants have been reduced, as the remaining inframarginal profit may not be sufficient (anymore) to invest in new coal-fired power plants. Hence, in the long term, a higher carbon price may have consequences for the type of investments in generation assets (see Sect. 7.5). Assets with lower carbon emissions (such as CCGC) or no carbon emissions (such as wind turbines) may become more attractive economically (see Sect. 7.5).

In the short term, a change in the dispatch of power plants occurs when the increase in the carbon price is sufficiently high to result in a change within the merit order. Whether this happens does not only depend on the level of the carbon price, but also on the spread between the prices of coal and natural gas, corrected for the thermal efficiencies of the respective power plants. The metrics to compare gas and coal plants which also account for the costs of carbon emissions are the so-called clean spark spread and clean dark spread (see Sect. 3.3.4).

When the coal-fired power plants and the CCGC plants change positions within the merit order, then another type of plant may become the marginal, price-setting plant (see Fig. 8.14). Such a change, which is called fuel switch, does not only result in a higher price of electricity, but also in a relatively lower production level by coal-fired power plants, resulting in a lower amount of carbon emissions.

Let's now turn to how subsidies for renewable energy affect the electricity market. In Chap. 3, we have seen that the supply by renewable energy sources as wind and solar power is positioned at the left side of merit order due to their lower marginal costs. As a result, when subsidizing renewable energy results in more renewable energy capacity, the merit order is shifted to the right, which has a

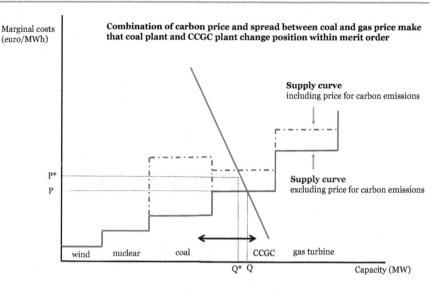

Fig. 8.14 Impact of a carbon price on electricity market (variant 2)

downward effect on the electricity price (see also Sect. 7.5.3). In a situation in which gas-fired power plants are the price-setting plants, like in Fig. 8.13, an increase in the supply by renewable sources may have as a consequence that gas-fired power plants are being replaced, while the coal-fired power plants can keep producing as they remain the inframarginal plants. Hence, the emission-reducing effect of renewable energy is limited when gas-fired generation forms the marginal plant as the extended production by renewable sources does not replace the coal-fired plants, but the less polluting gas-fired plants.

When subsidies for renewable energy are introduced in an electricity market which is also subject to an emissions trading scheme, the latter may foster the environmental effect by making coal-fired power plants the marginal plants by raising the costs of carbon emissions. In addition, higher carbon prices result in higher electricity prices which reduces the subsidy amounts needed for renewables as their market revenues (i.e. the private marginal benefits) are increased. Hence, these two effects refer to a positive interaction of emissions trading on the effectiveness of support schemes.

Support for renewables also affects the efficiency of emissions trading schemes. An increase in the supply by renewable energy reduces the production by other (fossil energy) sources, which results in lower carbon emissions in the electricity sector, but it does not necessarily affect the total carbon emissions by all participants in the trading scheme. This is the so-called waterbed effect of the emissions trading scheme. Reduction of carbon emissions, for example due to replacement of coal-fired power plants by renewable energy sources, results in a declining demand for or a growing supply of emission allowances and, hence, in a lower price of CO_2

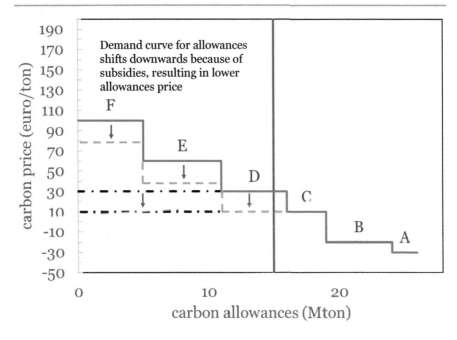

Fig. 8.15 Impact of subsidies for reduction options for allowances market

emission allowances (see Fig. 8.15). This occurs when the emissions trading scheme is a cap-and-trade system. In such a scheme, the environmental impact (i.e. the level of the total emissions) is determined by the cap and the costs incurred are determined by the price of the allowances, as we have seen in Sect. 8.5. When this waterbed effect occurs, reduction of emissions due to the promotion of renewable energy has, ceteris paribus, no environmental impact, but it only affects the costs incurred for the remaining emission reduction. Therefore, subsidizing renewable electricity indirectly also means subsidizing the other ETS participants as it reduces the price of carbon allowances. This also implies that because of the waterbed effects the productive efficiency of the emissions trading scheme is reduced when subsidies are given to particular reduction options (Löschel and Schenker 2017).

Any measures to increase the price for emissions of CO_2 by, for example, levying a CO_2 tax or by the introduction of a floor in the price will therefore yield no environmental impact if there is a cap-and-trade emission trading scheme. Such measures may lead nationally to a reduction of carbon emissions, but due to the functioning of the international emissions trading scheme, it will result in a lower price of the emission allowances, so that elsewhere in the system emissions will increase.

In the EU ETS, this waterbed effect will be partly undone by the abovementioned MSR and its cancellation policy as of 2023 (Perino 2018). When, for instance, subsidies for renewable energy result in a lower demand for allowances, the Total Number of Allowances in Circulation increases and when this number exceeds the threshold of 833 million, less allowances will be auctioned. In this way, the EU ETS is adapted in order to partly mitigate the waterbed effect, which implies that subsidies for renewables do have an emission-reducing effect.

8.7 Security of Supply Measures

8.7.1 Defining Security of Supply from Economic Perspective

As energy is a crucial commodity for many economic activities, security of energy supply is generally seen as an important social objective. The concept of 'security of supply' has, however, different connotations. The economic interpretation of security of energy supply differs from the physical or technical definition. In the latter sense, energy supply is defined as secure when its availability is always guaranteed, i.e. it is defined as an undisturbed supply. In the economic interpretation, however, it may be efficient to have sometimes very high prices and, hence, limited supply for consumers, or even interrupted supply. Very high energy prices can be efficient if they result from scarcity in the market. In such cases, scarcity prices are needed to realize a market equilibrium (in the short term) and to give incentives to economic agents to invest in more capacity (supply side response) or to substitute to other types of energy or commodities (demand side response), as was discussed in Chap. 7.

From this economic definition of security of supply follows that the primary regulation to secure energy supply is to improve the functioning of energy markets, which means that energy prices should be able to reflect actual market conditions and that all market participants should be able to respond to these prices. If energy prices which economic agents have to pay or receive deviate from the efficient market prices, the responses by these agents cannot be efficient. Factors which make that economic agents are not able to respond to changing prices are, among others, a limited access to infrastructure and the presence of bottlenecks in the infrastructure. Removing such constraints for the activities of market participants may, therefore, contribute to the security of energy supply.

Despite such regulatory measures to improve the functioning of energy markets, they may not be able to deliver the optimal security of supply. Economic agents may, for instance, fail to sufficiently take into account the negative effects of their energy consumption on security of supply. Because resources of fossil energy are non-renewable (i.e. they are limited in magnitude), consumption of these resources now implies that less (domestic) resources are available for future generations and that more energy has to be imported in the future (see Sect. 2.4.2). The former effect (i.e. depletion) needs not to be an external effect, as this effect is generally priced in energy markets. Because of the depletion of resources, resource owners have an incentive to postpone production to the future in order to realize a higher price. The Hotelling theory says that the price increases with the rate of interest (see Sect. 2.4.2). Hence, the fact that non-renewable resources are limited in size is not a market failure because this fact is acknowledged by market participants and, hence, priced by markets.

The fact that increased consumption may result in increasing imports in the future, however, may not be fully internalized in market prices. Increasing imports may result in higher political dependency, which reduces the independent decision-making power of a country. If this effect is not included in current prices, it is a negative externality.

8.7.2 Strategic Energy Reserves

Governments may want to implement measures to address the market failure of energy users who insufficiently pay attention to the political risks of consuming energy. An example of this type of regulation is the establishment of strategic energy reserves. These reserves can be seen as a kind of a collective insurance against the risk of insecure supply of energy. The costs of this insurance consist of the costs of storing the energy while the benefits consist of the reduction in economic damage in case of a disruption. A method to assess such insurance policies is to calculate the break-even frequency of a disruption: how often should a disruption occur (i.e. how often should there be benefits of the insurance) in order to recoup all costs? If this break-even frequency is lower than the likely frequency, then the policy is efficient.

The most prominent example of such reserves is the policy of the IEA countries to have sufficient oil in reserves for at least 90 days of net oil imports. In case of disruption in the supply of oil, the IEA countries may decide to release oil from these reserves.[3] Up to now, these strategic oil reserves have only been used three times: during the Gulf war in 1991, after the Hurricanes Katrina and Rita in Gulf of Mexico in 2005 and during the Libyan Civil War in 2011. In all these cases, there was a significant reduction in supply of oil due to an external supply shock which strongly raised the oil price.

Another example of strategic energy reserves consists of the strategic generation capacity in electricity markets. This capacity is meant for situations in which the electricity market is not able to clear because of limited flexibility on the supply and/or demand side. These reserves are generally operated by the transmission network operators to keep the electricity grid in balance (see Sect. 7.5).

In both examples, a public agency (governments and network operator, respectively) invests in extra capacity to be able to supply to the market in case of supply disruptions or inability of the market to find an equilibrium. These measures are meant to address market failures, but they can also have adverse effects on the functioning of markets. If market participants know that a public agency operates a strategic reserve which will be used in case of (extreme) scarcity, they may invest less in capacity themselves. This is the so-called crowding-out effect of public

---

[3]See https://www.iea.org/areas-of-work/ensuring-energy-security/oil-security.

investments. Insofar this effect occurs, the public investments do not result in extra security, but only result in higher costs for society.

Public investments in strategic reserves may also result in windfall profits for existing capacity operators. This happens if the public agency that is responsible for the strategic reserves purchases the required extra capacity from market participants that offer capacity that already is in operation. If these participants just sell their already existing capacity to the public agency, this may result in extra revenues for them without having any effect on the total amount of installed capacity.

Because of these potential adverse effects of strategic energy reserves, governments can also consider other types of regulatory measures to address the negative externality on security of supply. Other measures to address this negative externality are the implementation of a tax on energy consumption or the implementation of a cap on domestic production. Through a tax on consumption, consumers are incentivized to take more costs into account when making their consumption decision, while through a cap on domestic production, a country invests in keeping domestic reserves available for future times of scarcity.

Exercises

8.1 What makes carbon emissions a more complicated externality to solve than, e.g. emissions of small particles?

8.2 Why is pricing the negative externality generally more efficient than subsidizing the opposite measure?

8.3 How can an energy tax be designed such that it minimizes the risk of loss of competitiveness of internationally operating companies?

8.4 Which factors have to be taken into account when determining the level of a subsidy for renewable energy?

8.5 Which factors determine the price of CO_2 allowances in the primary allocation of an emissions trading scheme?

8.6 Which information can be inferred from the level of the carbon price in an emissions trading scheme?

References

Barrett, S. (2007). *Why cooperate? The incentive to supply global public goods*. Oxford University Press.
ECN (2016). *National energy outlook 2016*. Petten.
Hepburn, C., & Teytelboym, A. (2017). Reforming the EU ETS: Where are we now? In I. Parry, K. Pittel, & H. Vollebergh (Eds.), *Energy tax and regulatory policy in Europe*. CESifo Seminar Series.

Löschel, A., & Schenker, O. (2017). On the coherence of economic instruments: Climate, renewables, and energy efficiency policies. In I. Parry, K. Pittel, & H. Vollebergh (Eds.), *Energy tax and regulatory policy in Europe*. CESifo Seminar Series.

Nentjes, A., & Woerdman, E. (2012). Tradable permits versus tradable credits: A survey and analysis. *International Review of Environmental and Resource Economics, 6*(1), 1–78.

Perino, G. (2018). New EU ETS phase 4 rules temporarily puncture waterbed. *Nature Climate Change, 8*, 260–271.

Perman, R., Ma, Y., McGilvray, J., & Common, M. (1999). *Natural resource & environmental economics* (2nd ed.). Prentice Hall.

Sandmo, A. (2000). *The public economics of the environment*. New York: Oxford University Press.

Sorrel, S. (2009). The rebound effect: Definition and estimation. In J. Evans & L. C. Hunt (Eds.), *International handbook on the economics of energy*. Cheltenham: Edward Elgar.

Speck, S. (2013). Carbon taxation: Two decades of experience and future prospects. *Carbon Management, 4*(2), 171–183.

Tietenberg, T. H. (2006). Emissions trading; Principles and practice. Resources for the future (2nd ed.).

Market Power in Wholesale and Retail Energy Markets

9.1 Introduction

A fundamental characteristic of energy markets is their vulnerability to the presence of market power, which is that one or more suppliers can influence market prices. This vulnerability is related to a number of factors: low price sensitivity of demand, inflexibility of supply, limited abilities to store energy, and restricted capacities of the transportation network. These characteristics are discussed in Sect. 9.3. First, Sect. 9.2 discusses the general conditions and welfare consequences of the presence of firms using market power. Afterwards, Sect. 9.4 discusses how the presence and use of market power in energy markets can be monitored. Finally, Sect. 9.5 discusses a number of regulatory options to address the market failure of market power.

9.2 Conditions for and Consequences of Use of Market Power

9.2.1 Ability and Incentive to Behave Strategically

Market power is the ability of market parties to influence market outcomes by behaving strategically. This behaviour is called strategic because it is meant to realize an objective outside the firm. A firm with market power can, for instance, reduce its supply to the market in order to raise the equilibrium price. Before a firm behaves strategically, two conditions need to be satisfied: the firm needs to have both the ability and the incentive to do so. A firm may have the ability to raise the market price, but this may not be beneficial for that firm and, as a result, a (rationally behaving) firm is not going to use this ability.

The ability to influence market outcomes depends on abilities of other suppliers as well as users to respond to changing prices. When the other suppliers can

© Springer Nature Switzerland AG 2021
M. Mulder, *Regulation of Energy Markets*, Lecture Notes in Energy 80,
https://doi.org/10.1007/978-3-030-58319-4_9

respond easily by increasing their supply while the consumers can also respond quickly by switching to another supplier, a firm does not have the ability to influence market outcomes. After all, any action of a firm in that direction, for instance, by reducing its production, will be followed by responses of competitors or customers. Hence, the less the other market participants can respond, the stronger the ability of a firm to behave strategically.

The incentive to do so depends on the impact of changing market outcomes on the profit of the firm. This incentive generally depends on a trade-off related to strategic behaviour: will the extra profits resulting from a higher price be sufficient to compensate for the costs of the reduction in supply?

A firm with both the ability and incentive to behave strategically can consider two types of behaviour to influence market outcomes: physical as well as economic withholding. A reduction in supply is called physical withholding as the firm offers less commodities to the market. A firm with market power could also choose for economic withholding, which is that it does not reduce the volume of its offer to the markets, but that it raises its minimum required price.

Suppose a market is characterized by stepwise demand and supply functions, which reflect the marginal willingness-to-pay of buyers and the minimum required prices of sellers, respectively (see Fig. 9.1). The demand curve is relatively inelastic in the middle and very elastic at low demand volumes. Suppose further that one supplier (S1) has a relatively large production capacity with relatively low marginal costs. When this supplier would utilize all its resources to produce, it could take care of a large part of the total supply (i.e. $Q_{S1,A}$) to the market, which results in supply curve A. As a result, the market price can be close to its own marginal costs (P^*), resulting in a relatively small (operational) profit. This price is determined by the marginal costs of the next supplier in the merit order (S2, in this example). The equilibrium volume (Q^*) is only slightly higher than the output of supplier S1.

For supplier S1, capacity withholding may be beneficial. If this supplier reduces its supply from $Q_{S1,A}$ to $Q_{S1,B}$, the supply curve shifts to the left (supply curve B). If the other suppliers (S2–S5) do not behave strategically, but just respond to the market price and keep supplying on the basis of their marginal costs, P^{**} will be the new equilibrium market price, with supplier S5 being the new marginal, price-setting supplier.[1] The result of this strategic behaviour is not only that the market price is higher (P^{**} instead of P^*), but also that the equilibrium volume is lower (Q^{**} instead of Q^*). For supplier S1, this is beneficial when the extra profits due to the higher price over the remaining production (i.e. $Q_{S1,B}$) exceed the loss of profits due to lower production (i.e. from $Q_{S1,A}$ to $Q_{S1,B}$).

The ability of supplier S1 to behave strategically is determined by the marginal costs of the other producers (S2–S5) and the slope of the demand curve. In this example, the other supplies have higher marginal costs, which implies that they are not able to respond by offering the commodity at the same price, while at the same time, the demand is very inelastic at various parts of the demand curve.

---

[1] Actually, the equilibrium price can be a bit higher as this supplier can charge a price equal to the WTP of the marginal buyer, as no other supplier is able to offer the product at a lower price.

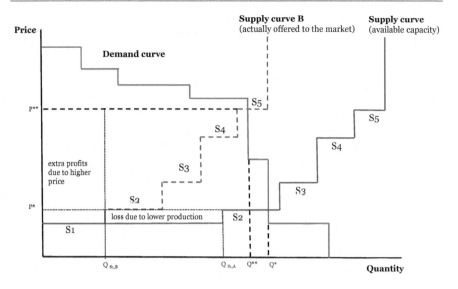

Fig. 9.1 Impact of physical withholding on profit of firm with market power

The incentive to behave strategically is determined by the trade-off between the two opposing effects of capacity withholding: a volume effect and a price effect. The first effect is that the total supply by the firm is lower than what it otherwise would have been. Without the use of strategic behaviour, supplier S1 would produce $Q_{S1,A}$ and almost supplying the whole market, but in case of physical withholding the production is reduced to $Q_{S1,B}$. Hence, the sacrifice (L) the firm makes by reducing its production is equal to the foregone (inframarginal) profits being the difference between the market price (p^*) and its marginal costs (mc):

$$L = (P^* - mc) * \left(Q_{S1,A} - Q_{S1,B}\right) \tag{9.1}$$

In order to make physical withholding beneficial for the firm, this sacrifice should be more than the compensation by the extra benefits (B) resulting from the higher price on the remaining production, which is equal to:

$$B = (p^{**} - p^*) * Q_{S1,B} \tag{9.2}$$

To formulate this more generally, a firm has an incentive to reduce its supply when the present value of the extra benefits (B) exceeds the present value of the foregone profits (L). As the extra profits may be smaller than the foregone profits, a firm with the ability to influence market outcomes does not have the incentive to do so. For instance, an electricity firm with only one power plant may have the ability to influence market prices because it is indispensable to satisfy total demand as all other generation plants are operating on full capacity, but a reduction of its production may, however, hardly be beneficial for this firm as the negative effect on the profits of a volume reduction may exceed the positive effect of higher prices. Hence, in order to benefit from a reduction in production, a firm needs to have sufficient so-called inframarginal capacity that remain producing and that, hence, can benefit from the higher prices.

When individual firms don't have the ability and the incentive to behave strategically individually, they may raise their ability as well as incentive by collaborating with each other, i.e. by concluding an agreement to jointly reduce production or increase the required prices. In terms of the above example, suppose that the supply capacity of S1 is distributed over a number of suppliers, then, these suppliers need to collaborate in order to have the same market power as in the case of one firm. When more firms in a market together reduce production, they share the costs of this strategic behaviour, while all firms enjoy the same benefits. When such a cartel is successful, the total production is less than what it would be in a competitive market. When the lower levels of producing energy result in lower environmental effects, such as carbon emissions, it is said that cartels of producers are the friends of the environmentalist (Perman et al. 1999). Nevertheless, cartels to influence market outcomes at the expense of consumers are strictly forbidden in countries with a competition policy (see also Sect. 3.4.2).

9.2.2   Welfare Consequences of Use of Market Power

When firms with the ability and the incentive to use market power act accordingly, several negative welfare effects result. A reduction in supply to the market (i.e. physical withholding) shifts the supply curve to the left, while an increase in the required minimum prices shifts the supply curve upwards (i.e. economic withholding). In both cases, the equilibrium price rises (see Fig. 9.2). Because of these higher prices, some consumers will not buy the product anymore although their willingness-to-pay exceeds the marginal costs of the supply. This loss of welfare is called the deadweight loss of the use of market power. This loss is indicated by the triangle ABC in the figure.

Those consumers that continue consuming in spite of the higher prices have to pay more for the products. Hence, this results in a redistribution of welfare from consumers to producers. This means that the consumer surplus decreases, while the producer surplus increases. Although the sum (i.e. the total welfare) remains the same, this redistribution of welfare is generally seen as a negative effect. This in particular holds when the suppliers are located in another country (i.e. the goods are imported), as then this transfer of welfare is equal to a loss of welfare for the importing country. Afterall, the importing country has to pay more to the exporting country for consuming the goods. The redistribution of welfare due to the higher prices resulting from the use of market power is often also seen as a negative effect when both suppliers and buyers are located within the same country. In that case, the total welfare of the country is not affected, but generally, competition authorities are directed at maximizing the welfare of consumers. The reason for this is that the

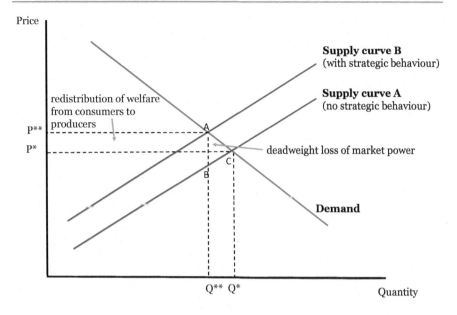

Fig. 9.2 Welfare effects of use of market power

ultimate objective of economic activities is to satisfy the wishes of consumers, not to maximize profits from production.[2] A shift of welfare to producers is, therefore, seen as a reduction in (consumer) welfare.

Besides these direct welfare effects of the use of market power, also effects on productive and dynamic efficiency may occur. A firm that does not face a competitive pressure to reduce the costs of its products or to improve the quality has less incentives to improve efficiency and to innovate (Motta 2004). In addition, when a market is dominated by a few firms, inefficient firms may be able to remain in business, while they would leave the market in competitive circumstances.

This does, however, not imply that the presence of firms with market power is always bad for welfare. In some cases, the ability to charge a bit higher prices is needed to compensate for the risks and costs of innovation. In highly competitive markets, firms don't make any profit, which may hinder them to invest in, for instance, research and development. Hence, when firms have to make high upfront costs in infrastructure or innovation, the market prices should be able to stay above the marginal cost level, otherwise these costs cannot be recouped, and consequently, firms will not make these investments. To what extent firms should have some market power to charge prices above the level of marginal costs is an

---

[2]This was first formulated by Adam Smith in his *An Inquiry into the Nature and Causes of the Wealth of Nations* (1776) as: 'Consumption is the sole end and purpose of all production; and the interest of the producer ought to be attended to only so far as it may be necessary for promoting that of the consumer'.

empirical question as this depends on the precise characteristics in the industry. In Sect. 9.4, we will discuss how to monitor and assess the presence of market power in energy markets. First, we discuss why energy markets are sensitive to the presence of market power.

9.3   Vulnerability of Energy Markets to Market Power

9.3.1   Inflexibility of Energy Supply

The ability of a firm to behave strategically depends on the options others have. If other suppliers can easily respond to an increase in price, the ability of a firm to influence market outcomes is limited. In the oil market, for instance, non-OPEC suppliers can only respond to physical withholding by OPEC members by utilizing their more expensive resources (see Sect. 3.3.2). As a result, the production of cheap oil from OPEC members, in particular Middle East countries, is replaced by more expensive oil from other regions. If other regions, not participating in the OPEC, would find new resources with low production costs, the ability of this cartel to influence the market outcomes would vanish.

The ability on the supply side to respond to price changes depends on the type of technologies, the flexibility to change production levels in the short term, and the time needed to expand production capacity. In the short term, this ability is generally more limited than in the long term. In electricity markets, for instance, the abilities of producers to expand production in the short term are constrained by the magnitude of the installed capacity and the technical characteristics of the power plants. Nuclear and coal-fired power plants, for instance, have to make so-called dynamic costs for ramping up and down their production, which makes it costly to quickly change the production level. As a result, even if the electricity price in the intraday market increases strongly, the production level of these plants cannot easily be raised, even if there is unused capacity left, because of these dynamic costs.

If all installed capacity in a market is fully utilized, each of the suppliers is able to reduce its own production without being afraid that its competitors will increase their production. In the short term, therefore, a market that is operating on full capacity contributes to the ability of individual suppliers to behave strategically. In the long run, however, firms are able to install new generation capacities which reduces the ability of other firms to behave strategically. This implies that in energy markets, market power is in particular a market failure that occurs in the short term, because the flexibility options of other suppliers and consumers are limited. This market failure may be addressed by the market itself in the long term, for instance through entrance of new firms which invest in new generation plants. Regulators, therefore, focus on the ability of these new firms to enter the market, for instance by regulating the access to the electricity grid (see Sect. 9.5.2).

Box 9.1 Physical withholding in the oil market

A clear example of the use of the market power in energy markets is offered by the oil market. The magnitude of the reserves in the Middle East region is huge. The proven reserves in this region are equal to about 50% of the global reserves and 23 years of current global oil consumption (see Fig. 9.3). As the marginal costs of oil production in this region are much lower than elsewhere (see Sect. 3.3.2), they could be the sole supplier to the global market. It is, however, not in the interest of these producers to produce as much as they can as this would reduce the market price and, hence, their profits significantly.

Because the producers in these regions are the only one having such low marginal costs fields, they can act strategically by applying the strategy of physically withholding. If all reserves in this region would be operated by one firm (country), then this agent could simply act unilaterally, but since the reserves are owned and operated by a number of firms and countries, they need to collaborate. Such collaboration to jointly act strategically is called a cartel. The latter exists in the oil market as Organization of Oil Producing Countries (OPEC): the members of this cartel make agreements with each other that all members restrict their production (often way below what they could produce) in order to raise the market price.

The ability of this cartel to control the total production of the cartel members is determined by market circumstances. When the market is tight and all members as well as the producers outside the cartel (such as US shale oil producers) are producing on full capacity, then cartel members have the ability to influence the market price. When, on the contrary, all producers are

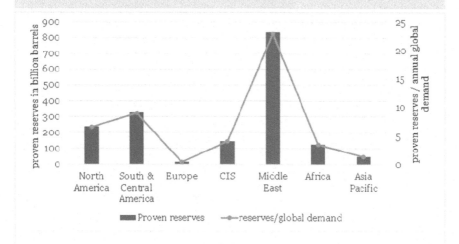

Fig. 9.3 Proven reserves per region, in billion barrels and in relation to global demand (2018). *Source* BP (2019)

producing below their capacity, then this is hardly possible because individual cartel members have the ability to cheat, i.e. to produce more instead of committing to a cartel agreement to reduce less. Because of this risk of cheating behaviour, the oil cartel is less effective in periods of a low utilization of production capacity. In order to reduce this risk, cartels like OPEC have implemented mechanisms to identify violations of agreements and to punish the violators (Nersesian 2016).

9.3.2 Inflexibility of Energy Demand

The ability of a firm to behave strategically does not only depend on the flexibility of competitors, but also on the flexibility on the demand side. If consumers are able to quickly respond to an increase in the price or reduction in the quality of a good, then firms would not have the option to influence market outcomes. Afterall, a strong response by consumers mitigates the effect of strategic supply behaviour on the prices.

The flexibility of consumers is determined by the presence and costs of alternatives and the costs they have to make to adapt to an alternative. The latter are called switching costs, which consist of costs for searching for and assessing of alternatives and the costs of implementing an alternative (see also Sect. 5.3). The size of these costs affects the price elasticity of demand for a commodity. The higher the switching costs, the less attractive it is for consumers to reduce current demand and to switch to an alternative. In such situations, the price elasticity is low, reflected by a steep demand curve.

If the demand curve is steep, a change in the supply curve, for instance, due to economic withholding, hardly affects total market volume but strongly affects the market price. This is indicated by Fig. 9.4 in which it is assumed that the marginal, price-setting supplier, located on the right part of the merit order, applies economic withholding which shifts that part of the supply curve upwards from A to B, and as a result, the market price increases strongly, from p^* to p^{**}.

From this follows that price-inelastic consumers contribute to the ability of firms to behave strategically and to make a profit by physically or economically withholding capacity. In such situations, the deadweight loss of this strategic behaviour is small as consumers keep consuming despite the high prices. The distributional effects are, on the contrary, large as the high prices result in a transfer of welfare from buyers to suppliers (see further on this topic Sect. 11.2).

In energy markets, the demand is generally inelastic and in particular in the short term (see Sect. 2.4.7). For instance, if consumers use gas for heating their houses, they can hardly respond to higher prices in the short term. The only short-term option they have is to accept a lower level of heating in their premises. All other

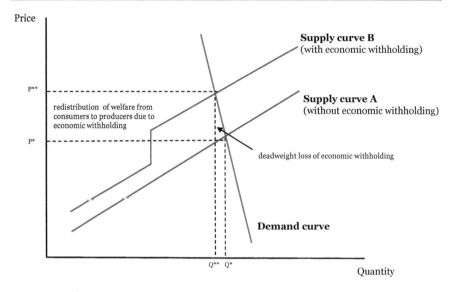

Fig. 9.4 Welfare effects of economic withholding in case of inelastic demand

options to reduce gas demand cost time to realize, such as insulation of houses and the substituting to another source of heating, such as heat pumps. This low flexibility of gas consumers implies that the producers can have the option to behave strategically during periods with low temperatures and high demand for heating.

9.3.3 Inability to Store Energy

The option for producers to behave strategically is negatively affected by the presence of storage facilities. If storage of an energy commodity is technically possible and economically feasible, such as in the markets for oil, coal and gas, then the storage can be used as an extra source of supply in times of high energy prices (see Box 9.2). To what extent the presence of storage contributes to the reduction of market power of suppliers depends on which parties operate the storages. If the storages are operated by firms who are also active as producers, then the storages can just be used as part of their strategic game.

Box 9.2 Three types of roles for storages in the gas market

In gas markets, storages play an important role because of the price inelastic demand for gas in the short term, the strong seasonality and weather dependence of gas demand, and the limited number of suppliers. The presence of storages affects the ability of producers to behave strategically. The roles for gas storages can be distinguished in commercial, balancing and securing supply.

The combination of a relatively strong dependence on external (weather) circumstances and the relatively low price sensitivity makes that a sudden increase in gas demand is hardly mitigated by the resulting increase in prices and has to be met by an increase in supply. As a result, gas prices can be volatile as well, which gives rise to opportunities for so-called intertemporal price arbitrage. Because of these opportunities, storages in gas market can be profitable investments (see Sect. 3.3.3). The presence of such commercial storages also limit the ability of producers to influence the market outcomes as in case of withholding of gas by suppliers, more gas can be released from storages.

Besides this commercial objective of storages, they can fulfil a role in real-time balancing the gas grid. As this grid needs to be in permanently balance, production and consumption need more or less to be equal (see Sect. 7.4). Differently from electricity grids, gas grids do have some buffer themselves as the pressure in a gas network may fluctuate within a bandwidth. This flexibility is called line pack. For larger deviations between production and consumption, other flexibility options are, however, needed, and storages are often used for this purpose.

Another reason to have storages in the gas market is related to the objective to secure the supply of gas. In particular when the supply of gas is coming from a small number of sources (e.g. countries), it may be efficient to invest in storages as a kind of insurance against interruptions in supply due to technical or geopolitical factors (Nersesian 2016). For commercial parties, such investments may be not profitable, but governments may stimulate such investments in order to be less dependent on foreign suppliers of gas (see further on this issue, Sect. 8.7).

Just as in the gas market, also in the oil market storages are used for commercial purposes as well as for security of supply reasons. In most countries, oil and products made from oil (like gasoline) are stored in order to reduce the dependence on short-term market circumstances or to benefit from intertemporal price differences. In addition, many countries store these energy carriers as a kind of insurance against supply shocks. The countries of the International Energy Agency (IEA), for instance, have organized, triggered by the oil crises in the 1970s, a system of strategic oil reserves which needs to be sufficient to cover 90 days of import of oil by these countries in the previous year. In Sect. 8.7 for the discussion of the economics of such security of supply measures.

In electricity markets, however, storage of the commodity is hardly possible, as is explained in Sect. 2.3.4. As the demand for electricity is also very price inelastic in the short term, while the supply is constrained by the magnitude of installed capacity which is technically available to produce, this market is in particular vulnerable to the risk of firms behaving strategically during hours of high demand. Through conversion of electricity in other energy carriers, like hydrogen, flexibility can be added to this market. As the costs of this conversion process are still fairly high, this flexibility is yet not very effective in removing the ability and the incentive of electricity producers to behave strategically. Another conversion technology that adds flexibility to the electricity market is pumped hydro, in which electricity is converted into gravitational potential energy by pumping water to a reservoir to a higher elevation level in mountains, while this energy can be converted back into electricity by releasing the water through turbines.

9.3.4 Presence of Constraints in Energy Networks

A source of market power which is typical for industries based on networks is the presence of network constraints. Transmission and distribution grids face technical limits, and if these limits are reached, market transactions are hindered. This is in particular relevant for electricity markets, because of the physical laws determining the flows of energy through conductors (see Sect. 2.4.4), the limited availability of storage facilities, and the relatively high costs of other flexibility sources (see e.g. Léautier 2018). If, for instance, the infrastructure for importing electricity into a country is fully utilized, then it is not possible for importing firms to increase their imports. Hence, their ability to respond to market prices is limited as well, which increases the ability of domestic producers to behave strategically. These type of bottlenecks in networks may occur between so-called market zones in electricity markets and between entry–exit zones in gas markets (see further on this issue Sect. 7.3.3).

Network bottlenecks within these zones do not influence the market transactions and, hence, do not influence the ability of market players to use market power. Within the market zones in electricity markets and entry–exit zones in gas markets, the markets are cleared based on the supply and demand from these zones and the available import from and export to other zones. Network constraints within these zones do not play a role, as both systems are based on the idea that any bottleneck within the zone has to be solved by the network operator and that market parties may assume that such bottlenecks do not exist. Insofar these bottlenecks do exist, network operators can solve them through congestion management (see Box 7.3).

The presence of such congestion management schemes in zonal electricity markets, however, may trigger strategic behaviour if some producers are able to create local bottlenecks by offering electricity in the spot market in the expectation that this supply cannot be generated, but that it will be asked to increase generation by plants located in the non-congested part of the network for which they ask a price above the marginal costs. Hence, if the number of firms operating in a market zone is limited, they may have an influence on the creation of internal grid bottlenecks.

9.4   Monitoring Market Power in Energy Markets

Because of the adverse welfare effects when firms use their market power to influence market outcomes, regulators may want to monitor the outcomes, structure and behaviour in markets in order to detect and address such strategic behaviour in a timely and adequate way. This monitoring activity is part of general competition policy (see Sect. 3.4.1), but it also plays a role in the regulation of energy markets as it contributes to the understanding how these markets function and to what extent additional regulatory measures are required (Léautier 2018). If the presence of market power is due to structural characteristics in the market, such as lack of network capacity, regulators can take measures directed at these characteristics, which is discussed further in Sect. 9.5. We first discuss a number of indicators which can be used to monitor the presence of market power in energy markets. It will become clear that none of these indicators presents the full picture regarding the intensity of competition, and therefore, they have to be used together in order to be able to adequately assess this (see Creti and Fontini 2019).

9.4.1   Market Outcomes

The first type of indicators refers to the outcomes of the market. In case of perfect competition, which is used as benchmark for the analysis of market failures (see Sect. 4.2.4), market prices are equal to the marginal costs of the marginal supplier. In such an equilibrium situation, it is not possible to improve the allocation of a commodity. A measure to assess market prices from this perspective is the Lerner index (LI), which measures the markup of the market price (p) on top of the marginal costs (mc) of a firm (i) in relation to the market price:

$$LI_i = \frac{p - mc_i}{p} \tag{9.3}$$

In case of perfect competition, the Lerner index is equal to 0 as the price is equal to the marginal costs. The higher the Lerner index, the more the market price is above the marginal costs, which may indicate the presence of strategic behaviour of a firm with market power. Although this measure is theoretically the best measure to assess market prices, in practice it is complicated to use because of the information required to determine the marginal costs. In electricity markets, for instance, the marginal costs not only consist of the direct variable costs needed to fuel a turbine, but it also includes dynamic costs, which are the costs of ramping up and down a power plant.

Another problem in applying the Lerner index is that one has to control for the occurrence of scarcity prices. Suppose the market situation in an electricity market is as shown in Fig. 9.5. As the demand curve intersects the supply curve on the vertical part on the right, the generation capacity is fully utilized and a scarcity price results which is way above the marginal (operating) costs of all plants. For each

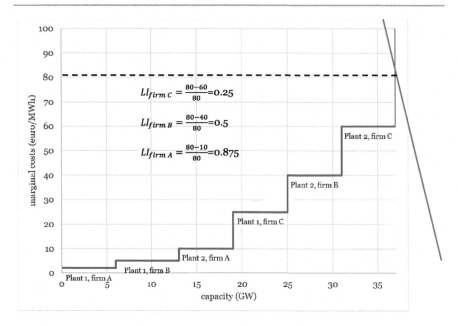

Fig. 9.5 Lerner index per firm in electricity market based on short-term marginal costs

firm, the Lerner index can be calculated by using the market price (which is, in this example, 80 euro/MWh) and the marginal costs of each firm, which is equal to the highest marginal operational costs of all plants utilized by a firm. For firms A, B and C, these highest costs are 10, 40 and 60, respectively, which results in a Lerner index for these three firms of 0.875, 0.5 and 0.25, respectively.

Hence, in times of scarcity, the Lerner index can be significantly above 0. When calculating the Lerner index in times of scarcity, the marginal costs need also to include the costs of extending the capacity, as was discussed in Sect. 7.5. As investments in new capacity are only profitable when the price is above these long-run marginal costs for a longer period of time, this Lerner index needs also to be assessed over such a longer period and not on a short-term basis. This makes that the application of the Lerner index in electricity market is complicated. Nevertheless, estimating the Lerner index can be a useful first step in monitoring the intensity of competition in a market and to measure to what extent the price exceeds the marginal costs (Mulder 2015).

Another type of measure to monitor the outcome of a market refers to the spread across the prices charged by the various suppliers. This spread is a measure for the price dispersion. In a perfectly competitive market, no supplier is able to charge a price above the (long-term) marginal costs, as a higher price would immediately be followed by customers switching to other suppliers. Hence, if in such a market the commodities are homogeneous, the prices charged by suppliers should be equal to each other. This means that in such a market, there is no price dispersion.

When, however, competition is hampered by a lack of information or the presence of search and switching costs, some suppliers may be able to charge higher prices than competitors in order to maximize their profits. Hence, in less competitive markets, price dispersion can be significant. However, when there is no competition at all because of the presence of a monopolist or a very effective cartel agreement among suppliers, then all prices are equal as well. Hence, there exists a non-monotonic relationship between price dispersion and intensity of competition (see Fig. 9.6). This implies that a change in price dispersion may be related to a change in the intensity of competition, but the direction of this relationship is not unambiguous and can only be determined by also looking at other aspects of the market.

The intensity of competition in a market is also reflected in the degree output prices are related to input prices. In a perfectly competitive market, the output prices are directly linked to the marginal costs, and, hence, to the prices of inputs. This relationship is the so-called pass-through rate. In a competitive market, an absolute change in the (marginal) input prices is translated into a similar absolute change in the output prices. In less competitive markets, however, the pass-through rate can be different. The reason for this is that in such markets, the output prices are only partly related to the marginal costs. In less competitive markets, the output prices

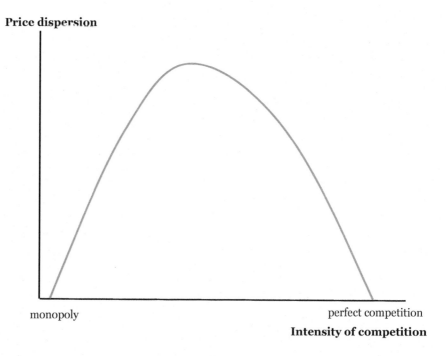

Price dispersion

monopoly   perfect competition

Intensity of competition

Fig. 9.6 Relationship between price dispersion (i.e. spread between highest and lowest prices) and intensity of competition

are set in such a way that the profits of suppliers are maximized, while in competitive markets, output prices are fully exogenous to individual suppliers and, therefore, only linked to the marginal costs of the marginal supplier. As said in Sect. 6.2.1, firms in competitive markets are, therefore, defined as price takers, while firms in less competitive markets can be seen as price setters.

In the extreme case of a monopoly, the pass-through rate is determined by the shape of the demand curve. Suppose in a market, the inverse demand curve is a linear downward sloping curve, while the marginal costs are constant (see Fig. 9.7). In such a market, the pass-through rate of an increase in marginal costs is 50%, while in the other extreme case of a perfectly competitive market, the pass-through rate is 100%. When the demand curve has a different shape, for instance convex to the origin, the pass-through rate of a monopolist will be different (see RBB 2014).

In a competitive setting, the relative pass-through rate may deviate from 100% if the output price is also related to other costs components. This relative rate is also called pass-through elasticity and measures the percentage change in output prices in response to a percentage change in costs. This ratio will be below one if the competitive price is based on the sum of costs and a margin. This is, for instance, relevant in retail energy markets, where retailers base the price of the energy for their customers on the energy price they have to pay in the wholesale market plus a margin to cover the costs of the retailer.

The pass-through rate may also be different between upward and downward changes in input prices, resulting in a so-called asymmetric cost pass-through. When all firms in an industry are operating on their capacity constraint, no firm has an incentive to pass on decreases in input prices, as other firms are not able to increase their production. Increases in industry-wide input prices, on the contrary, will be fully translated into higher output prices as all firms face the same increase

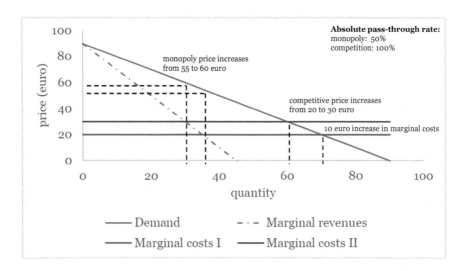

Fig. 9.7 Pass-through of costs in monopoly and perfect competition: numeric example

in marginal costs. Another factor why firms can be less inclined to reduce output prices when their input prices decrease is that consumers may be less active in searching when output prices are low, which means that the intensity of competition reduces when prices go down. This phenomenon of asymmetric cost pass-through is called rockets-and-feathers and was, for instance, found in the UK energy retail market (Ofgem 2011). Hence, when the pass-through rate is below 100%, this may be a sign of less intense competition.

9.4.2 Market Structure

Besides analysing the outcomes of market processes, regulators also tend to look at the structure of markets and the relative positions of market parties. If individual suppliers have a large market share, they may be able to behave strategically, although even in a market with only two suppliers, competition can be intense (see Sect. 4.2.4). Market-structure indicators can, however, be useful for a first screening of the presence of market power.

To get an idea about the market structure, a regulator can look at structural measures like the Herfindahl–Hirschman index (HHI), which measures the sum of the square of the market shares (ms) of all firms (i):

$$HHI = \sum_i ms_i^2 \tag{9.4}$$

As a rule of thumb, some regulators state that an HHI below 1200 indicates a competitive market and an HHI above 1800 a concentrated market.[3] Another structural measure which is often used is the C3 method, which just measures the joint market share of the three largest firms.

Although using measures for the degree of concentration in the market can be useful to get a first idea about the functioning of the market, it is important to realize that a concentrated market in itself is not necessarily bad for market outcomes. In some markets, a concentrated structure may be efficient if economies of scale or scope exist in production or when firms have to make extensive costs for research and development which can only be recouped when the firms have a high market share which enables them to charge prices which are above the short-term marginal costs.

Nevertheless, a concentrated market may facilitate higher market prices. This is, for instance, the case in markets with capacity constraints in production, such as in electricity markets. When the competition in such a market is characterized by so-called Cournot competition,[4] the market structure (HHI) is related to the market

---

[3]The HHI can be calculated through measures for market shares expressed as percentages or as perunages (i.e. percentage divided by 100). In the former case, the HHI varies between 0 and 10,000, in the latter between 0 and 1.

[4]Cournot competition is where suppliers compete in the number of quantities they offer to the market instead of the prices they ask.

outcome, measured by the weighted average Lerner index (LI), with market shares (m) of firms (i) as weights, as well as the price elasticity of demand (ε):

$$LI = \sum_{i} m_i LI_i = \frac{HHI}{\varepsilon} \qquad (9.5)$$

This formula indicates that there is a relationship between the degree of market concentration and the degree of market power (Motta 2004). In addition, it also says that given a level of concentration, the prices will more strongly exceed the marginal costs, the lower the price elasticity of demand. This ratio shows that for the presence of above-competitive prices, the price sensitivity of demand is crucial. A market-structure measure that not only focuses on the supply side, but also includes information on the demand side may, therefore, be better able to assess the ability of firms to behave strategically.

A measure regarding the oil market which includes information on both the supply and the demand side is the relative call on OPEC (rCoO). The absolute call on OPEC (aCoO) calculates to what extent production from OPEC countries is needed based on the difference between the total global oil demand (D_{global}) and the supply from non-OPEC countries ($S_{non\text{-}OPEC}$). The relative call on OPEC relates this difference to the actual production by OPEC countries (S_{OPEC}):

$$rCoO = \frac{aCoO}{S_{OPEC}} = \frac{D_{global} - S_{non-OPEC}}{S_{OPEC}} \qquad (9.6)$$

If this ratio is close to 1, then all OPEC production is needed to satisfy the residual demand which cannot be supplied by non-OPEC producers. In such situations, if one member of the OPEC reduces its production, the oil price has to rise in order to give an incentive to consumers to use less oil because the supply is not able to meet demand anymore. Such a situation occurred, for instance, in 2010 when the oil market was tight and the oil price increased, while in 2018, the actual OPEC production was significantly higher than needed to satisfy the residual demand (see Fig. 9.8). In this analysis, non-OPEC producers are seen as competitive, non-strategic suppliers which only respond to the market price.

If the rCoP ratio is way below 1, then OPEC produces significantly more than what is needed to satisfy current demand. As a result, a reduction in OPEC production hardly influences the market price, and it will only influence the amount of oil that is stored. In such situations of oversupply, it is difficult for the OPEC as a cartel to let all members stick to a production agreement, because a country that reduces its production will not see higher prices (i.e. the price effect is negligible), while it does face lower profits due to the lower production levels (i.e. the volume effect can be large). This situation occurred in spring 2020 in the beginning of the Corona crisis which resulted in a strong reduction of global oil demand, while the producers did not reduce their supply to the market in the same pace. As a result, the oil prices declined strongly.

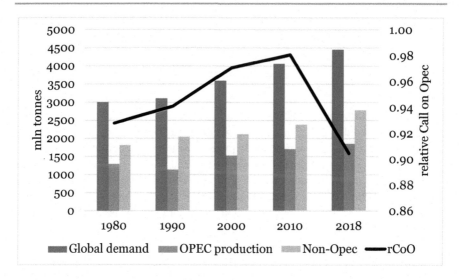

Fig. 9.8 Market power of OPEC measured through relative call on OPEC. *Source* IEA, Oil Information 2019

A measure regarding the electricity market which also includes information on both supply and demand side is the Residual Supply Index (RSI). This measure calculates to what extent a firm (i) is needed to satisfy demand. This index is calculated as the aggregate of the generation capacities (CAP) of all other suppliers ($j \neq i$) in the market and the total demand (D):

$$RSI_i = \frac{\sum_{j \neq i} CAP_j}{D} \tag{9.7}$$

Box 9.3 Calculating the RSI in an electricity market

Suppose the supply side of an electricity market consists of three firms operating six different plant types each having different marginal generation costs. The installed capacity per plant type and per firm and the related marginal costs are as follows:

Plant type	Firm	Capacity (GW)	Marginal costs (euro/MWh)
1	A	3	2
2	B	5	5
3	A	3	10
4	C	5	25
5	B	4	40
6	C	2	60

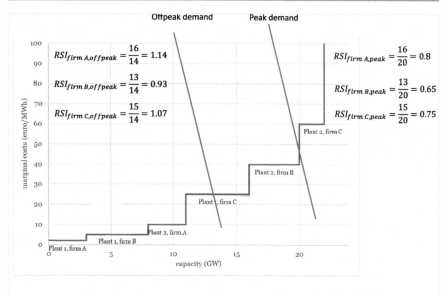

Fig. 9.9 RSI per electricity firm during off-peak and peak hours

Suppose further that the load fluctuates between 14 GWh during off-peak hours and 20 GWh during peak hours. Using these numbers, the pivotality of each firm can be calculated for these two demand situations (see Fig. 9.9). During off-peak hours, only firm B is pivotal, as the total capacity of firm A and C is not sufficient to meet total off-peak demand. During peak hours, all firms are pivotal, as for each firm holds that the demand is larger than the aggregated capacity of the other two firms.

If this ratio is above 1, then firm i is not needed to satisfy demand, and consequently, it doesn't have any ability to raise the price. If such a firm would reduce its production, other firms could easily increase their production. However, if the RSI is below 1, then firm i is indispensable to generate sufficient electricity for demand, which gives the firm the ability to behave strategically. This ability of a firm may change from hour to hour and day to day, as the demand for electricity is highly volatile, while also the capacity of generators is not constant. This latter holds in particular for renewable generation capacity, such as wind turbines and solar PV, which is dependent on weather circumstances and timing of the day. For instance, during hours when the demand is high (peak hours), while the generation by renewable sources is low as result of unfavourable weather conditions, the firms operating conventional capacity may have the ability to charge higher prices (see also Box 9.3).

9.4.3   Behaviour of Market Parties

The behaviour of market parties is also informative about the intensity of competition among suppliers. In a competitive market, suppliers do not withhold capacity, which means that they offer all products where the marginal costs are below or equal to the market price. A non-strategic electricity producer that can produce electricity at 50 euro/MWh will supply this electricity to the market when the market price is higher or equal to that level. If a firm does not utilize this capacity, then this may be a sign of withholding capacity, either through submitting a higher bid (i.e. required price) to the power exchange or by not submitting a bid at all. The application of this rule in monitoring of actual dispatch of electricity plants is, however, a bit more complicated because of the presence of dynamic costs of ramping up and down a power plant. In addition, power producers may also decide not to offer capacity of a particular plant in the spot market (day-ahead or intraday) in order to have some reserve capacity in order to prevent imbalance costs (see Sect. 7.4).

Another indication of a presence of market power is a lack of entrance of new firms, even when market prices are relatively high and the incumbent firms are making high profits. The absence of entrance can be due to entry barriers like a limited access to resources or transport capacity, but also be due to the need to make large upfront investments. When such barriers hinder the entrance of new firms, existing (incumbent) suppliers are more able to raise market prices.

A key factor behind the ability of incumbents to use market power is the flexibility of consumers. If they can easily switch to another supplier, the market power of an individual supplier is limited. Therefore, switching behaviour of consumers can be used as a measure for the intensity of competition, in particular in retail markets. Figure 9.10 shows, for instance, the number of households monthly switching from electricity contract in the UK. Until 2013, these switches mainly occur between contracts from the same supplier, but since then more households switch to a contract of another supplier, which is an indication of more intense competition among retailers.

There is, however, not a linear relationship between degree of switching and the intensity of competition. If in a market, barriers for switching disappear, it becomes easier for consumers to switch to another supplier which results in more intense competition. This process, however, results in more equal prices and a lower price dispersion. When prices are more or less equal, consumers have less incentive to switch. Hence, in a perfectly competitive market with no switching barriers, prices may become more equal, which results in less switching behaviour. This means that the inverse U-curve that describes the relationship between price dispersion and intensity of competition (see Fig. 9.6) is also relevant to describe the relationship between degree of switching and intensity of competition.

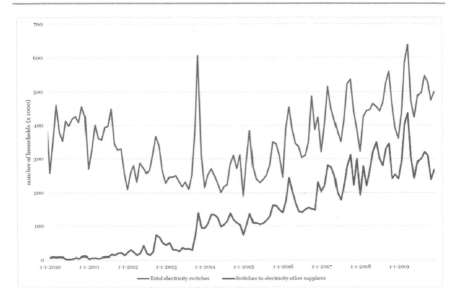

Fig. 9.10 Number of households switching from electricity contract, UK, per month in 2010–2019. *Source* Ofgem, https://www.ofgem.gov.uk/data-portal/retail-market-indicators

---

9.5 Regulatory Measures to Address Market Power in Energy Markets

9.5.1 Introduction

Because the use of market power to influence market outcomes may result in allocative and productive inefficiencies while it also may redistribute welfare from consumers to producers, societies may want to reduce the ability and incentive of market players to do so. While competition authorities are directed at the actually realized behaviour of market participants, through merger control and antitrust, regulators focus on the design of markets in order to prevent anticompetitive behaviour. As the presence of market power can be due to several characteristics of markets, regulatory measures have to be directed at those factors which are relevant in particular circumstances. The regulatory measures to address market power can be distinguished in removing entry barriers on the supply side, increasing the flexibility of consumers, fostering storage and other flexibility options, as well as removing network bottlenecks.

9.5.2 Removing Entry Barriers

One potential source of market power of incumbent firms is that competing firms have limited abilities to extend their production capacity or even to enter the market. As regulators cannot do much to reduce the time required to build a new installation, they may have a role in reducing the administrative costs for such an investment. Firms often require all kind of permissions to build a new installation because of other reasons, such as environmental concerns or safety for the direct neighbourhood. Sometimes, however, such permissions are motivated by the wish to protect incumbent producers from competitive entry.[5] Changing such permission rules may, therefore, contribute to the functioning of markets.

Having access to energy markets depends strongly on having access to the crucial infrastructure. Without a connection to the gas or electricity grid, new firms cannot enter the markets. Because access to the energy infrastructure is essential to obtain competitive energy markets, regulators generally force so-called third-party access (TPA), which means that other firms (and consumers) are allowed to make use of the infrastructure. This TPA also prescribes that the conditions for access should be reasonable, including the tariffs which have to be paid, which was discussed in Chap. 6. In addition, because the management of the critical infrastructure should treat all users of that infrastructure equally, regulators in many countries require that the network manager is not allowed to be active himself in the same business. This is called unbundling of network and commercial activities, which was discussed in Sect. 3.4.2.

9.5.3 Increasing Demand Flexibility

A necessary condition for firms to have market power is that consumers are not flexible, meaning that they cannot easily change their behaviour when the prices of good increase or the quality of that good decreases. When consumers are rather inflexible, as they generally are in energy markets, regulators can consider measures to foster their ability to make other choices. These choices may refer to the consumption of the energy as a commodity, but also to the conditions under which the energy is consumed. The first response refers to how much the energy consumption can be adapted when the price of energy changes. In Sect. 2.4.7, we have seen that this price elasticity is low, in particular in the short term. This low ability to change the consumption pattern in response to prices is related to the way consumers receive information on actual use and how they are charged for energy use. In most countries, residential electricity consumers are charged on the basis of a profile (which is an average, assumed consumption pattern), while they are not informed about the actual profile of their consumption and the actual corresponding market prices. By providing consumers with real-time information about their energy use

---

[5]In such cases regulation (or in general, policy) is captured by interest groups, in particular the incumbent industry.

as well as enabling retailers to offer contracts based on spot-market prices, the sensitivity of residential consumers to price changes increases. This will increase the price elasticity of demand in the wholesale energy market, which reduces the ability of suppliers in this market to behave strategically.

In addition to regulatory measures to provide consumers with more actual information on their consumption, it is also important to give consumers options to choose among competing suppliers. If consumers cannot easily oversee the market and all competing bids and if they have to make high switching costs in order to go to another supplier or contract, their current supplier may benefit from this low flexibility of their customers by charging higher prices or offering lower quality of services. Hence, consumer flexibility is fostered when the market becomes more transparent and when it is easier to change from contract and supplier (see further on this issue Sect. 5.3).

9.5.4 Fostering Storage and Other Flexible Sources

The presence of storages and the costs of using them strongly influence the ability of producers to behave strategically. Storages are not a net supplier over a longer period of time, which means that they can only play a role in addressing temporary shortages in the market. In the gas market, for instance, where demand is strongly related to outside temperature, storages are filled when the gas demand and hence, the gas prices are relatively low (i.e. during summer periods) and used when the gas demand and gas prices are relatively high (i.e. during cold winter periods). The presence of such storages limits the abilities of producers to behave strategically during cold winter periods. Therefore, regulators may want to foster access to storages under reasonable conditions, just as the access to the transportation infrastructure is regulated. Such a regulation of access to storage is in particular relevant when the amount of storages is limited and when extending the number of storages is complicated. This regulation of the access to storages may consist of imposing capacity allocation mechanisms (CAM) and congestion management mechanisms. The former includes rules how the capacity of a storage facility is allocated to potential users, while the latter refers to how capacity is reallocated in case of congestion.

Flexibility in the market does not only come from storages, as it can also be supplied by producers. For instance, in gas markets, some fields have the ability to quickly adjust production levels from hour to hour. Such fields are referred to as swing suppliers. This ability gives the operator of the field the advantage to vary the output level according to market circumstances. The Groningen gas field in the Netherlands, for instance, used to be a major swing supplier to the European gas market. The supply from this field peaked in winter periods and was relatively low in the summer, which implies that it was primarily used to meet peak demand during winter time when the gas price is higher. From the perspective of a regulator who wants to address market power, the presence of such a field also creates a risk for competition, because of its unique characteristics which enable the operator to profit from higher prices during cold winter periods, while it could also have the option to behave strategically. Because of this risk, a regulator may decide to make the flexibility services provided by such fields subject to regulatory constraints, such a maximum price that can be charged for the gas delivered as flexibility source.

Although in electricity markets it is hardly possible to store electricity itself, by converting electricity into another energy carrier, flexible sources can be utilized. An example of such a source is pumped hydropower. When electricity prices are low, electricity is used to pump water to reservoirs at higher altitudes, while the water is released again in order to turn on a turbine and produce electricity when the electricity price is high. Another flexibility option in electricity market is to convert electricity into hydrogen through electrolysis during periods of low prices and produce electricity again through fuel cells when the prices are high. The presence of such flexibility options removes the ability of electricity producer to behave strategically. Hence, effective regulatory measures to address this ability may also include the promotion of such technologies.

9.5.5   Removing Network Bottlenecks

Firms and consumers do not only need access to the essential infrastructure, it is also important to them that the infrastructure has sufficient capacity to meet all demand. If the capacity is fully utilized, this may result in congestion which is that the users of infrastructure are constrained in their production or consumption, as was discussed in Sect. 7.3. Besides the direct loss of welfare caused by the reduction in market transactions, congestions may also reduce the intensity of competition or even enable some firms to behave strategically. Several regulatory measures can be taken to foster the availability of network infrastructure for users.

A basic measure is to give all energy producers and users the right to use the network and to give the operators of the network the obligation to connect all producers and consumers who want to make use of the infrastructure. This access should be realized at so-called reasonable tariffs, as is discussed in Chap. 6. Moreover, the connection should be realized within a short period of time. Regulators may require that the network has always sufficient capacity to transport the energy at any time. This means that the network has to be designed to meet peak demand, which implies that the operators should prevent situations of congestion. In line with this, the regulator may require that the network operators offer the service to transport energy without bordering the network users with concerns about network capacity.

In the European gas markets, this is realized through the so-called entry–exit systems (see Sect. 3.3.3). The users of the gas network only have to indicate how much capacity they need to inject or withdraw gas from the network and how much they really are going to inject and/or withdrawn at specific periods of time. The first type of bookings is called capacity bookings and generally refers to a longer period of time (say a quarter or a year), while the latter is called transport nominations. Once the gas has been injected into the network, the producers (or traders) do not need to make any further bookings for the required capacity during transport. The network operator has the full responsibility to transport and to realize the capacity which is required. In the electricity market, a similar system exists. Network users may behave as if the network is a copper plate, in which no barriers for transport exist. Sometimes, however, specific parts within the network are congested, meaning that the capacity of a part of the network (such as a substation) is not sufficient to handle the current which is put on the line. For such situations, a regulator may want that the network operator applies a kind of congestion management, as was discussed in Sect. 7.3.

In this respect, one needs to distinguish two types of congestion: physical and contractual congestion. Physical congestion refers to a lack of physical capacity, which can only be solved by either reducing the supply and/or demand or increasing the grid capacity. In energy markets, network congestion often is characterized by contractual congestion, which is that market participants have booked all capacity through their capacity nominations, while the actual infrastructure is not completely used. Although contractual congestion does not create any risk for the physical stability of the energy transport, it may hinder the entrance of other players to the market. When all capacity is booked by incumbent producers, new entrants may not be able to enter the market. A regulatory solution to address this market barrier is to impose rules that unused capacity, i.e. booked capacity that is not nominated for transport, will be taken away by the network operator. Such regulator measures can have the form of Use-It-Or-Loose-It (UIOLI) clauses or Use-It-Or-Sell-It (UIOSI) clauses. Through such measures, grid capacity can be made available for other players which may not only reduce the market power of incumbent suppliers, but also result in more benefits resulting from trade.

Exercises

9.1 What is the difference between ability and incentive to use market power?

9.2 What is difference between economic and physical withholding?

9.3 How does the presence of storage affect market power?

9.4 Why is the Lerner index not a perfect measure in electricity markets?

9.5 How can demand flexibility be stimulated in energy markets?

References

BP. (2019). *Statistical review of world energy* (68th ed.).
Creti, A., & Fontini, F. (2019). *Economics of electricity: Markets, competition and rules.* Cambridge University Press.
Léautier, T.-O. (2018). *Imperfect markets and imperfect regulation; An introduction to the microeconomics and political economy of power markets.* Cambridge, MA: The MIT Press.

Motta, M. (2004). *Competition policy; Theory and practice.* Cambridge, MA: Cambridge University Press.

Mulder, M. (2015). Competition in the Dutch electricity wholesale market: An empirical analysis over 2006–2011. *The Energy Journal, 36*(2), 1–28.

Mulder, M., & Willems, B. (2019). The Dutch retail electricity market. *Energy Policy, 127,* 228–239.

Nersesian, R. L. (2016). *Energy economics; Markets, history and policy.* London/New York: Routledge.

Ofgem. (2011). Do energy bills respond faster to rising costs than falling costs? London, 21 March.

Perman, R. Y., Ma, J. M., & Common, M. (1999). *Natural resource & environmental economics.* England: Pearson Education Limited.

RBB Economics. (2014). Cost pass-through: Theory, measurement, and potential policy implications; A report prepared for the Office of Fair Trading, February.

International Restrictions on Trade in Energy

10

10.1 Introduction

When international trade in energy is restricted, several negative welfare effects may occur. Removing trade barriers, therefore, can contribute to the functioning of energy markets. This chapter first discusses the negative effects of international trade barriers on productive efficiency (i.e. costs per unit of production), allocative efficiency (i.e. intensity of competition and international differences in prices) and security of supply (Sect. 10.3). Then, Sect. 10.4 discusses what kind of international trade barriers exist in energy markets and by what kind of regulatory measures these barriers may be reduced. Among others, attention is paid to both network extension and a higher utilization of existing infrastructure. Finally, this Chapter analyses the implications of an improved international integration of markets for the effectiveness of national energy policies (Sect. 10.5).

10.2 Theory of Integration of Markets

10.2.1 Market Outcomes Without Integration

When regional markets are not connected with markets in other regions, they have to be cleared in every region separately. As a result, the outcomes of the separate regional markets may differ as the outcomes only depend on the specific circumstances in a region. Suppose there are two countries (A and B) in which the same type of commodity is produced as well as consumed. Suppose further that the demand (D) for a commodity in both countries A and B has the same relation to the price (p):

$$D_{A,B}(p) = 90 - p \tag{10.1}$$

© Springer Nature Switzerland AG 2021
M. Mulder, *Regulation of Energy Markets*, Lecture Notes in Energy 80,
https://doi.org/10.1007/978-3-030-58319-4_10

The countries differ, however, in their ability to produce the commodity. Suppose that producing the commodity in country A is more expensive than in country B, resulting in the following two supply functions:

$$S_A(p) = \frac{1}{2}p \tag{10.2}$$

$$S_B(p) = \frac{3}{2}p \tag{10.3}$$

The specification of the supply curve of country A means that the supply increases by 50% of the increase in the price, which is equal to saying that the marginal costs of supplying the commodity increase twice as much as the increase in the quantity as the inverse supply function is equal to p = 2q. In country B, the inverse supply function is p = 2/3q, meaning that the marginal costs of supplying the commodity increase with 2/3 of an increase in quantity. Hence, producing the commodity in country B is less expensive.

The equilibrium outcomes can be determined by equating the demand (D) and supply (S) curves for both countries A and B (see also Sect. 4.2.3):

$$Country\,A : 90 - p = \frac{1}{2}p \overset{yields}{\rightarrow} p = 60, q = 30 \tag{10.4}$$

$$Country\,B : 90 - p = \frac{3}{2}p \overset{yields}{\rightarrow} p = 36, q = 54 \tag{10.5}$$

The equilibrium price in country A is higher than in country B (i.e. 60 vs. 36 euro per unit), while the equilibrium output is lower in that country (i.e. 30 vs. 54). In other words, because country B has less expensive technologies, i.e. with lower marginal costs, the market clears at a lower price and a higher volume than in country A (Fig. 10.1).

Because of the absence of any integration between these two markets as well as the difference in production costs, consumers in country A have to pay a much higher price for the same commodity than those in country B. Consumers from country A, therefore, would prefer to buy the commodity in country B, which is only possible, of course, when a connection is realized between the two regions.

10.2.2   Market Outcomes with Full Integration

The most extreme form of connecting markets is realizing a full merger without any remaining cross-border bottlenecks. When markets merge into one integrated market, the demand and supply curves are merged as well. The integrated curves can be determined as the sum of the curves of the individual markets because, at a given price, producers and consumers in both countries supply or demand a specific

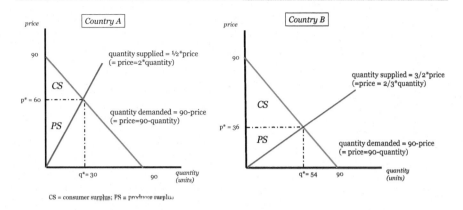

Fig. 10.1 Market outcomes in situation without market integration

number of commodities. The aggregated demand curve and supply curve are as follows:

$$S(p) = S_A(p) + S_B(p) = 2p \tag{10.6}$$

$$D(p) = D_A(p) + D_B(p) = 180 - 2p \tag{10.7}$$

The equilibrium outcome of this integrated market can be determined by equating these aggregated demand and supply curves:

$$S(p) = D(p) \overset{yields}{\rightarrow} 180 = 4p \overset{yields}{\rightarrow} p = 45, q = 90 \tag{10.8}$$

The equilibrium price and quantity in the integrated market are 45 and 90 euro per unit, respectively (see Fig. 10.2). The amount consumed in both countries is the same, as both countries have the same demand function (i.e. the quantity consumed in both countries is 45 units). The production, however, differs between these countries. In country A, the production is 22.5 units, while country B produces 67.5 units.[1] Country A produces less than without the market merger, because it has relatively expensive production technologies. As the domestic production in this country is below the domestic consumption, this country has to import. Country B produces more than what is domestically consumed and exports the difference. As a result, country B is exporting (67.5 − 45=) 22.5 units, while country A imports these (45 − 22.5=) 22.5 units.

From this example, it appears that (an unconstrained) market integration results in equal prices between countries. From an economic perspective, therefore, market integration is defined as a situation in which traders are able to arbitrage between

---

[1]Production in country A is based on the supply curve of 1/2* price which gives 22.5 units, and production in country B is based on the supply curve of 3/2* price which gives 67.5 units.

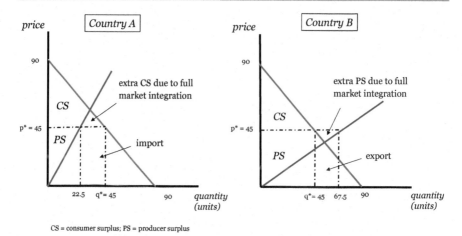

CS = consumer surplus; PS = producer surplus

Fig. 10.2 Market outcomes in situation with full market integration

countries resulting in equal prices. The relationship between market integration and regional price differences is called the law of one price, which says that in the absence of transportation constraints and (marginal) costs, prices in the related countries are equal. This law of one price is defined as follows:

$$p^A < p^B + t^{B,A}; q^{B,A} = 0 \tag{10.9}$$

$$p^A = p^B + t^{B,A}; q^{B,A} \leq Q^{B,A} \tag{10.10}$$

$$p^A > p^B + t^{B,A}; q^{B,A} = Q^{B,A} \tag{10.11}$$

This economic law says that the commodity flows from market B to market A ($q^{B,A}$) will be zero if the domestic price (p^B) plus the transport costs ($t^{B,A}$) exceeds the price in market A (p^A). If the commodity can freely flow between both markets (i.e. the flow is below the capacity constraint, $Q^{B,A}$), the price difference between both markets will be equal to the costs of transportation. Price differences can only remain higher than the transportation costs when the cross-border capacity is below the volume needed to equalize prices.

Note, however, that market integration is not equal to realizing direct connections between countries, as the integration of prices can also be realized through trade with other countries (see Box 10.1). Another factor which may contribute to harmonization of international prices is the presence of a common factor that affects the markets in the various countries. An example of such a common factor in the gas market is the outside temperature: during cold winter days, gas demand for heating will be high in a number of countries, which may raise gas prices everywhere.

The above example also shows that all countries benefit from international trade. Country A benefits because consumers can consume more and pay lower prices resulting in a higher consumer surplus (see Table 10.1). Country B benefits because producers can produce more and receive higher prices resulting in a higher producer surplus. However, producers in country A produce less, while they also receive a lower price which results in lower profits. In addition, consumers in country B consume less, while they also have to pay more. Their decline in welfare is, however, more than compensated by an increase in welfare of the other consumers and producers who consume and produce more. Hence, market integration (and, in general, international trade) not only results in higher overall welfare, but it also redistributes welfare.

10.2.3 Market Outcomes with Limited Integration

Above, we have assumed that there is an abundant capacity for international trade resulting in full market integration. In electricity and gas markets, this would imply that the capacity of the cross-border network is more than sufficient to transport the import/export flows. Often, however, this capacity is restricted from time to time, as discussed in Sect. 7.3. Suppose in the above example, the magnitude of the cross-border capacity (C) is 10 units. This implies that the maximum import (i.e. the difference between the domestic consumption and production) in the country with the highest marginal costs (i.e. A) cannot be higher than C (i.e. 10), while the export from country B faces the same constraint, which gives the following restrictions on international trade:

$$D_A(p) - S_A(p) \leq C \tag{10.12}$$

$$S_B(p) - D_B(p) \leq C \tag{10.13}$$

The implication of the presence of a constraint on trade is that the markets are merged up to the capacity of the constraint. Because of the constraint, the equilibrium outcomes have to be determined per market:

$$D_A(p) - S_A(p) = C \overset{yields}{\rightarrow} (90 - p) - \frac{1}{2}p = 10 \overset{yields}{\rightarrow} p = 53.33, q = 36.67 \tag{10.14}$$

$$S_B(p) - D_B(p) = C \overset{yields}{\rightarrow} \frac{3}{2}p - (90 - p) = 10 \overset{yields}{\rightarrow} p = 40, q = 50 \tag{10.15}$$

In country A, the domestic consumption in the market equilibrium (36.67) is 10 units higher than the domestic production (26.67), while in country B, the domestic consumption (50) is 10 units lower than the domestic production (60) (see Fig. 10.3). This implies that in country A the price must be lower than compared to

Table 10.1 Numeric welfare effects of integrating markets in several variants

	No integration		Full integration		Limited integration	
	Country A	Country B	Country A	Country B	Country A	Country B
Production (quantity)	30	54	22.5	67.5	26.67	60
Consumption (quantity)	30	54	45	45	36.67	50
Import (quantity)	0	0	22.5	0	10	0
Export (quantity)	0	0	0	22.5	0	10
Price (euro/unit)	60	36	45	45	53.33	40
Producer surplus (euro)	$(30 * 60)/2 = 900$	$(54 * 36)/2 = 972$	$(22.5 * 45)/2 = 506.25$	$(67.5 * 45)/2 = 1518.75$	$(26.67 * 53.33)/2 = 711.15$	$(60 * 40)/2 = 1200$
Consumer surplus (euro)	$(90 - 60) * 30/2 = 450$	$(90 - 36) * 54/2 = 1458$	$(90 - 45) * 45/2 = 1012.5$	$(90 - 45) * 45/2 = 1012.5$	$(90 - 53.33) * 36.67/2 = 672.34$	$(90 - 40) * 50/2 = 1250$
Total welfare (euro)	1350	2430	1518.75	2531.25	1383.50	2450
Congestion rents (euro)	n.a	n.a	0	0	$(53.33 - 40) * 10 = 133.33$	
Congestion costs (euro)	n.a	n.a	0	0	$(1518.75 + 2531.25) - (1383.50 + 2450) = 216.50$	

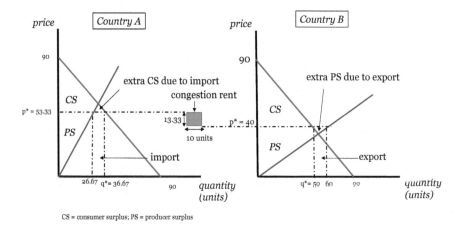

CS = consumer surplus; PS = producer surplus

Fig. 10.3 Market outcomes in situation with limited market integration

the situation of autarky (without any international trade), while in country B the price must be higher due to the export, but this effect is smaller than in the situation of full market integration.

Because of the presence of price differences between both countries, there is a remaining benefit for traders. In the equilibrium situation, traders can import energy in country B at the price of 40 euro/unit and sell it in country A at the price of 53.33 euro/unit. This benefit is called the congestion rent, which is the profit resulting from the presence of congestion. The size of this rent is 10 units times the price difference $(53.33 - 40) = 13.33$ euro/unit $= 133.33$ euro.

Traders can only realize the congestion rent themselves if the cross-border capacity would be freely available to them. If this capacity is, however, operated by a network operator, this operator may charge a price for using the capacity (see also Box 7.2). As long as this price is smaller than the price difference between the two countries, traders still have an incentive to buy in the lowest priced country and export it to the highest price country. Consequently, the network operator is able to charge a price which is almost equal to the price difference. Hence, this operator captures the congestion rent.

When price differences between markets remain, they are not only congestion rents, but also congestion costs. The congestion costs consist of the loss of welfare because of the congestion. These costs are equal to the difference in the welfare realized in case of full market integration and the welfare realized when then there is a limited integration. In this example, the size of these congestion costs is 216.50 euro.

Box 10.1 Interconnecting markets is not equal to integrating markets

Although market integration is strongly linked to interconnecting markets, these two concepts do not have the same meaning. Two countries may be connected with each other through cross-border transport capacity, but the markets do not need to be perfectly integrated. And the other way around, countries may be perfectly integrated, while they are not connected to each other. The former situation happens when prices in two neighbouring countries remain different despite the presence of interconnection capacity. This occurs when this capacity is not sufficient to facilitate all traded required to realize equal prices in both markets, like in the example in Sect. 10.2.3. Hence, market integration is defined as the situation in which market prices are equal or are not more different than the transportation costs.

Hence, equalization of prices only occurs when market parties are able to import from the lower-cost country and export to the higher-cost country until the prices are equal to the transportation costs. From this follows that market integration is facilitated by increasing cross-border transport capacity, but it does not necessarily depend on this. Market integration may also happen when there is a third country which is connected to both other countries. For instance, if countries B and C don't have a direct connection with each other, but both are connected with country A, then traders may be able to realize full market integration between all three countries, provided that the transport capacities between the countries A and B and between countries A and C are sufficient to facilitate all the trade that is needed to equalize prices (see Fig. 10.4).

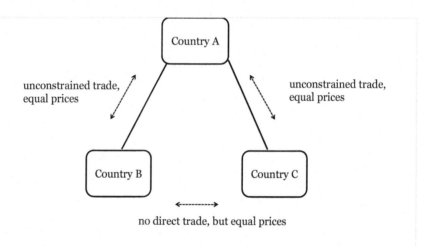

Fig. 10.4 Market integration between two countries without direct interconnection

10.3 Economic Effects of International Trade

10.3.1 Introduction

While coal and oil markets have been global markets for decades, gas and electricity markets have been hindered by cross-border barriers for many years. These barriers are both physically, i.e. in terms of technical transport capacity, and contractual, i.e. in terms of the availability of existing capacity for market participants. The presence of network constraints may give market players the ability to influence market outcomes. Hence, removing cross-border network bottlenecks may facilitate competition as it reduces the ability of players to behave strategically. In addition, integration of markets may have more benefits as it may also result in higher productive efficiency and higher security of supply (Giesbertz and Mulder 2008; ACER 2013). In addition, integration of markets may reduce the costs of producing a commodity, while the latter means that a market may become less vulnerable to shocks. In this Section, we briefly discuss the economic reasons to remove barriers for international trade.

10.3.2 Effects on Productive Efficiency

A major source of the potential welfare contribution of market integration refers to the benefits in terms of productive efficiency (Dutton and Lockwood 2017). Without interconnections, each country produces electricity that is needed to satisfy demand using the domestic generation capacity even if the marginal costs are high. In the presence of interconnections, however, countries with high marginal costs and, hence, high electricity prices will import the electricity from other countries that can use technologies with lower marginal generation costs, as we have seen in the previous section. Raising interconnection capacity between two markets, therefore, may result in lower overall costs of production when it enables low-cost producers to export a commodity to a neighbouring market where the marginal production costs (and price) are higher.

A precondition for having this benefit of international trade is, of course, that the countries differ in their abilities to produce a commodity. In the case of electricity, some countries have the option to generate electricity through hydropower plants (e.g. Norway), while others may have better access to primary energy sources such as coal (e.g. Germany) or natural gas (e.g. the Netherlands) (see Box 7.8). Such differences in production costs partly result from differences in natural circumstances. Costs differences may also be the result of historical choices regarding the production portfolio. A country that has extensively invested in nuclear power plants, for instance, may have a competitive advantage at times of high gas or coal prices over countries that have more invested in fossil-fuel power plants.

A country that is mainly using hydropower plants, for instance, is able to generate electricity at low marginal costs when the water reservoirs are full of water after a period of rain fall (because the opportunity costs of using water are negligible), but when these reservoirs are almost empty the electricity price will be high because of scarce water resources (and high opportunity costs of water use). In the latter period, consumers in this country may want to import electricity from other countries with lower prices, while in the former period, producers may want to export in order to realize higher prices. To what extent this import and export take place also depends on the marginal generation costs in other countries. The marginal costs in countries with generation portfolios based on fossil-fuel power plants are strongly related to the prices of natural gas or coal, which implies that the electricity prices in these countries fluctuate with these fuel prices (see Sect. 3.3.4). Hence, when markets with different generation portfolios are connected to each other, the incentives to trade fluctuate with changes in weather circumstances and changes in relative fuel prices.

Box 10.2 Spillover effects of local energy transition policies

Regional spillover effects of energy policies do not only occur between, but also within countries. Empowered by the decentralized character of renewable energy in combination with financial support schemes (see Sect. 8.4), some local communities try to become independent from the international electricity markets. This preference for independent renewable energy systems is connected with preferences of the concerned agents for physically realizing the energy transition themselves without depending on existing market systems. This tendency to be independent affects, however, all parts of an electricity system as it reduces the demand response potential in the market, the financial basis for maintaining the networks as well as the basis for taxing energy.

Such spillover effects to other parts of a market are normal phenomena, and all market participants will continuously have to adjust to changing circumstances. These spillover effects imply, however, a rise in the costs for other market participants. The pursuit of local systems where all energy is renewably generated as much as possible in the own region raises the costs of the energy transition, though, while the volume of renewable energy generated nationally does not necessarily increase. After all, the smaller the region that has to generate the renewable energy, the fewer its possibilities to choose the most efficient options. The costs of local solutions, therefore, are contrary to the benefits of international integration of markets, as the latter process generally results in a higher productive efficiency, more competition while it also increases security of supply.

Another factor why international trade may contribute to productive efficiency is the presence of differences in load patterns between countries (see e.g. Schavemaker and Sluis 2017). With restricted possibilities for international trade, countries need a higher domestic production capacity in order to deal with their peak load. Hence, countries are more forced to utilize plants with relatively high marginal

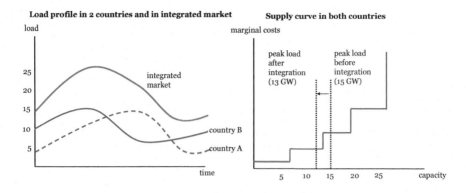

Fig. 10.5 Impact of market integration on load profile and dispatch of power plants

costs. In case of interconnections, however, these plants may not be needed if the peak load of the integrated market is relatively lower than the peak load in the separate countries. This can be illustrated by the following example.

Assume there are two countries (A and B) with a load profile as shown by Fig. 10.5. The peak load has the same value in both countries (say 15 GWh/h), but they are not at the same moment of time. The lowest values of the load profile are also at different moments, while they are also at a different level (4 GWh/h in country A and 7 GWh/h in country B). Without market integration, each country needs to have sufficient generation capacity to meet the peaks in load.

By integrating the markets, the peak load of the integrated markets is less than twice the peaks of the individual countries (in this example 26 GWh/h), because these peaks do not occur at the same time. Hence, the load in the integrated market has a lower volatility what also follows from the ratio between the highest and lowest load level during a period. In country A, this ratio is (15/4=) 3.75 and in country B (15/7=) 2.14, but in the integrated market this ratio has declined to 26/13 (=2.0).

The lower peak-load levels are relevant for the productive efficiency of both markets. Suppose both markets have the same generation portfolio (i.e. the same merit orders), then each market will produce 50% of the total production in the integrated market. Because of the decrease in the peak load in the integrated market, the peak-load production per country decreases as well: in this example from (30/2=) 15 GWh/h to (26/2=) 13 GWh/h, which implies that the power plants with the higher marginal costs do not need to be dispatched anymore. As a result, the marginal costs in the integrated system are lower than in the individual non-integrated markets. This implies that integrating countries with an equal generation portfolio but with a different load profile may result in less volatile energy prices. Because of this mechanism, market integration also affects investment decisions. In an integrated market, less investments in peak capacity are needed because the load-duration curve is flatter (see Sect. 7.5).

Table 10.2 Impact of integrating markets with different load profiles on RSI

	Country A	Country B	Integrated market
Number of firms and installed capacity per firm	2 firms, each 10 GW	2 firms, each 10 GW	4 firms, each 10 GW
Peak load	15 GWh/h	15 GWh/h	26 GWh/h
RSI per firm, during peak demand	10/15 = 0.66	10/15 = 0.66	30/26 = 1.15

10.3.3 Effects on Competition

Market integration may also have a positive effect on competition when there are more market participants in a larger market, which makes separate producers less indispensable. Suppose that in the above example, in both country A and B the electricity is generated by two firms each having the same level of generation capacity (say 10 GW each). During peak-load hours (when load increases to 15 GWh/h) in the situation without integration, the RSI of each firm is 0.66, which means that each firm is pivotal and has the ability to act strategically (see Table 10.2). By integrating both markets, the integrated peak-load becomes 26 GWh/h. The RSI of each firm is now 1.15,[2] which means that no firm is pivotal anymore. As the utilization of the generation capacity decreases and the electricity prices decrease as we have seen above, however, there will be less investments and, hence, less capacity after a number of years (see also Sect. 7.5). Consequently, the RSI in the integrated market may decrease again. This shows that market integration may have short-term positive effects on competition, but in the long term this effect may be neutralized by responses on firm level.

In addition, market integration makes it also harder to collude. In case no firm is pivotal and no one is able to act strategically, firms can try to do this by colluding with other firms. This collusion can be done without any explicit agreement (so-called tacit collusion), but by monitoring each other's behaviour and giving signals that a firm is prepared to coordinate their production decisions. Such a behaviour is legally not forbidden, but may harm the efficiency of markets. Such a tacit collusion is more complex when there are more firms supplying to the market.

The coordination of behaviour can also be done explicitly by concluding contracts (so-called cartel agreements). In most countries, however, such agreements are strictly forbidden by competition law (see Sect. 3.4). The ability to collude explicitly decreases when the number of firms is higher. This is related to the fact that a higher number of firms in a market makes it more difficult to reach an agreement on how the coordination should be implemented, while it is also easier for individual firms to cheat, i.e. not to stick to an agreement. Hence, because market integration increases the number of firms, it reduces the ability for strategic behaviour through coordinated conduct. As an indication for this ability,

[2]RSI of a firm is calculated as the aggregated capacity of all other players (i.e. 30) over total market demand (i.e. 26). See Sect. 9.4.2.

competition authorities look at structural measures. In the above example, the HHI of the non-integrated markets is 5000 (which indicates a highly concentrated market), but in the integrated market the HHI has declined to 2500 (which is still an indication of a concentrated market).[3]

10.3.4 Effects on Security of Supply

Integration of markets may also contribute to a higher degree of supply security in the short term, because a larger system reduces the chance that supply and demand shocks occur simultaneously in a large part of the market, while at the same time there are more options for flexibility. If in country A, 10% of the installed generation capacity has to be taken out of production because of a sudden technical outage, and this supply shock refers to only 5% of the installed capacity in the integrated market. Hence, market integration reduces the impact of a particular shock to the market. In the long term, however, market parties will adapt their volume of installed capacity because of the lower prices, which mitigates this effect.

In addition, market integration may increase the flexibility of the market to respond to a shock as more flexibility options as well as market participants are present that can react, such as other generation units that can increase production or consumers than can reduce their consumption (see Box 10.3).

Box 10.3 Resilience of integrated gas markets to unexpected shocks

The European gas market has witnessed a number of supply shocks. In January 2009 and February 2012, for instance, the supply of Russian gas through the Ukraine was interrupted during a number of days in the winter periods when the outside temperatures were very low, and hence, the demand for gas in Europe was high. The disruption in the transit route to Western Europe in 2009 resulted in a significant loss of gas flows destinated for the southern German gas market (NCG). The decline in supply to this market in a period of very low temperatures immediately raised the day-ahead price from 23 to 27 euro/MWh (see Fig. 10.6). The reduced supply was replaced by a combination of storage withdrawals, increased flows from Russia through the Yamal-Pipeline transiting Belarus and Poland, and higher cross-border flows from the Netherlands. These extra flows from the Netherlands were largely sourced from the United Kingdom through the interconnector, driven by increased price differentials between the UK gas spot market (NBP, Net Balancing Point) and the continental gas hubs. The increased export from the UK resulted in a relatively strong decline of the British storage levels. This increased demand for gas in the UK raised the spot price at NBP. Because of

---

[3]As both firms have the same generation portfolio, the market share of each firm in the non-integrated market is 50%, resulting in an HHI of $2 \times 50\%^2 = 5000$. In an integrated market, the market share of each firm is 0.25, resulting in an HHI of $4 \times 25\% = 2500$. See Sect. 9.4.2.

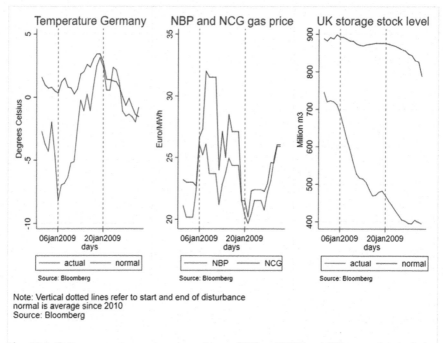

Fig. 10.6 Daily average temperature, gas prices on NBP and NCG and UK storage levels during Ukraine gas supply crisis, January 2009. *Source* Bartelet and Mulder (2020)

these responses to the supply shock, no forced terminations of gas were needed to help the market to find new equilibrium solutions. When the supply of natural gas from Russia through the Ukraine was restored, the gas prices in Europe returned back to normal levels as well. This episode showed that the gas market was able to handle a sudden shock in supply.

Moreover, in a larger regional market, generally a larger variety of production techniques is used, which means that the total supply is less dependent on a particular source, which reduces the risk that a mayor part of the supply will be disturbed.

On the other hand, integrating markets also implies that each market may be affected when in one market a disruption in supply occurs, while in the extreme situation of isolated markets, markets are only sensitive to what happens in the own region. Although the sensitivity to risks in isolated markets may seem to be smaller, these markets need to take more measures in order to prevent negative effects of a disturbance as the effects are not mitigated by being connected to other markets.

10.4   Regulatory Measures to Foster International Integration

10.4.1   Introduction

In order to realize the welfare-increasing effects of integrating markets, regulators can take a number of measures. The first measure to consider, because it is most efficient, is to increase the availability and utilization of the existing cross-border capacity. This is discussed in Sect. 10.4.2. When the existing capacity is highly available to market participants and fully utilized, it may be efficient to extent this capacity, which is discussed in Sect. 10.4.3. Generally, such investments in extending the cross-border capacity are done by the operators operating the grid on both sides of the border. Sometimes, however, these network operators may not want to make such investments because they perceive the risks of such investments as to high. In such cases, commercial investors may be willing to step in. This would create other regulatory problems, as commercial investors generally want to have exemption from regulation. This topic is discussed in Sect. 10.4.4.

10.4.2   Virtual Extension of Cross-Border Capacity

In energy markets based on zones, such as the entry–exit schemes in gas markets and the market zones in European electricity markets, traders have free access to inter-zonal capacity, but for the international trade they have to cross the market area borders. This means that additional arrangements have to be made to get access to the foreign market. In gas markets, firms used to book the capacity for cross-border trade a long period in advance. This capacity was allocated on the basis of First-Come-First-Served (FCFS), which means that the available capacity was allocated among potential users in the order of the requests. As in the past, the gas market was dominated by only a few firms, these (incumbent) firms acquired all capacity. As a result, the cross-border capacity was not available for other firms (including new entrants) when they wanted to import or export. This even happened when the incumbent firms that acquired the access rights to the capacity did not always fully use (i.e. nominate) the capacity. Although physically there was sufficient transport capacity available, virtually the capacity was congested. This type of congestion is called contractual congestion. This type of congestion on borders also occurred in European electricity markets.

Regulators can address this issue in two ways. One way is to impose rules on the use of booked capacity which will not be nominated. An example of such a rule is the Use-It-Or-Lose-It (UIOLI) clause, which enables the network operator to reallocate capacity that is booked in advance, but that not will be used. Another type of measure is to impose an obligation on market parties that they have to resell booked capacity when they are not going to use is (i.e. a Use-It-Or-Sell-It (UIOSI) clause). Regulators can facilitate market parties to resell capacity by creating a secondary market for transport capacity.

The other, more direct way to addressing the issue of contractual congestion is to change the way of primary allocation. Instead of using the FCFS method, network operators can sell the access to the cross-border capacity through open auctions in which all potential market parties can join. This opens the market to new entrants as well-existing players that want to use more cross-border capacity.

In electricity markets, the cross-border capacity issue is increasingly addressed by so-called market coupling, which is also called implicit auctioning of cross-zonal capacity. This means that the spot markets for the commodity of electricity are coupled with the market for capacity of transport (see Fig. 10.7). Traders do not need to book cross-border capacity separately for the commodity trading, as the operators of the spot markets solve this by using the latest information on expected flows received from the operator of the high-voltage network. This information refers to the expected utilization of bilateral connection lines in relation to the available capacity of those lines (i.e. the available capacity for trade). An advanced form of market coupling is so-called flow-based market coupling in which the allocation of capacity of the various border points is based on an estimation how the current flows within the electricity network, using the insights from the physical (Kirchhoff's) law regarding the paths of current through a meshed network (see Sects. 2.2.5 and 2.4.4). The difference with just market coupling is that in the flow-based system, all cross-border points are other relevant network components are simultaneously included in the analysis. This system has been introduced in the Northwest-European electricity market, enlarging the available capacity for international trade.

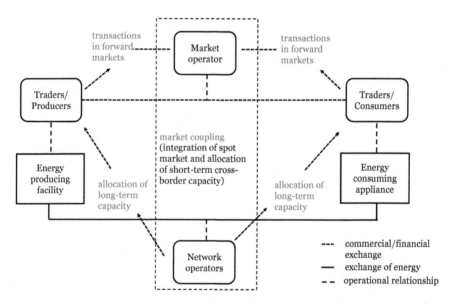

Fig. 10.7 Relationship between allocation of cross-border capacity and commercial transactions

This systems of market coupling hold, however, only for short-term markets. For the markets where long-term products, like year-ahead contracts are traded, market parties still have to buy capacity for that period. The allocation of this long-term capacity is allocated through auctioning.

10.4.3 Physical Extension of Grid Capacity

By virtual extension of cross-border capacity, this capacity remains the same in physical sense but is used more efficiently. This regulatory measure is relatively efficient, as there are no investments needed in the grid itself. The impact of this virtual extension is constrained, however, by the existing physical capacity of the grid. In some cases, however, it may be efficient to extent the grid capacity, both within a market zone and between market zones. The former is relevant for international trade when congestions within a market zone hinder cross-border flows, which effect results from the fact that a constraint somewhere in a meshed power network, affect flows everywhere (see Sect. 2.4.4).

The value of investing in grid capacity in order to facilitate international trade is determined by the expected impact on differences in power prices between the respective markets. The higher the price differences, the more traders are prepared to pay for getting access to the cross-border capacity and the more they want to utilize the line, as was discussed in Sect. 10.2. The size of the price differences can be seen as the shadow price for the cross-border capacity, i.e. this is the price traders are prepared to pay to have one unit more capacity. An investment in an additional unit of cross-border capacity is efficient when the present value of the future congestion costs exceeds the unit costs of the investments. The economic interpretation of this is that an investment in grid capacity is efficient when it results in a higher productive efficiency (i.e. lower costs per unit of production of energy), higher allocative efficiency (i.e. more intense competition and, hence, lower prices for energy users) or higher security of supply (i.e. less risk of very high prices or interrupted supply).

In a regulated environment, the profits for the network operator do not depend on the expected cross-border price differences, as the revenues for the network operator are maximized. In such a regulated environment, the decision to invest in cross-border capacity should be based on a social cost–benefit analysis as was shown by the stylized examples of Sect. 10.2. If the outcome of this analysis is positive, then the regulator can allow the network operator to realize the investment.

When the extension of the grid capacity has been realized, the investment value can be added to the Regulated Asset Base (RAB), which raises the level of allowed revenues (see Sect. 6.7.1). This regulatory reimbursement of efficient investments gives an incentive to network operator to search for those investments that contribute to solving congestions.

When the network operator auctions the (long-term) capacity and, hence, realizes more revenues than what is allowed, this surplus of revenues can be transferred to network users by lowering future network tariffs. Consequently, network users benefit from any remaining congestion rents, while the network operator receives regulatory incentives to mitigate congestion costs by investing in grid capacity.

10.4.4   Public or Merchant Investment?

Although it may be clear that increasing the physical cross-border capacity may contribute to the efficiency of markets, it not always evident what type of agent should realize this investment. In principle, the highest welfare improvement is realized when the investment is done by a public investor (i.e. the network operator) that maximizes social welfare, which is equal to the sum of consumer and producer surplus. In theory, the welfare-maximizing design of an investment is where the marginal social benefits equal the marginal social costs of the investment.

There are, however, several factors which may make that the incumbent investors (i.e. the network operators) are not able to realize this level of investments in interconnections. These investors do not always have the incentives to make these investments (for instance, because of lack of unbundling with commercial activities in generation), regulators may not be able to commit to a long-term policy regarding allowed revenues, while there may also be coordination problems between the regulators of the markets on both sides of the interconnection (Rubino and Cuomo 2015). Therefore, merchant investors may be an alternative to realize more, welfare-improving interconnections.

The major difference between a merchant investment and a similar investment done by an incumbent network operator is that for a merchant investor the objective is to maximize the congestion profits resulting from using the line. As a result, a merchant investor does not want to remove all congestion as then the price differences between the two markets will vanish, and hence, no congestion rents can be earned anymore. Consequently, merchant investors have an incentive to invest less than what would be socially optimal (Sereno and Efthimiadis 2018). In addition, merchant investors may also have an incentive to depart from the optimal design of the investment as they maximize their private profits coming from congesting revenues instead of the benefits for the full network, including effects on reliability. Indirect effects of a particular interconnection on the availability of capacity in other parts of the network, resulting from the physical (Kirchhoff's) laws, may be neglected, just as the impact on reliability of the grid. Nevertheless, the merchant interconnections may create significant welfare effects (Doucet et al. 2013; Gerbaulet and Weber 2018).

Moreover, merchant investments in interconnections may result in an uneven distribution of welfare resulting from the interconnection, i.e. merchant investors may acquire a major part of the welfare improvement of the interconnector if no additional (regulatory) conditions would be imposed on the merchant investment. Merchant investors, just as network operators, sell the interconnection capacity to traders (or players in the electricity market) and the maximum price these traders are prepared to pay is (theoretically) determined by the (expected) differences in prices between both markets, as was shown in Sect. 10.2. The more the merchant investor is able to charge a fee for using the line which is close to this willingness-to-pay, the more the welfare improvement resulting from the interconnector is realized by the merchant investor. This is the fundamental economic effect that the owner of the scarce resource gets the scarcity (congestion) rent, the so-called Ricardian rent.

This distributional effect of the welfare of an interconnection can be formulated, as indicated in Sect. 10.2, in terms of the microeconomic concepts of consumer surplus (CS), producer surplus (PS) and the surplus for the scarce resource, in this case the congestion rent (CR). These economic concepts are, however, not really useful for policy discussions in practice. The concepts CS, PS and CR are theoretical concepts, not saying much about which groups in society are receiving which part of the improvement in welfare. The CS refer to the surplus realized by all users of the commodity, which means in the case of energy that the 'consumers' mainly consist of large energy-intensive industries, possibly partly owned by a group of international investors. For PS holds more or less the same: the 'producers' are everyone who produces electricity, so including residential prosumers and small-scale wind parks. The CR can be acquired by a commercial investor, but this player can be a firm as any other.

Hence, one cannot conclude that because of the fact the 'merchant takes it all' having no interconnector is equal to having a merchant investment. Such a conclusion is based on translating the theoretical economic concepts CS, PS and CR to the decision-making in practice without paying attention to the players (firms, people) who are playing the role of consumer, producer or network operator. A merchant investment that generates a positive overall welfare is always beneficial for society, even if the merchant investor acquires this welfare completely.

A regulator may allow a merchant investor to build a line in a network that is operated by a network operator and that, hence, the tariffs for using the merchant line will be regulated (see Box 10.4 on possible consequences for the allowed compensation for the costs of capital).

Box 10.4 Risks and returns for a merchant investor under regulation

When an investment in a regulated network is realized, the regulator determines the required compensation for the costs. The central parameter in the determination of the compensation for the capital costs is the WACC (see Sect. 6.5). To what extent, should the estimate for the WACC be different in case a merchant investor realizes the investment? In order to answer this question, the various components of the WACC formula have to be analysed. The basic components are the (1) costs of equity, (2) costs of debt and (3) the gearing.

1. The risks for the providers of equity are measured through the so-called equity premium, which is the compensation required by investors as compensation for the non-diversifiable risk. This premium depends on the market-risk premium and the equity beta. The market-risk premium is

completely independent of the characteristics of individual projects, so it does not matter whether a project is conducted by a TSO or a merchant investor, the market-risk premium remains the same. The equity beta, however, is to some extent related to the characteristics of the investor. The equity beta depends on the asset beta, which describes how the risk of a project is related to the market risk, and the gearing, which is the ratio debt/equity in the project. A merchant investor may face higher risks because it may be less well equipped to minimize the risks caused by general economic circumstances (i.e. the market risk) than a TSO operating the full electricity grid, resulting in a higher asset beta.

2. The risks for the providers of debt are translated into the so-called debt premium: the extra compensation above the risk-free interest rate which is needed to compensate providers of debt for the extra risks incurred when they provide debt capital to the project. The level of the risk premium does not only depend on the characteristics of an investment project, but also on the financial strength and the type of ownership of the firm. If a regulator bases the risk premium on the financial rating of firms which are state owned and mainly active in a regulated business, this premium may be too low for a firm that faces a higher financial risk purely because of its type of ownership and character of (revenues from) other activities.

3. In addition, a merchant investor may be less able to attract debt than a TSO operating in a regulated environment and backed by the state as owner, which implies that the share of equity is higher (i.e. gearing is lower). A lower gearing implies that the asset risk is allocated over a higher amount of equity, which lowers the equity beta.

Concluding, when a regulator decides to allow a non-TSO company (i.e. a commercial investor) to build and operate a cross-border connection, then it may also need to adapt the allowed rate-of-return on capital (i.e. the WACC). *Source* Mulder (2018).

10.5 International Spillover Effects of National Energy Policies

10.5.1 International Dimension of Energy Policies

Integration of markets can positively affect the functioning of markets, in particular the productive efficiency, the intensity of competition and the security of supply, as we have seen in Sect. 10.3. In addition to this, market integration may also have an effect on the effectiveness of national energy policies. Because of connecting and integrating markets, national policy interventions are not limited to the national market, but spread over the entire integrated region. This holds also for energy transition policies.

Energy transition policies in Europe are highly nationally oriented, but they result from European policy. Under the EU Renewable Energy Directive, there are binding objectives for the portion of renewable energy both on an EU and on a Member State level. In the EU as a whole, for instance, the portion of renewable energy in the total energy consumption must be 20% in 2020 and 32% in 2030, with different objectives for the separate Member States. The Netherlands, for instance, was required to generate at least 14% renewable energy in 2020, but this percentage was between 30 and 49% for example for the Nordic countries (EU 2018). These differences in obligations across Member States result from differences in starting positions and their possibilities to promote renewable energy. Although the Member States are obliged to realize the renewable energy objectives, they are free to choose how they translate these obligations both in terms of tightening the objectives and the choice of policy instruments. Consequently, the EU countries implement a variety of policy instruments to foster the share of renewable energy, such as different types of support schemes and command-and-control measures as a forced closure of coal-fired power plants (see Chap. 8). These different types of policy measures may affect the energy markets in other countries.

10.5.2   National Policies in International Electricity Markets

Because energy markets are increasingly integrated, national measures to influence the composition of the domestic energy sector have cross-border effects. The increase in German wind power, for instance, has not only resulted in lower domestic prices, but also lower prices in neighbouring countries (see e.g. Mulder and Scholtens 2016). This reduction of the Dutch power prices, for instance, as a result of the German policies to promote renewable energy, has made subsidising renewable energy in the Netherlands more expensive, as the subsidies cover the difference between the costs of renewable energy (i.e. the LCOE) and the power prices (see Sect. 8.4). In addition, when the volume of the cross-border capacity is limited, the policy to stimulate renewable energy in one country may lead to larger price differentials with neighbouring countries, as occurred between Germany and France (Keppler 2017).

National measures to make the generation by, for instance, coal-fired power stations more expensive (e.g. by introducing a tax on coal) or even impossible (by mandatory closure) may, as a matter of fact, stimulate production by coal-fired power stations in the neighbouring countries. Such national measures shift the merit order of the domestic production upwards (in the case of coal tax) or to the left (in the case of closure), making other, foreign power plants more competitive. Therefore, a closure of power plants in one country will lead, ceteris paribus, to more generation with coal-fired power stations in the neighbouring countries and a consequent increase in their CO_2 emissions (Mulder and Zeng 2018). This does not only mean that the domestic environmental impact is partly undone, but also that the neighbouring countries will have to incur more costs to realize their energy transition objectives.

Apart from these cross-border price effects of differences in national energy policies, there is also a cross-border effect on the deployment of power stations. When wind turbines generate more electricity in a country, they reduce the domestic conventional production (because of the merit-order effect), but this may raise the conventional production in neighbouring countries. This paradoxical outcome is the result of the fact that the power supplied by wind turbines in a country must find its way through the network to the customers and this may not only result in local bottlenecks (as discussed in Sect. 7.3), but it may also reduce the available capacity for cross-border trade. This effect results from the physical (Kirchhoff's) laws determining how energy flows within an electricity grid (see Box 10.5). Energy flows that transit another country are called transit flows when the source (i.e. the producing country) and the sink (i.e. the consuming country) are different. When both source and sink are within the same country, these flows are called loop flows (Schavemaker and Sluis 2017). Network operators managing the cross-border grid have to keep sufficient capacity available for these kinds of flows. Therefore, they reserve a part of the technical capacity for this purpose, limiting the available capacity for trade. It appears that at many European borders, the cross-border capacity available for the market is less than 50% of the technical capacity, leaving much to gain in terms of efficiency (ACER-CEER 2019). Because of this mechanism, a higher production of renewable electricity in one country may have as a result that the neighbouring countries can import less, which may result in larger price differences.

Box 10.5 Allocation of technical capacity in zonal electricity markets

The capacity that is available for exchange between zones in electricity markets is not simply equal to the technical capacity. This is due to the physical laws that determine the flows of electricity (see Sect. 2.4.4). The capacity that can be made available on a particular border is also affected by the use of other components of the grid as electric energy always searches for the route of the lowest resistance. Network operators have, therefore, to reserve a part of the technical capacity on the border for the so-called transit and loop flows in order to prevent overloading of network components. Besides this margin for transit and loop flows, network operators typically reserve some capacity for emergency (contingency) situations. The remaining capacity can be used for the commercial exchange of electricity. This capacity is called the available cross-zonal capacity.

When the size of the capacity that is available for commercial trade has been determined, the next issue is to allocate this capacity as efficiently as possible. The first step in this allocation process is to allocate the capacity over various types of contracts: long-term, short-term and balancing (this step is called 'splitting'). In the next step, the capacity for the various types of contracts is allocated over users. The long-term capacity is generally auctioned, while the capacity for the short-term is increasingly integrated with the organization of the short-term (i.e. spot) markets. This integration is called market coupling. Through the introduction of (flow-based) market coupling, in which the trade in electricity is integrated with the allocation of cross-border capacity, the efficiency of capacity usage has increased strongly over the past years.

Given the increased international integration of electricity markets, these examples show that national energy transition policies must be coordinated to minimize their costs. The same applies to measures to improve the operation of the electricity market, such as through the introduction of capacity mechanisms as discussed in Sect. 7.5.5 (ACER 2013). After all, the introduction of a capacity mechanism in one country may lead to externalities in the neighbouring countries, such as lower electricity prices and, consequently, reduced investments in production capacity (Cramton et al. 2013).

Exercises

10.1 When two countries integrate, this may result in lower prices due to productive and allocative efficiencies. Discuss how this may affect the intensity of competition in the long term.

10.2 How does integration of two neighbouring electricity markets with different generation portfolios having different system-marginal costs affect the welfare of consumers and producers in both countries?

10.3 How can market integration contribute to stabilizing electricity prices?

10.4 How does extending the cross-border capacity between two countries affect the international spillover effects of national renewable energy policies in one country?

References

ACER. (2013). Capacity remuneration mechanisms and the internal market for electricity, 30 July.
ACER-CEER. (2019). Annual report on the monitoring of the internal electricity and natural gas markets in 2018. Ljubliana/Brussels.
Bartelet, H., & Mulder, M. (2020). Natural gas markets in the European Union: Testing resilience. *Economics of Energy & Environmental Policy, 9*(1).
Cramton, P., Ockenfels, A., & Stoft, S. (2013). Capacity market fundamentals.
Doucet, J., Kleit, A., & Fikirdanis, S. (2013). Valuing electricity transmission: The case of Alberta. *Energy Economics, 36,* 396–404.

Dutton, J., & Lockwood, M. (2017). Ideas, institutions and interests in the politics of cross-border electricity interconnection: Greenlink, Britain and Ireland. *Energy Policy, 105,* 375–385.

EU (2018). Directive 2018/2001 of the European Parliament and the Council on the promotion of the use of energy from renewable sources, 11 December.

Gerbaulet, C., & Weber, A. (2018). When regulators do not agree: Are merchant interconnections an option? Insights from an analysis of options for network expansion in the Baltic Sea region. *Energy Policy, 117,* 228–246.

Giesbertz, P., & Mulder, M. (2008). The economics of interconnection: The case of the Northwest European electricity market. IAEE Energy Forum, second quarter.

Keppler (2017). Rationales for capacity remuneration mechanisms: security of supply externalities and asymmetric investment incentives. *Energy Policy, 105,* 562–570.

Mulder, M. (2018). Merchant investments in interconnections in the European electricity market. CEER Policy Papers no. 4, September.

Mulder, M., & Scholtens, B. (2016). A plant-level analysis of the spill-over effects of the German Energiewende. *Applied Energy, 183,* 1259–1271.

Mulder, M., & Zeng, Y. (2018). Exploring interaction effects of climate policies: a model analysis of the power market. *Resource and Energy Economics, 54,* 165–185.

Rubino, A., & Cuomo, M. (2015). A regulatory assessment of the Electricity Transmission Investment in the EU. *Energy Policy, 85,* 464–474.

Schavemaker, P., & van der Sluis, L. (2017). *Electrical power system essentials* (2nd ed.). New York: Wiley.

Sereno, L., & Efthimiadis, T. (2018). Capacity constraints, transmission investments and incentive schemes. *Energy Policy, 119,* 8–27.

Distributional Effects and Equity Concerns in Energy Markets

11.1 Introduction

Well-functioning markets result, by definition, in efficient market outcomes, but this does not mean that the outcomes are always socially accepted as the optimal outcomes. This chapter shows that there are many possible efficient market outcomes depending on the initial allocation of endowments. In Sect. 11.2, it will become clear that the concepts of allocative efficiency and distribution have different meanings. The other two sections of this chapter discuss regulatory measures to redistribute welfare. Section 11.3 addresses the question how resource rents can be redistributed within society, while Sect. 11.4 discusses the issues of energy poverty and fairness and what types of regulatory measures can be used to make the distribution of welfare more aligned with social objectives.

11.2 Theory: Allocative Efficiency and Distribution

To understand the differences in the concepts of allocative efficiency and distribution and how they are related to each other, the so-called Edgeworth box is very useful (Varian 2003). In this box, the indifference curves of two economic agents (say, I and II) for the consumption of two different goods (say, A and B) are depicted. An indifference curve shows all combinations of products which give an economic agent the same level of utility and as a result that agent is indifferent, i.e. it does not have a preference regarding each of the combinations, as is explained in Sect. 4.2.1. The horizontal axis in the Edgeworth box shows the quantity of product A and the vertical axis the quantity of product B (see Fig. 11.1). It is assumed that the length of both curves indicates the total number of these products that is available in the economy. In this example, the total number of available units of both product types is 10. This implies that if, for instance, agent I consumes 6 units

© Springer Nature Switzerland AG 2021
M. Mulder, *Regulation of Energy Markets*, Lecture Notes in Energy 80,
https://doi.org/10.1007/978-3-030-58319-4_11

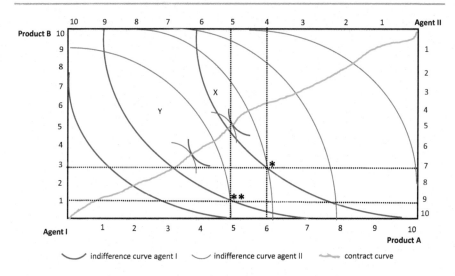

Fig. 11.1 Edgeworth box: efficient allocation for various distributions of resources

of product A, only 4 units are left for agent II. As the total number of products is constrained in this economy, there is a limited set of feasible allocations.

The indifference curves for agent I are depicted from the perspective of the angle on the left-down side of the box (the red lines), while the indifference curves for agent II are shown from the perspective of the angle on the right-top side (the blue lines). The longer the distance from an indifference curve to the respective origins, the higher the utility of the agents as more of one product is consumed without a reduction in the consumption of the other.

Suppose the initial allocation of the products is indicated in the box by *, in which agent I consumes 6 units of product A and 3 units of product B, while agent 2 consumes 4 units of product A and 7 units of product B. This allocation is not efficient, as both economic agents could arrive at a higher indifference curve by having a different mix of products. The possible options of reallocating the products to arrive at a higher efficiency are indicated by area X, which shows all combinations of the allocation of the products in which both agents enjoy a higher utility as these positions are further away from both origins than the indifference curves where * is located on. The same analysis can be done for any other initial allocation of the products. If the initial allocation of the products is, for instance, depicted by **, then the range Y between the two relevant indifference curves gives all options where trade would be beneficial for both agents.

In all these cases, the utility of both agents can increase when they enter in an exchange of products. From the areas X and Y, it is clear that there are many possible solutions in which both agents can enjoy an improvement. Within this range of options for realizing a higher utility for both agents, there is only subset of options which are called Pareto efficient, which are situations where the utility of some agents can only be further improved by reducing the utility of other agents. The set of Pareto efficient allocation consists of these allocations of the products where the indifference curves do not cross each other, but are tangent. In the figure, only one tangent situation is depicted, but there are many more in both the areas X and Y. In fact, a curve can be drawn from the origin of agent I to the origin of agent II which shows all Pareto efficient allocation of products. This curve reflecting all Pareto efficient allocations is called the contract curve.

From this theoretical analysis, we learn that for any given initial allocation of products, such as depicted by * or ** in the figure, a number of Pareto efficient reallocations exists, and that for all feasible initial allocation of products, a much larger set of Pareto efficient reallocations exists. Hence, allocative efficiency of markets is an important concept to analyse the functioning of markets, but it does not describe the full picture. Allocative efficiency only refers to a subset of potential efficient outcomes of the economic process. When the distribution of endowments is different, another subset of Pareto efficient allocations exists.

Societies may have preferences regarding the initial allocation of products as well as the distribution of the results of economic transactions which may call for additional regulation, even if markets function efficiently. For the assessment of alternative distributions of welfare, we need the concept of the social-welfare function. This function describes the welfare of a society as depending on the welfare of the individual economic agents. This function can be formulated in different ways. One way to do this is the so-called Benthamite welfare function, which defines the welfare of a society as the weighted sum of the utilities of the individual economic agents (Varian 2003). The weights depend on the societal or ethical debate in society on how to value the relative welfare positions of the individual agents. When the weights have been determined, the optimal allocation of resources can be chosen from the set of Pareto efficient allocations. Hence, which position on the contract curve of Fig. 11.1 maximizes social welfare depends on the societal debate on the relative weights for the various agents' utilities. If a society, for instance, attaches a higher value to welfare of agent I, then allocations more to the right top of the contract curve will be preferred.

This societal debate on the optimal distribution of welfare is very relevant in energy markets for two reasons. The first reason is that natural energy resources are distributed unevenly across the globe and economic agents, which means that there is an uneven distribution of revenues from the production of these resources as well. This may result in large differences in economic welfare. The other reason is that energy products are generally viewed as basic commodities which should be available at reasonable conditions to everyone, but this is not always the case. Below, we will discuss more into detail what these differences in economic welfare explain, and what regulators can do to address them.

11.3 Distribution of Resource Rents

11.3.1 Theory

In competitive markets, no supplier is able to make supra-normal profits in the long run. By supra-normal profits are meant profits which exceed the compensation required to recoup the costs of capital. In the short term, the revenues can exceed this level due to constraints on the supply side, but in well-functioning markets the resulting extra profits will attract investments to generate new supply, either from the existing suppliers and/or from entrants. This extension of supply will bring down prices, and hence, revenues. The opposite type of mechanism occurs when the demand level is strongly below the supply constraints. In such situations, prices are set by the short-run marginal costs, which may force inefficient firms, i.e. firms with marginal costs above the market price, to leave the market. As a result, the total capacity to supply reduces which may create scarcity in the market and, hence, higher product prices in the future.

As a consequence of this adaptation mechanism within markets, as was also shown in the discussion of investments in electricity markets in Sect. 7.5, in the long term the prices are just sufficient to give the suppliers a profit which is precisely needed to recoup their costs of capital. As this profit is needed to recoup the cost of capital, this profit cannot be seen as real profit. Therefore, this profit is just the accounting profit (π_a) as calculated as the difference between revenues (R) and operational costs (C_o).

$$\pi_a = R - C_o \tag{11.1}$$

In competitive markets, firms are not able to make more profit in the long run, which means that they are not able to realize a so-called economic profit (π_e). The economic profit is the profit which remains after subtracting the costs of capital (C_c) from the accounting profit:

$$\pi_e = \pi_a - C_c \tag{11.2}$$

Economic profits are also called rents which can be seen as income which are not related to a sacrifice (i.e. opportunity costs). Rents can only occur when there are factors present which hinder that new firms enter the market or that other firms can increase their supply, which are the mechanisms that make that in well-functioning markets rents cannot occur for a long period of time.

One of the crucial factors which facilitate the realization of rents is the presence of differences in natural characteristics. The classical example is a farmer who owns a productive piece of land, while other farms in his region only have less productive pieces of land. The former farmer is able to make more profit because it is making less costs but receiving the same price for the agricultural products as the other farmers in his region. This example was first presented by the economist Ricardo and that's why this rent is also called Ricardian rent. An example from energy systems is presence of oil rents resulting from the uneven spatial distribution of oil resources and the differences in production costs (see Box 11.1).

Another factor which may result in rents is a legal rule that give, for instance, a licence to a limited number of firms to conduct a specific activity. If a licensing scheme does not have the option to give additional licences to other players, the firms which have received a licence are more or less protected from competitive threats from entrants. Note that this protection from competition and the resulting ability to realize economic profits might be the objective of the licensing scheme. This holds in particular for licences to sell products from innovation, which are meant to give the innovating companies certainty about the revenues from the innovation. Hence, in this case a licensing scheme is necessary to address the positive externality of innovation (see Sect. 8.2).

The fundamental factor that explains the occurrence of economic rents is that the owner of a scarce resource is able to capture all the profits in a market. All other suppliers in a supply chain only receive compensation for the costs they make because of the competitive process they are subject to. When the owner of a scarce resource is protected from competition, he is able to make economic profit provided there is demand for the commodity.

This mechanism is illustrated by Fig. 11.3 which shows the occurrence of economic rents in two market situations. In both situations, there are three types of producers (A, B and C) using different types of technologies to produce a commodity. Producer A has the lowest costs (and, hence, the lowest required price), and producer C the highest. When demand exceeds the capacity of the producers A and B, then the marginal costs of producer C will set the price (market situation I). As a result, producers A and B will realize marginal revenues (i.e. a price) that exceed their marginal costs. If these marginal costs refer to the long-run marginal costs (i.e. including costs of capital), then the difference between the price and the marginal costs creates an economic profit or scarcity rent. In this case, such rents are also called inframarginal rents.

Box 11.1 Rents in the oil market

Because of the uneven spatial distribution of oil resources and the large differences in the marginal costs of producing oil, the inframarginal producers are able to make huge resource rents. This ability is increased in times when these producers are effective in their cartel behaviour to restrict production and, hence, raise the oil prices (see Sect. 9.2.1). In such periods, as occurred in the periods 1973/85 and 2003/2008, the oil market was tight and the OPEC cartel successful. This resulted in large rents from oil production, which constituted major parts of the GDP (10–50%) of these countries (see Fig. 11.2).

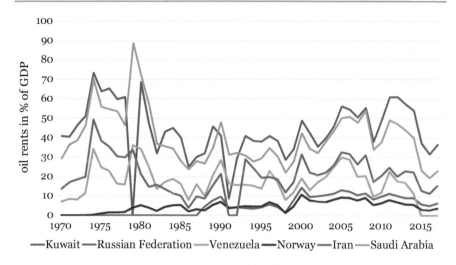

Fig. 11.2 Oil rents in % of GDP, per country. *Source* Worldbank

Such rents imply that investors are making a higher return on their investment than what is possible elsewhere, which makes that the existing companies have an incentive to increase their production and that other companies would like to enter the market. Suppose that this is only possible for producers using technology B, but not for producers using technology A (market situation II) (see Fig. 11.3). Consequently, the capacity of producer B will increase up to the point that the prices have decreased until the level of the long-run marginal costs. In that situation, this producer will not realize an economic profit anymore. Producer A, however, is still making an economic profit, because the capacity of his technology is constrained, while its long-run costs are below the costs of producer B. From this stylized example, it follows that (a) the occurrence of scarcity rents is an important incentive for investors to extend production capacity or to enter a market, as was also shown in Sect. 7.5, and (b) that these scarcity rents continue to exist when producers with the lowest costs cannot extent their capacity or are able to prevent someone is doing this (i.e. by organizing a cartel) (see also Conrad 1999).

11.3.2 Regulation

When scarcity rents are related to the use of natural resources, they are called resource rents. The magnitude of the available reserves is limited in the short term, which may result in positive economic profits. In the longer term, these economic profits trigger new investments in exploration for new resources and development of new reserves (see Sect.2.4.2). Generally, these reserves have higher marginal costs, which mean that the reserves with the lowest marginal costs still remain able to realize a positive economic profit.

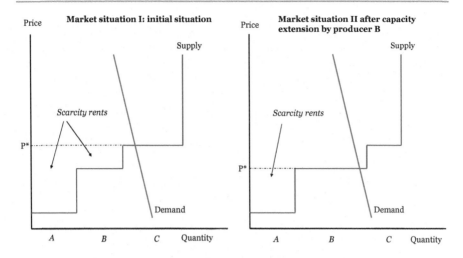

Fig. 11.3 Occurrence of scarcity (or: inframarginal) rents in markets in two situations

The presence of positive economic rents is not a market failure in itself, but just an outcome of the economic process. Governments (i.e. societies) may, however, believe that society should benefit from the profits resulting from the exploitation of natural resources instead of the (private) companies that produce the resources. The general idea behind this is that the society possesses the natural resources and, hence, that they should also benefit from the revenues from it. The regulatory question now is how to let society benefit from the economic profits resulting from the exploitation of natural resources (Daniel et al. 2010). The regulatory solutions can be distinguished in three types: ownership, taxation and financial regulation (see Table 11.1).

Table 11.1 Regulatory options to redistribute resource rents

Regulatory measure	Method	Risks
Public ownership	Resource rents are directly earned by public entity	Less incentives for productive and dynamic efficiency
Resource tax	Realized resource rents are transferred to resource owner (i.e. state) through a specific taxation in the course of the exploitation	When not appropriately designed, the resource tax may hinder efficient exploitation projects or leave resource rents with the investors
Auctioning	Expected resource rents are transferred to the resource owner (i.e. state) at the start of the exploitation	Auction price is too high because of the winners curse (too optimistic about revenues and costs), making a successful exploitation not possible

The most direct way to let the society benefit from the resource rents is to let public companies, owned by the state, exploit the natural resources. The revenues from this exploitation then immediately go to the state which may use them for the public interest. This regulatory option has, however, as a caveat, which is that publicly owned companies may not be that efficient (see also Sect. 3.4.2). If a company has the licence to conduct an activity, without facing any competitive pressure, this company has no incentive to increase its productive efficiency. Because of the limited incentives to become more efficient, the production costs are higher and, hence, the economic profits smaller than what would be the case when the company faces competition. The incentives to become more efficient increase when a publicly owned company becomes privatized, as then the shareholders can put pressure on the management to increase its efficiency. Although a public shareholder may also try to give stronger incentives, this pressure is less powerful as the public shareholder does not have an outside option. Private shareholders always have the option to sell their shares which affects the share price and, hence, the capability of the company to attract capital for new investments. Because of this mechanism, the management of a privately owned company has stronger incentives to listen to the wishes of shareholders than the management of a publicly owned company. This effect may, however, be partly neutralized if there exists a well-functioning labour market for managers which give them an incentive to perform well.

When a private company receives the licence to exploit a natural resource, the problem for the regulator is how to transfer the economic profits coming from this exploitation to the society without removing the incentives for companies to invest in exploration, development and production of resources. The regulatory instrument to solve this problem is implementing a resource tax, which is a tax on the economic profit from natural resources. Resource tax has to be distinguished from corporate income tax. The latter refers to the tax on the return on capital, which is a part of the costs of capital (as was discussed in Sect. 6.5). Corporate income tax has to be paid by any company that realizes a positive accounting profit, depending on the precise design of the tax scheme. In theory, the resource tax is a tax on top of the corporate income tax and refers only to the scarcity rents (see e.g. Nakhle 2009). In practice, a resource tax can be differentiated across different types of reserves, taking into account the differences in costs per unit of the commodity. By including such a differentiation, the resource tax can be calibrated to capture a large part of the (estimated) level of scarcity rents.

While a resource tax is a regulatory measure to transfer (a part of) *realized* resource rents to society, a regulatory measure to transfer the *expected* resource rents to society is auctioning of licences to exploit a natural resource. One of the advantages of this regulatory option is that competition is established among a group of potential companies that have the interest to exploit the reserves. This regulatory option is also called competition for the market (see also Sect. 6.2.3). In this case, companies are invited to submit bids for exploitation, while the bids include the price the companies want to pay to the resource owner (typically, the state). When the auction is competitive, this resulting price is equal to the expected resource rent, because in a situation of perfect competition no firm can make a supra-normal profit. Another advantage of the auctioning system is that the government can make use of the information and knowledge available among market parties over the expected production and costs.

Auctioning has, however, also some caveats. One caveat is that the winner may overestimate the future production volume and/or commodity prices or underestimated the costs of exploitation, and as a result, the winner pays too much and may go bankrupt or at least not being able to continue the exploitation of the resources. This risk is called the winners curse. The winning bid in an auction can also appear to be too low when the resource prices increase more than expected during the time of the auction. This will result in a positive economic profit realized by the company that has won the licence for exploitation.

Hence, auctioning is an efficient method for allocating a licence for production, but it goes with serious risks. Because of these risks, many countries prefer to use public ownership of the exploiting company or to allow privately owned companies in combination with resources taxes as instruments to redistribute the resource rents to society.

11.4 Energy Poverty and Fairness

11.4.1 Theory

Another distributional issue in competitive well-functioning markets is that commodity prices may be highly volatile and frequently reach high levels. Such high levels are sometimes needed to trigger economic agents to respond to a tight market situation. Electricity producers, for instance, are only going to invest in new production plants, when the electricity wholesale prices are sufficiently high for a sufficiently long period to enable them to recoup the investment costs. As we have seen in Sect. 7.5, the occurrence of scarcity prices belongs to the essential characteristics of energy-only markets. High prices in times of scarcity also give incentives to consumers to look for cheaper alternatives for a specific activity or use of energy, to temporarily postpone their consumption (i.e. load shifting) or to reduce their consumption.

Hence, when market prices are able to fluctuate unrestrictedly, the responses by suppliers and consumers can be efficient, which means that those responses will happen with the lowest opportunity costs. Those producers and consumers who can change their behaviour at lower costs then the increase in energy costs, will do so, while others will keep producing and consuming in the same way. As a result, the groups of consumers who are not adapting their behaviour in response to an increase in the energy prices will face (much) higher costs of using energy. These higher costs are equal to the higher revenues for the producers. Hence, the change in prices has distributional consequences. This effect is depicted by Fig. 11.4, showing the distributional effects of a rise in prices due to a reduction in the available production capacity which makes that the demand exceeds the remaining capacity.

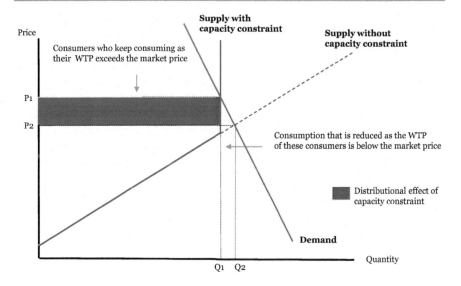

Fig. 11.4 Distributional effects from scarcity prices

If the demand curve intersects the supply curve at an output level which exceeds the available capacity, this output level cannot be realized. If the capacity is limited in the short term, as it generally is, the only way to find a market equilibrium is to have higher prices. These higher prices will give incentives to some consumers to change their consumption, through one or more of the mechanisms described above. In theory, those consumers with a WTP which is below the market price will stop consuming, while all other consumers with a WTP which exceeds the market price keep consuming. The latter group continue to consume although they have to pay more for the same commodity. The extra price they have to pay more because of the capacity constraint is the difference between the scarcity price (P1, which is the price where demand and supply find an equilibrium given the capacity constraint) and the unconstrained market price (p2). The latter price can be assumed to be higher than the marginal costs of the constrained supply, as the marginal costs will increase when the volume of output increases. The extra price these consumers have to pay is extra revenues for the producers. The red area in Fig. 11.4 shows this distributional effect of the occurrence of scarcity prices.

Although the possibility of high prices forms a precondition for efficient markets, in order to have incentives for both producers and consumers to respond to changing market circumstances, it is important to realize that high (scarcity) prices automatically result in a distribution of welfare. It is generally considered, at least by economists, that these distributional effects belong to essential characteristics of markets, without the need to correct them. There is, however, one exception on this rule, and this refers to the position of residential consumers and in particular those consumers with the lowest incomes in a society.

The fact that energy is a basic commodity that is needed by everyone implies that consumers generally don't have much options to adapt their energy consumption when the prices increase. Using the above theoretical concepts, many consumers have an (implied) WTP for energy which is very high, and as a result, they just keep consuming energy even if the price of it increases strongly. As a consequence, rising prices result in higher energy costs for many consumers, in particular those consumers which the lowest incomes (see Box 11.2).

Box 11.2 Sensitivity to changes in energy price varies across income groups

The sensitivity of households to rising energy prices depends on many factors, such as the degree of insulation of houses and the type of usage of energy, such as for cooling and heating. A major factor appears to be the income level. Because energy is a basic commodity, its consumption does not linearly increase with income. As a result, the higher the income level, on average, the lower the share of energy expenditures in total household expenditures (see Fig. 11.5). Therefore, rising energy prices in particularly affect lower income groups (see also Bhattacharyya 2019).

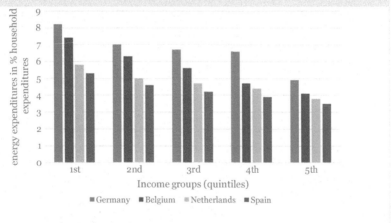

Fig. 11.5 Energy expenditures in % of household expenditures in number of EU countries, 2015. *Source* EU Energy Poverty Observatory

---

Residential consumers are, however, not fully exposed to the changes in the energy prices, as the residential end-user bill is not only determined by changes in the wholesale energy prices, but also by the margin charged by the retailer and governmental energy taxes. Suppose a retailer operates in a perfectly competitive market and fully passes on changes in the wholesale price (p_w) to the retail price (p_r) (i.e. $dp_r = dp_w$) (see Sect. 9.4.3). The relative sensitivity of the retail price to the changes in the wholesale price also depends on the margin (m_r) the retailer asks and the taxes per unit of energy (t_e):

$$\frac{dp_r}{p_r} = \frac{dp_w}{p_w + m_r + t_e} \qquad (11.3)$$

The higher the retail margin and the higher the taxes on energy, the less end-use prices for consumers are affected in relative sense to changes in wholesale prices. If, for instance, the joint share of the margin and taxes in the final residential energy price is 50%, then an increase in the wholesale price of 10% results in an increase of the residential price of 5%. Hence, the higher the taxes on energy, the less consumer prices are affected by changing wholesale prices.

In addition to this tax effect, retailers generally also offer products with fixed prices for a long period of time, which make that consumers are not exposed to short-term fluctuations in energy wholesale prices (see Box 5.3). Although these changes are reflected in retail prices in the long term, the use of long-term contracts mitigates the price volatility and the sensitivity of residential consumers for short-term changes in the wholesale market.

Box 11.3 Behavioural-economic and ethical criteria for defining fairness of dynamic network tariffs

In a zonal electricity system, network users may assume that there is unlimited transport capacity within the market zone (see Sect. 7.3.3). As a result, network users do not pay attention to the degree of utilization of the distribution infrastructure when they want to consume or produce electricity. As a result, the operator of the grid has to handle intra-zonal congestions when they arise. This can be done through ex-post congestion management in which transactions concluded in the forward wholesale market are corrected afterwards. Another option is to make the grid tariffs dynamic and location specific. Typically, the network tariffs in a zonal market are primarily meant to compensate the network operator for its operational and capital costs, and as a result, these tariffs are only related to the capacity of connections and the magnitude of the network use (see Sect. 6.6). By adding a dynamic and/or locational component to these tariffs, they may also give incentives to network users to pay attention to the occurrence of congestion.

However, tariffs that fluctuate in time and across locations may have unfavourable distributional consequences. The relevant question is this respect is to what extent residential consumers feel that such tariff structures are fair. The classical economic approach is not able to provide an answer to this question as it only has tools to assess optimal efficiency, but not to determine the optimal distribution. From behavioural-economics, however, a number of criteria regarding fairness of tariffs can be inferred (Neuteleers et al. 2017). In order to be fair, tariffs should be related to costs, predictable, and not result in exploitation of customers. Also from ethics, a number of criteria can be formulated. One of these criteria is equality, which means in this case that all users should be treated in the same way. Another ethical principle is that the intervention (i.e. the dynamic tariffs) should not result in more inequality.

Using these criteria, one can conclude that dynamic tariffs are not necessarily unfair. In order to make them fair, the tariff levels should be related to costs, predictable and be applied in a transparent manner, treating all network users in the various types of conditions in a similar way.

Besides the impact of changing prices on the ability of consumers to use energy, there is also the related issue of fairness. Although high prices during scarcity hours may be efficient and consumers may also be able to pay these high prices, people may perceive it as not fair when they have to pay very high prices only because the commodity has become scarcer. This may be relevant, for instance, when network operators want to introduce dynamic network tariffs in order to give network users incentives to use the network optimally (see Sect. 7.3). In order to assess the fairness of such dynamic network tariffs, insights from behaviour economics can be useful (see Box 11.3).

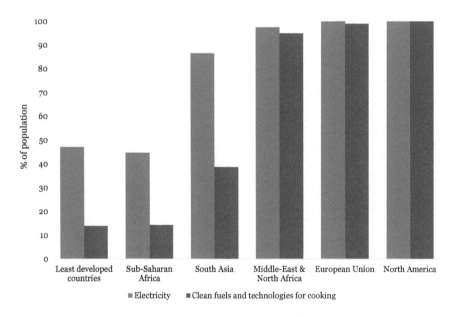

Fig. 11.6 Access to electricity and clean fuels in number of global regions, 2016 (in % of population). *Source* Worldbank: https://data.worldbank.org/

11.4.2 Regulation

In particular for consumers with low incomes, societies may believe that a strong increase in the energy bill is not fair. Therefore, governments may want to assure that everyone has access to energy at reasonable prices. In particular, they may want to prevent that low-income groups face financial difficulties or interruptions in their energy supply because of developments in energy markets and energy policies. When access to infrastructure has been realized, as is the case in most developed countries (see Fig. 11.6), the main issue regarding energy poverty is related to the price levels (González-Equino 2015). Even in developed countries, like, the European countries, a significant number of households experience difficulties in paying their energy bills (see Fig. 11.7).

Regulators have a number of options to address distributional effects of high energy prices. These can be distinguished in interventions in energy markets, specific measures regarding energy consumers and general income policies.

An example of an intervention in energy markets in order to prevent that consumers face a too strong increase in their energy bill is the implementation of a cap on the energy prices retailers may charge. If this cap is related to the cost increase of retailers, then it may still result in a high variability when the wholesale prices increase strongly. When the cap on retail prices is not related to the wholesale prices the retailers have to pay, but only to what regulators believe is reasonable for consumers to pay, then such a policy may have serious negative effects for the

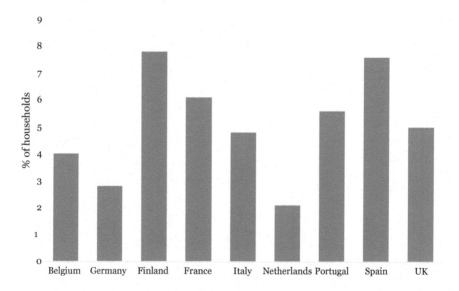

Fig. 11.7 Presence of energy poverty in number of European countries, 2015 (% of households with arrears on energy bills). *Source* EU Energy Poverty Observatory: https://www.energypoverty. eu/

Table 11.2 Regulatory options to address distributional effects of high energy prices

Regulatory measure	Method	Risks
Intervening in energy prices	Imposing rules on the level and/or increase in retail energy prices	Negative financial effects for retailers which may distort the reliability of supply Reduced incentives for consumers to contribute to stabilization of market
Compensating energy consumers	Giving subsidies for energy saving measures or exemptions in energy tax	Subsidies levels can be higher than needed Subsidies may encourage inefficient measures or be neutralized by rebound effect
General income support	Giving financial support through income tax or other schemes not related to energy consumption	Support may be given to consumers who are less affected by increase in energy prices

reliability of energy supply. Retailers may have to pay higher prices for their inputs in the wholesale markets, but if they are constrained in to what extent they pass on these prices to the end-users, they may face financial difficulties themselves (Table 11.2).

Such caps on retail prices may not only have adverse financial effects for market participants, they may also create inefficiencies. For instances, if consumers are protected from higher prices, they may continue to consume energy to the same extent, even when the market is tight and the real value of the energy is above the willingness-to-pay of some consumers. Because of the lack of responses by consumers resulting from a cap on retail prices, other market participants have to respond more strongly in order to obtain a market equilibrium. For instance, other types of consumers (e.g. from industry) will reduce their energy use more than what they would do when the wholesale prices increase less because of a higher participation of residential consumers. Hence, when some part of the demand (such as from residential consumers) is protected from high prices, the prices for other parts of the demand have to increase even higher in order to let the demand equalize supply. If also other parts of the demand are not able to respond, then the market cannot find an equilibrium which may result in physical disruptions or a black out of the energy system. Consequently, interventions in prices in order to protect consumers from high prices set the reliability of the system at a risk.

Instead of applying this kind of heavy-handed regulation directed at the prices in energy markets, regulators can take measures which are specifically directed at the group of low-income consumers. This group can be defined by using a metric like the percentage of energy expenditures of total income. A regulator can help such groups of consumers by providing support in reducing their energy consumption, for instance by giving subsidies for energy saving measures, by giving them exemptions on the tax on energy or by applying so-called social grid tariffs for particular income groups (see Creti and Fontini 2019). The advantage of this type of regulatory measure is that it can be well directed at the group of consumers that is most affected by high energy prices, while it also directly reduces their sensitivity to the high prices.

The third group of regulatory measures is not directly related to energy consumption, but is meant to give general income support to those groups that face financial difficulties because of, among others, rising energy prices. This support can be given through, for instance, a general reduction in tax tariffs for the lowest income groups, higher exemptions for income tax or giving them tax credits (i.e. reduction in income tax). The caveat of this type of measure is that it may be difficult to target it at the group of residential households that is strongly affected by high energy prices. For instance, if the price of gas strongly increases, this in particular affects those consumers that use gas, for instance to heat their premises, while consumers that use other methods for heating may be not be affected at all by such a price increase, but they would benefit from the general income support.

Exercises

11.1 What is the difference between allocation and distribution?

11.2 Explain why an efficient allocation is not necessarily equal to the optimal allocation.

11.3 How can resource rents be redistributed without negatively affecting incentives for producers of natural resources?

11.4 How can a regulator implement a tax on energy use of residential households in order to incentivize them to improve energy efficiency without lowering their net income?

11.5 Under which conditions can peak-load prices for residential consumers be viewed as fair?

References

Bhattacharyya, S. C. (2019). *Energy economics; Concepts, issues, markets and governance* (2nd ed.). Berlin: Springer.
Conrad, J. M. (1999). *Resource economics*. Cambridge University Press.
Creti, A., & Fontini, F. (2019). *Economics of electricity: Markets, competition and rules.* Cambridge University Press.
Daniel, P., Keen, M., & McPherson, C. (2010). *The taxation of petroleum and minerals; principles, problems and practice.* New York: International Monetary Fund, Routledge.
González-Equino, M. (2015). Energy poverty: An overview. *Renewable and Sustainable Energy Reviews, 47*, 377–385.

Nakhle, C. (2009). Petroleum taxation. In J. Evans & L. C. Hunt (Ed.), *International handbook on the economics of energy*. Cheltenham: Edward Elgar.

Neuteleers, S., Mulder, M., & Hindriks, F. (2017). Assessing fairness of dynamic grid tariffs. *Energy Policy, 108*, 111–120.

Varian, H. R. (2003). *Intermediate microeconomics* (6th ed.). New York/London: W.W. Norton & Company.

Regulating Energy Markets: Concluding Remarks

12

12.1 Introduction

This textbook has analysed the characteristics of energy markets and how the functioning of these markets can be improved through regulation. This analysis has departed from the public-interest approach which provides the analytical framework in order to search for optimal regulatory measures to address market failures. This final chapter summarizes the lessons learned regarding the key conditions for well-functioning energy markets and by what kind of regulatory measures governments may help to improve the functioning of these markets. Section 12.2 discusses the fundamental components of regulating energy markets and realizing energy transition, while Sect. 12.3 discusses the economic criteria for defining the optimal regulatory measures. Section 12.4 concludes this textbook by applying the lessons learned to the analysis of the optimal policies to promote energy transition.

12.2 Components of Regulating Energy Markets

Although energy markets, just as any other market, facilitate the exchange of commodities between producers and consumers, there are a number of characteristics of these markets which make the functioning of this exchange more challenging. The most fundamental of these characteristics is the fact that all trades in gas, electricity and heat are based on the presence of a physical infrastructure for transmission and distribution. Because of this characteristic, these markets can only function when there is an appropriate regulation of balancing responsibility within a network area (so-called entry–exit zones in gas markets and market zones in (European) electricity markets) as well as an efficient allocation of grid capacity between these areas (Fig. 12.1). In particular for electricity markets, it is also crucial to have a regulation that assures the adequacy of production capacity.

© Springer Nature Switzerland AG 2021
M. Mulder, *Regulation of Energy Markets*, Lecture Notes in Energy 80,
https://doi.org/10.1007/978-3-030-58319-4_12

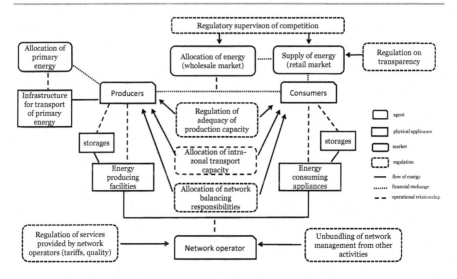

Fig. 12.1 Basic elements of regulating energy markets

In addition, because the infrastructure for transmission and distribution is generally characterized by natural monopolies, regulation is needed to guarantee all market participants access to this infrastructure against reasonable conditions, while the operator of the networks also needs to have regulatory incentives to improve and/or maintain the quality of the services provided. Because of the crucial role of the network infrastructure, it is also important that the operator does not have any commercial interest in activities that make use of the infrastructure, such as production and trade, which can be realized through unbundling of network operation from the other activities in the energy chain.

Another characteristic of energy markets which calls for specific regulation follows from the fact that the demand for energy is generally price inelastic, while the production is generally constrained in the short term and storage is not always possible. As a result, energy markets are vulnerable to the abuse of market power. That's why monitoring energy markets on the presence and use of market power is also a necessary component of regulation of energy markets, just as taking appropriate actions to reduce the risk that market outcomes are distorted because of some participants having market power. To these appropriate actions belong the removal of barriers for trade between market areas as that reduces the pivotality of individual producers, while it increases the substitution possibilities for consumers. As a result of such regulatory measures, wholesale markets for gas and electricity have increasingly become international markets in which differences in prices between market zones occur less frequently and are smaller in magnitude.

Because energy is a commodity that everyone needs, but not all consumers are capable to respond adequately to changing market circumstances, regulation may be required to protect those consumers who are most vulnerable to exploitation by retailers. This regulation includes, among others, rules that foster the transparency in retail markets.

All these regulations are needed to improve allocative efficiency in the wholesale and retail markets as well as in the use of networks. Moreover, regulatory measures are needed to give incentives to network operators to operate productively efficiently and to realize an optimal quality of services.

On top of these basic regulations for the functioning of markets and use of networks, specific regulation is required to solve environmental market failures. As long energy use is mainly based on fossil resources, it results in carbon emissions having negative effects on the climate. In order to tackle this negative environmental effect, regulators have several options to intervene in energy markets (see Fig. 12.2). The majority of these measures are meant to influence the decisions taken by individual producers and consumers. By implementing a tax on the use of (fossil) energy, giving subsidies for producing renewable energy, implementing energy-efficiency measures or introducing an emissions trading scheme, market participants receive financial incentives to make other choices regarding their use of energy. In each of these measures, these participants are still free to use fossil energy, but the costs of those choices are increased. By command-and-control measures, however, like a forced closure of coal-fired power plants or standards on the energy efficiency of appliances, the regulator directly prescribes what market participants have to do. These types of measures can be more effective, but generally, they are also less efficient because of the information asymmetry between regulator and regulated agents.

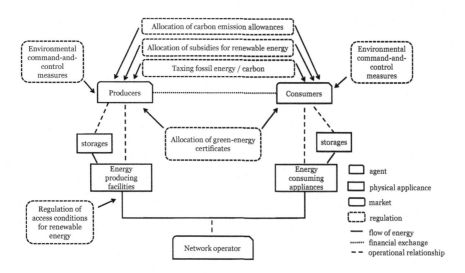

Fig. 12.2 Regulating energy transition and climate policy

Other types of regulatory measures to promote the transition towards renewable energy are the establishment of systems for green-energy certificates, which enables consumers to pay a (market) premium for energy that is produced in a renewable way. In addition to this, regulators may also give priority access to renewable energy sources when the transportation infrastructure is congested.

12.3 Economic Criteria for Optimal Regulation

Based on the lessons learned in the previous chapters, we are able to formulate a number of criteria for an optimal regulation (see Table 12.1). The starting point for formulating these criteria consists of the general conditions for well-functioning markets and the various types of market failures as discussed in Sect. 4.2. The basic requirement is that the decisions in markets are taken by many different decision-making units (i.e. economic agents) and that everyone only pursues its

Table 12.1 Criteria for optimal regulation

Condition well-functioning market	Market failure	Regulation	Condition for regulation
All effects of decisions of economic agents are taken into account	External effects	Internalizing external effects by • Subsidies • Taxes • Standards • Trading schemes	• Related to value of external effect • Preventing windfall profits • Financeability constraint
No possibility to behave strategically	Market power	• Regulation of natural monopolies • Removing network bottlenecks • Increasing demand flexibility	• Incentives for productive efficiency • Financeability constraint • Allocatively efficient tariffs • Efficient use of infrastructure
All agents have full information on the offers made by other agents	Information asymmetry	• Promoting transparency • Imposing rules regarding contracts • Implementing schemes for certificates	Giving incentives to suppliers to be transparent without taking over their role
No transaction costs	Barriers to enter markets	• Reducing entry barriers, promoting transparency	Preventing inefficient entrance by setting access conditions at efficient level

own interest while the market takes care of the coordination of their wishes. Such a decentralized organization of economies results in the highest possible welfare, provided that all conditions for such an organization are satisfied.

One such condition for a well-functioning market is that all market participants take the effects of their decisions into account. When this is not the case, activities by these participants cause external effects (negative or positive). The appropriate regulation to address this market failure is to internalize these effects through a financial regulatory instrument. The most efficient way to internalize these effects is to introduce prices, taxes or subsidies which are efficient, which means that the prices reflect the marginal valuation and costs of the effects. In addition, these internalization measures should also be related to a constraint on windfall profits, which means that the valuation of the external effect should not be higher than the actual amount needed to compensate for extra costs. When the regulator pursues specific objectives, such as regarding a specific percentage of renewable energy, then the financeability constraint needs to be taken into account, which means that the financial regulatory instrument should be chosen such that all the (efficient) costs to realize that objective can be covered.

Another condition for well-functioning markets is that market participants are not able or don't have the incentive to behave strategically in order to influence market outcomes. This situation occurs in case of natural monopolies. The appropriate regulation is to give the operator incentives to operate efficiently and to restrict the tariffs the operator may ask. These regulated tariffs should be both allocatively efficient and also generate sufficient revenues to the operator to compensate for the fixed costs. The opposite of the latter requirement is that the regulated tariffs should not exceed the amount that is needed in order to prevent that the regulation results in windfall profits and, hence, too high costs for those who pay the tariffs (i.e. the network users).

Another criterion refers to the ability of market participants to enter into transactions. In order to be able to do so, they need information on product characteristics and outcomes of trade. A key condition for well-functioning markets is that market parties have the same type of information. If consumers cannot have the same information on product qualities as producers have, that market will result in suboptimal volumes. If such information asymmetry exists, this should be addressed. Related to this point on information availability is the general criterion that transaction costs should be low, otherwise market parties will be hindered to make efficient deals.

Market transactions may also become distorted if some parties can be held up by others, which may exist in the case of dedicated investments. Parties will not make such investments if they do not have certainty before they make the investment decision on the revenues after the investment has been realized.

Market processes result in profits, and the distribution of these profits (rents) can be an important consideration for regulators. Microeconomic theory cannot be used to formulate criteria for optimal distribution, as the assessment of distribution depends on the preferences in society on what is (not) fair. From behavioural economics, however, we can derive some general principles on fairness. One of these principles is that people/firms should 'deserve' profits, meaning that they should have done something to be entitle to it. Hence, windfall profits are generally seen as unfair.

12.4   Economic Principles for an Efficient Energy Transition

The above criteria are related to the functioning of energy markets in general. A specific criterion for the assessment of regulation is its cost effectiveness. This criterion says that the policy objectives should be realized at lowest possible costs. Regulation may, however, also fail to deliver or be more costly than the benefits of regulation. Therefore, the last criterion is that regulatory measures should only be implemented until the point where the marginal benefits of these measures are above or equal to the marginal costs. This criterion is also relevant when governments pursue specific policy targets, such as regarding energy transition. To keep the costs of energy transition as low as possible and, thus, to increase the chances of successfully realizing the energy transition, it is important to uphold the basic economic principles of the organization of the energy market to the greatest possible extent. Experiences with the design of the energy markets in the last few decades have shown how the efficiency in these markets has gradually improved by applying several of these principles (see Table 12.2).

One basic principle of a well-organized market is that scarce resources must be allocated based on market prices. Market mechanisms can also be used to allocate, for instance, scarce subsidy funds. In order to realize the largest amount of renewable energy per unit of subsidy, support schemes should be set up in a generic way, i.e. independent of the technology that is used and where participants have to compete for a subsidy. When interventions are not based on market prices, inefficiencies may result. Interventions in energy markets, such as allowing network operators to invest in storage themselves, may suppress the role of the scarcity prices in incentivizing market participants. Consequently, such an intervention may negatively affect the efficiency of investments made by market participants in storage and generating capacity.

Another basic principle is that the responsibility for paying the costs of energy transition must be taken by those who have caused the costs or can influence them. This is the so-called cost-causation principle. As far as electricity grids are concerned, this implies that programme responsibility for keeping own portfolios in balance is a vital component of the incentives required to stimulate participants to produce (or consume) the electricity which they have sold (or bought) in the forward markets as well as that the market participants minimize the costs of creating imbalances. Another example of an application of this principle is that network users have to pay for using the network at particular moments of time or at particular places.

Table 12.2 Economic principles for realizing energy transition objectives in an efficient way

Economic principle	Examples	Efficiency effects of examples
Allocation of scarce resources through market mechanisms	• Allocation of subsidies for renewable energy through auctioning • Emissions trading schemes	• Renewable energy projects with lowest LCOE receive the subsidy • Emission reductions are realized by the methods having the lowest costs
Cost-causation principle	Costs have to be paid by those agents who have caused them, e.g. • Costs of system imbalance • Costs of network usage • External costs, like emissions of CO_2	• System users have incentives to prevent imbalances • Network users take the costs of using the network at particular moments of time and locations in account when making their investment and production decisions • When users of fossil energy are confronted with the external costs, they have incentive to explore efficient alternatives
Decentralized organization	• Investments in electricity generation plants • Dispatch (i.e. actual use) of these plants • Demand response by energy users	When investment, production and demand-response decisions are taking by independent decision-makers, there are stronger incentives to do this as efficiently as possible
Transparency and freedom	• Freedom to make use of networks • Freedom to enter into international trade	The more options market participants can consider, the lower their costs will be
Coordination only when required	• Networks are semi-public goods, which cannot be offered by a market • Hold-up problem in case of investing in new infrastructures (e.g. district heating, hydrogen) • International coordination of renewable energy policies	• Reliability of services can be offered at lower costs • Realization of welfare-improving investments which cannot be realized without coordination • Mitigation of international spillover effects of national energy policies

Even though energy transition is generally seen as something which can only succeed when all economic agents, including government agencies and energy business, closely cooperate, it remains important to maintain the principle of the decentralized organization of energy markets. This means that the decisions are made by separate parties which operate as autonomously as possible. Market prices cannot fulfil their driving and informing role well unless they result from independent decentralized decisions made by suppliers and consumers and all participants are exposed to these market prices. This increases the chance of successful innovations because, ultimately, the best initiatives will emerge from a multitude of more or less successful ones.

Cooperation or coordination between market participants in determining their investments or disinvestments in generating capacity decreases the efficiency of the system. Cooperation can generate welfare gains only where there are coordination problems which the market cannot solve, for instance, when new infrastructures are built, such as in heating or hydrogen networks, and others are phased out, such as gas networks.

Local and national governments may want to take over the role of market participants when they feel that the energy transition process is too slow. It may be that the market is less dynamic in taking the energy transition on board than governments would want them to, but it is important to realize that this doesn't necessarily imply that the market fails. Usually, the reason for market participants not to proactively implement green-energy projects is that the costs exceed their private or commercial interests. When local or national governments want to stimulate these market participants to take action in this respect, it is likely more efficient to change the market participants' costs or revenues.

Market participants can only be stimulated to contribute to achieving social objectives where competition is possible. This is not the case with electricity grids. A completely independent operation of these networks is, therefore, crucial. Such an independent operation is at risk when network operators add tasks to their role as network operator which can be carried out by market participants as well. While storage of electricity, for instance, may help to prevent that network congestions to occur, this task can also be carried out by market participants. To stimulate market participants to invest in storage where this is efficient, dynamic network tariffs may be helpful, based on the principle that scarce network capacity can be priced. An alternative option in case of grid congestions is that network operators request market participants to provide flexible capacity in a certain part of the network.

Furthermore, international integration of energy markets also helps to keep the costs of energy transition as low as possible. Even though local renewable energy initiatives are appealing to some, local solutions are, by definition, more expensive than solutions which come to the fore in an international system. Therefore, government support for local initiatives is not a logical thing to do when the ambition is to realize the energy transition at the lowest costs for society. International coordination of national energy transition policies is also required because energy markets are international markets, so that domestic interventions in the energy systems generally have cross-border implications, which may lead to higher costs to realize the energy transition. This means that domestic environmental effects can leak away to other countries, even without emissions trading.

Finally, in order to reduce the emissions of CO_2 it may not be effective to promote the share of renewable energy within the energy sector. In a system with emissions trading this measure need not to result in less emissions, because of the waterbed effect, as it only reduces the prices of emission allowances. In such a setup, subsidies for renewable energy can thus be seen as subsidies for all participants in the emissions trading system.

In order to reduce emissions, the most efficient and effective way is to reduce the number of allowances available for market participants. This reduction can be achieved by various types of interventions: lowering the number of allowances in the primary allocation, replacing energy use outside the trading scheme by energy use that belongs to the system (e.g. replacing gasoline cars by full-electric cars) or just buying allowances and, subsequently, cancelling them without using them. All these actions increase the scarcity in the emission allowances market, which leads to an increase in the CO_2 price and requires the participants of the trading scheme to do more about emission reduction. From the above follows that, although governments need to intervene strongly in energy markets in order to realize the societal objectives regarding carbon emissions, the realization of these objectives can be most efficiently done by applying the basic components of well-functioning energy markets. Hence, implementing energy transition policies is not at odds with the concept of liberalized markets based on free decentralized decision-making, as it is sometimes said, but the latter is able to facilitate an efficient realization of energy transition objectives.

Exercises

12.1 What are the regulatory requirements to foster the reliability of energy systems?

12.2 Why is applying the cost-causation principle a condition for an efficient energy transition?

12.3 Why is a decentralized organization of energy systems preferable from an economic perspective?

12.4 Why do local solutions of energy transition generally result in higher costs?

12.5 Why is realizing climate-policy objectives through markets based on decentralized decision-making more efficient than through central coordination?

Appendix 1: Definitions

1 tonne of oil is 7.33 barrel and has an energy content of about 42 GJ or 11.7 MWh 1 tonne of oil equivalent (toe) is defined as the energy content of 1 tonne of crude oil.

1 tonne of coal equivalent (tce) is defined as 7 million kilocalories which is equal to 29.3 GJ or 8.2 MWh. Hence, one tce is equal to 0.7 ton of oil equivalent (toe).

1000 m³ of natural gas is set equal to 35.17 GJ or 9.8 MWh of energy.

1000 kg. of hydrogen is equal to about 126 GJ or 35 MWh of energy.

1 GJ is equal to 0.28 MWh and 1 MWh = 3.6 GJ.

For more detailed information on conversion factors, see for instance IEA (2019), Oil Information.

Units:

To express the amount of energy, the International System of Units is used. The most commonly used units are the following:

Base 10	Prefix (symbol)	Prefix name	English name	Decimal
10^{15}	P	peta	quadrillion	1,000,000,000,000,000
10^{12}	T	tera	trillion	1,000,000,000,000
10^{9}	G	giga	billion	1,000,000,000
10^{6}	M	mega	million	1,000,000
10^{3}	k	kilo	thousand	1000
10^{0}				1

© Springer Nature Switzerland AG 2021
M. Mulder, *Regulation of Energy Markets*, Lecture Notes in Energy 80,
https://doi.org/10.1007/978-3-030-58319-4

Appendix 2: Answers on Exercises

Chapter 1

1.1 Because energy prices may fluctuate strongly, which may in particular affect those consumers who spend a relatively large portion of the income on energy.

1.2 No centralized coordination, decisions on production and consumption are made by individual firms and consumers; coordination is done through markets.

1.3 Because it analyses what would be the best regulation from a public-interest perspective; it does not explain the actual regulation.

1.4 Energy markets are based on the idea of decentralized decision-making (e.g. free entrance, consumer choice), while energy transition consists of policies to influence the decisions of firms and consumers.

1.5 Energy intensity is the energy use per unit of income, while energy efficiency refers to how much energy is used to produce a particular output.

Chapter 2

2.1 When power plants using natural gas as fuel, such as a CCGT, are the price-setting power plants.

2.2 Hydrogen can be produced through Steam Methane Reforming, which uses gas as input, and through electrolysis, which uses electricity and water as input.

2.3 Energy transition policies have as a consequence that the demand for fossil energy will reduce in the future which will reduce the future price of fossil energy. As the resources of fossil energy are limited in size, i.e. they are non-renewable, it is rational for the producers of this energy to advance the production.

2.4 LCOE of solar park is 104.5 euro/MWh.

2.5 Electricity can be converted into other energy carriers, for instance, through electrolysis or pumped hydro.

© Springer Nature Switzerland AG 2021
M. Mulder, *Regulation of Energy Markets*, Lecture Notes in Energy 80,
https://doi.org/10.1007/978-3-030-58319-4

Chapter 3

3.1 Liberalization refers to the introduction of decentralized decision-making, while privatization (only) refers to the ownership of assets/companies (which changes from public in private ownership).

3.2 Because exchanges only offer a set of standardized products (regarding timing), while on bilateral and OTC markets in principle every type of contract can be traded.

3.3 When suppliers want to reduce their future price risk and, therefore, they are prepared to accept a lower price than the expected spot price in the future.

3.4 A liquid market is a market where an individual transaction has no impact on the market price.

3.5 If imbalance is negative, this means that demand exceeds production, which is an indication the oil market is tight. The spot price will rise which is an incentive for owners of oil that is stored to release this.

3.6 This is only relevant to them for getting entry and/or exit rights. As import and export are seen as entry and exit, for international trade capacity booking is relevant for market parties.

3.7 The horizontal movement of the merit order due to changes in (expected) production of the renewable production and the resulting impact on the electricity price.

3.8 The ranking of all hours in a year from highest to lowest load levels.

3.9 The separation of the management of the transportation and distribution networks from other parts of the supply chain.

3.10 This is the regulation implemented by the regulator on the basis of the primary legislation, such as national energy laws.

Chapter 4

4.1 A positive externality means that an economic activity results in benefits which are not taken into account by the decision-maker as these benefits are not priced in a market. As a result, the optimal size of that activity will be less than when these benefits are taken into account.

4.2 (a) This can be solved by: $3q = 50 - 2q$, which results in: $q = 10$ and $p = 30$.

(b) If the firm does not include all costs, the market equilibrium can be found by: $2q = 50 - 2q$, which results in $q = 12.5$ and $p = 25$. Note that the equilibrium output is 12.5, which is possible in markets where the unit of the commodity is, for instance, MWh and, in principle, any quantity of a commodity can be exchanged.

(c) As the equilibrium price is lower, while the quantity is higher, this shows that in case of a negative externality consumers consume too much because prices do not reflect all costs.

4.3 Economies of scale in transport of electricity, gas and heat, resulting in a natural monopoly, economies of scope in transport and supply of energy resulting in costs advantages, and periods of high demand when all producers operate on their capacity constraints.

4.4 This will be the case when potential investors in the district-heating system are uncertain about their negotiation position towards future users (producers, users) once they have built the system. If they cannot conclude contracts with these users in advance, they may not be willing to invest.

4.5 Variants refer to the definition of the project alternatives (i.e. specification of policy measures) while scenarios refer to the assumptions on the external circumstances.

Chapter 5

5.1 Response by suppliers to a problem of information asymmetry, in order to give consumers certainty about the information they provide.

5.2 If consumers cannot trust information on the source, they are not willing to pay a higher price for commodities which are more expensive.

5.3 In mass balancing there is a strict relationship between the trade in certificates and the flows of energy, while this is not the case in a book-and-claim approach.

5.4 This raises the search and switching costs, which gives retailers more options to charge higher prices.

5.5 Retailers make more profit on non-savvy consumers which enables them to compete more fiercely in the market for savvy consumers offering them lower prices.

5.6 That the wholesale price behaves differently than assumed when a retail contract is concluded.

5.7 A low switching rate may be due to high search and switching costs which hinder consumers to switch to another supplier, but a low rate may also occur when these costs are very low and, as a result, competition among retailers is fierce, resulting in equal prices which give no incentive anymore to consumers to switch.

Chapter 6

6.1 A firm can have a natural monopoly while the average costs increase with production volume.

6.2 When competition is possible, i.e. more firms could operate on the market under similar costs.

6.3 Adverse selection, as this refers to information on the precise characteristics of the firm.

6.4 The marginal impact of changes in costs on profits. If this impact is −1 (or: −100%), the firms see the full consequences of any change in its costs, which gives it a strong incentive to reduce costs.

6.5 Tightness refers to the profit level a firm can realize in normal (average) circumstances, while incentive power refers to the marginal effect of costs (effort) on profit.

6.6 Regulation of tariffs based on total costs.

6.7 Productivity is ratio output/input while efficiency is the productivity of a firm in comparison with benchmark (technological frontier).

6.8 $p = c * fs$.

6.9 Incentive power = 0.

Chapter 7

7.1 There may be insufficient capacity for transport, and as a consequence, a network operator may need to switch off producers or consumers (involuntary load shedding).

7.2 Congestion management in zonal pricing; in a nodal pricing scheme with prices which vary between nodes in the network depending on the presence of constraints.

7.3 Pressure in gas network or voltage in electricity grid may become too high or too low, making it impossible for network users to use the energy.

7.4 Organization of balancing system which incentives network users to be in balance themselves or at least to see financial consequences of imbalances.

7.5 Total blackout as load can be much higher than production capacity.

7.6 By arranging reserve capacity and to set a price if the market is not able to do this. This price should be based on the VoLL.

Chapter 8

8.1 Carbon emissions have a global and long-term effect, while small particles only affect the environment in the direct neighbourhood, so in the latter case, agents experience the negative outcomes themselves more directly.

8.2 In case of subsidies, the effect is partly neutralized by the rebound effect. Subsidies also may result in free riders resulting in windfall profits. Subsidies also generally give less flexibility to participants, so it suffers more from the information asymmetry between regulator and firms and, as a result, it easily results in productive inefficiencies.

8.3 By using a relative benchmark as basis for the tax, or by reducing corporate income tax.

8.4 Social marginal benefits (i.e. the value of negative externality), LCOE of investment and other (private) benefits (revenues).

8.5 The number of allowances (i.e. the cap) and the marginal abatement cost curves of the participants.
8.6 How costly it is (on the margin) to realize the emission target.

Chapter 9

9.1 The ability says to what extent a firm is able to influence the market price; the incentive says to what extent the profits of the firm will increase by doing so.
9.2 Economic withholding means that suppliers require a price which exceeds their marginal costs in order to raise the market price. Physical withholding means that suppliers offer less products to the market in order to realize the same.
9.3 More storage means that producing firms face more competition, provided the storages are owned by other firms.
9.4 Firms need to make profits in order to recoup investments, so one has to control for that. In addition, the marginal costs also depend on dynamic costs for ramping up and down.
9.5 Demand flexibility can be fostered by enabling consumers to switch more easily to other retailers by increasing transparency about the market, and reducing costs for searching and switching.

Chapter 10

10.1 The productive and allocative efficiencies resulting from market integrating result in lower output prices, which affect the incentives for investments in generation capacity. The lower prices reduce the inframarginal profits, which may make some investments not profitable anymore. As a result, a short-term improvement in competition, indicated through the RSI, can vanish in the long term as the capacity in the market is reduced. Nevertheless, when the number of suppliers increase due to the market increasing, it may become more difficult for them to behave strategically.
10.2 When different system-marginal costs result in different electricity prices, market integration enables price arbitrage. If the cross-border transport capacity is not constrained, this price arbitrage results in full price equalization. Consumers in the country with the high prices initially benefit because they will have lower prices, while consumers in the other country will see higher prices. Producers in the country with the high prices initially lose because of the decline in the price, while producers in the other country will benefit because they experience an increase in their inframarginal profits as well as an increase in production levels. The overall welfare increases, although there are both losers and winners.
10.3 Integrating markets makes markets less vulnerable to shocks in demand or supply as there are more options to respond to these shocks. A failure in a power plant, for instance, can be more easily neutralized in a larger market with more

plants being able to respond and more consumers who can reduce their consumption. Hence, integration increases flexibility and, as a result, reduces the volatility of prices.

10.4 Without market integration, national energy policies cannot have a direct effect on energy markets in neighbouring markets. In case of market integration, however, the effects of national policies will also affect energy systems in neighbouring markets. If a country stimulates its renewable electricity production, for instance, its domestic electricity price goes down, which triggers international trade: export from this country will result in lower prices in the importing countries. As a consequence, for instance, in that country more subsidies are required to stimulate renewable electricity generation insofar these subsidies are based on the feed-in-premium system.

Chapter 11

11.1 Allocation refers to how scarce commodities are used, while distribution refers to how commodities (or income) are spread over economic agents.

11.2 The optimal allocation depends on how society values the distribution of resources (e.g. income) among agents. For every initial distribution of resources, several efficient allocations exist.

11.3 By designing a resource tax which only captures the profits above the profit which is required by investors given the risks of the project.

11.4 This can be done by, for instance, combining the tax on energy use with a lump see subsidy or reduction in labour income tax. In such a design, incentives to improve energy efficiency have increased while households do not to experience a reduction in net income.

11.5 When the scheme for peak-load pricing is transparent, peak prices can be foreseen while the revenues are used to reduce the congestion causing the peak-load prices.

Chapter 12

12.1 Regulation of balancing responsibility, production adequacy, operational security and capacity of networks.

12.2 Because it gives incentives to all agents who cause costs to the energy system (such as regarding balancing or network capacity) to take these costs into account when they are considering to produce or consume energy.

12.3 Because of information advantages, each agent can consider more relevant aspects for any decision than a centralized agent, provided that a coordination mechanism is in place, such as a well-functioning market.

12.4 In local energy systems, all the required flexibility to deal with the volatility in production and load have to be dealt with locally, such as through storages, while in an international system, more options for providing flexibility are available.

12.5 In decentralized systems with coordination through markets, more options will be used, which results in a higher productive and dynamic efficiency, as economic agents have an incentive to reduce costs, while there is a higher likelihood that new innovations will be realized.

Index

Printed by Printforce, the Netherlands